AF245505

# Date Palm Biotechnology Protocols Volume I

## Tissue Culture Applications

Edited by

## Jameel M. Al-Khayri

*Department of Agricultural Biotechnology*
*King Faisal University*
*Al-Hassa, SA*

## S. Mohan Jain

*Agricultural Sciences*
*University of Helsinki*
*Helsinki, FL*

## Dennis V. Johnson

*Cincinnati, OH, USA*

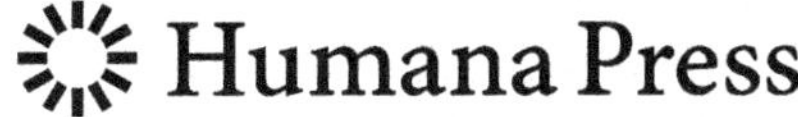 Humana Press

*Editors*
Jameel M. Al-Khayri
Department of Agricultural
 Biotechnology
King Faisal University
Al-Hassa, SA

S. Mohan Jain
Agricultural Sciences
University of Helsinki
Helsinki, FL

Dennis V. Johnson
Cincinnati, OH, USA

ISSN 1064-3745        ISSN 1940-6029  (electronic)
Methods in Molecular Biology
ISBN 978-1-4939-7155-8        ISBN 978-1-4939-7156-5  (eBook)
DOI 10.1007/978-1-4939-7156-5

Library of Congress Control Number: 2017945286



This Humana Press imprint is published by Springer Nature
The registered company is Springer Science+Business Media LLC
The registered company address is: 233 Spring Street, New York, NY 10013, U.S.A.

# Preface

Biotechnology is playing an essential role in sustaining sufficient food supplies challenged by the burgeoning human population and current global environmental changes. Development of plant biotechnology has reached advanced stages of applications in several important food crops. Palm fruit crops, particularly those grown predominantly in developing countries, are attracting increasing research attention in response to greater funding allocated for plant biotechnology research in these countries. Cultivation of date palm (*Phoenix dactylifera* L.), an economically important tree species grown in the arid and semiarid regions of the Middle East and North Africa, has expanded to Australia, Southern Africa, South America, Mexico and the southwestern USA. In addition to its high nutritive value and versatile tree by-products, dates have diverse medicinal properties due to a high content of bioactive compounds including carotenoids, polyphenols, tannins, and sterols. These compounds are known for antioxidant activity, cholesterol-lowering properties, and chemoprevention which may inhibit cancer, diabetes, and cardiovascular diseases.

Our earlier book entitled *Date Palm Biotechnology* provided a comprehensive coverage on various biotechnological aspects in relation to date palm. This book *Date Palm Biotechnology Protocols* is intended to supplement the previous volume and to provide precise stepwise protocols in the field of date palm biotechnology. Materials, equipment, methods, and analysis are detailed for easy adoption by novices and substantiated with relevant references for specialists desiring to indulge in new biotechnological applications.

The book consists of two volumes, *Vol 1: Tissue Culture Applications* and *Vol 2: Germplasm Conservation and Molecular Breeding*. This first volume comprises 27 chapters grouped in six parts: Part I Adventitious Organogenesis; Part II Somatic Embryogenesis; Part III Contamination, Hyperhydricity, and Acclimatization; Part IV Cell Suspension, Protoplast, and Bioreactors; Part V Genetic Transformation; and Part VI Secondary Metabolites and Abiotic Stress. Each protocol chapter starts with a brief introduction relevant to the topic methodology and then lists the necessary materials, including chemical and equipment, as well as reagent preparation, followed by detailed reproducible procedural steps supported with observational notes and illustrative photographs. The chapters in this book are authored and reviewed by prominent specialists demonstrating distinct research contributions to date palm biotechnology, invited from industry, universities, and research institutes.

This two-volume book is a valuable resource to students, researchers, scientists, commercial producers, consultants, and policymakers interested in agriculture or plant sciences particularly in date palm biotechnology. This protocol manual is highly recommended for teaching advanced level undergraduate as well as graduate courses in date palm biotechnology or other relevant courses.

We would like to express our sincere gratitude for the generosity and excellent cooperation of the chapter authors of these two volumes. In total, 125 scientists, representing 21 countries, have provided their detailed knowledge of the biotechnological protocols being employed to improve the date palm and enhance its contribution to the world's food needs. These two scientific books represent the epitome of international cooperation, transcending the various differences, which exist in the world, and emphasizing the common solutions to feeding the burgeoning human population of the planet we all share.

*Al-Hassa, SA*                                                                                    *Jameel M. Al-Khayri*
*Helsinki, FL*                                                                                        *S. Mohan Jain*
*Cincinnati, OH, USA*                                                                        *Dennis V. Johnson*

# Contents

## Part III   Contamination, Hyperhydricity and Acclimatization



## Part IV   Cell Suspension, Protoplast and Bioreactors



## Part V   Genetic Transformation

Part VI   Secondary Metabolites and Abiotic Stress

# Contributors


LARBI ABAHMANE • *Plant Biotechnology Laboratory, INRA, Marrakech Regional Center of Agricultural Research, Marrakech, Morocco*

ABEER H.I. ABDEL-KARIM • *Central Laboratory of Date Palm Research and Development, Agricultural Research Center, Giza, Egypt*

OLA H. ABDELBAR • *Department of Agricultural Botany, Faculty of Agriculture, Ain Shams University, Cairo, Egypt*

MANSOUR A. ABOHATEM • *Department of Biology, Faculty of Education and Languages, Amran University, Amran, Yemen; Plant Tissue Culture Laboratory, Public Corporation for Agricultural Services, Ministry of Agriculture and Irrigation, Sanáa, Yemen*

ALI M. AL-ALI • *Plant Tissue Culture and Genetic Engineering Laboratory, National Agriculture and Animal Resources Research Center, Ministry of Environment, Water and Agriculture, Riyadh, Saudi Arabia*

MAKKI A. AL-KHAFAJI • *Department of Horticulture and Landscape Gardening, College of Agriculture, University of Baghdad, Baghdad, Iraq*

ABD ULMONEEM H. AL-KHAMEES • *Plant Tissue Culture and Genetic Engineering Laboratory, National Agriculture and Animal Resources Research Center, Ministry of Environment, Water and Agriculture, Riyadh, Saudi Arabia*

ABDULLATIF A. AL-KHATEEB • *Department of Agricultural Biotechnology, College of Agriculture and Food Sciences, King Faisal University, Al-Hassa, Saudi Arabia*

SULIMAN A. AL-KHATEEB • *Department of Environment and Natural Resources, College of Agriculture and Food Sciences, King Faisal University, Al-Hassa, Saudi Arabia*

JAMEEL M. AL-KHAYRI • *Department of Agricultural Biotechnology, College of Agriculture and Food Sciences, King Faisal University, Al-Hassa, Saudi Arabia*

MAI A. ALLAM • *Genetic Engineering and Biotechnology Division, Department of Plant Biotechnology, National Research Center, Giza, Egypt*

ABDULMINAM H.A. ALMUSAWI • *Cell and Biotechnology Research Unit, College of Science, Basra University, Basra, Iraq*

SAMI O. AL-OTAIBI • *Plant Tissue Culture and Genetic Engineering Laboratory, National Agriculture and Animal Resources Research Center, Ministry of Environment, Water and Agriculture, Riyadh, Saudi Arabia*

ANSAM M.S. ALSHANAW • *Date Palm Research Centre, Basra University, Basra, Iraq*

SULTAN A. AL-SULAIMAN • *Plant Tissue Culture and Genetic Engineering Laboratory, National Agriculture and Animal Resources Research Center, Ministry of Environment, Water and Agriculture, Riyadh, Saudi Arabia*

HUSSAIN A. ALWAEL • *Department of Agricultural Biotechnology, College of Agriculture and Food Sciences, King Faisal University, Al-Hassa, Saudi Arabia*

MEGAHED H. AMMAR • *Plant Production Department, College of Food and Agriculture Sciences, King Saud University, Riyadh, Saudi Arabia*

KAZEM ARZANI • *Department of Horticulture Science, Faculty of Agriculture, Tarbiat Modares University, Tehran, Iran*

MOHMMED BAAZIZ • *Laboratoire de Biotechnologies-Biochimie, Valorisation et Protection des Plantes, Faculté des Sciences Semlalia, Université Cadi Ayyad, Marrakech, Morocco*

SALEH M. BADER • *State Board of Agricultural Research, Ministry of Agriculture, Baghdad, Iraq*

YAZID BAKIL • *Plant Tissue Culture Laboratory, Public Corporation for Agricultural Services, Ministry of Agriculture and Irrigation, Sanáa, Yemen*

EMNA BAKLOUTI • *Laboratory of Plant Biotechnology, Faculty of Sciences of Sfax, University of Sfax, Sfax, Tunisia*

CHOKRI BAYOUDH • *Centre Régional de Recherches en Horticulture et Agriculture Biologique, Chott Mariem, Tunisia*

NAZIM BOUFIS • *Laboratoire des Ressources Génétiques et Biotechnologies, Ecole Nationale Supérieure Agronomique (ENSA, ES1603), Algiers, Algeria; Division Biotechnologies et Amélioration des Plantes, Institut National de la Recherche Agronomique d'Algérie, Algiers, Algeria*

MOHSEN BRAJEH • *Department of Agronomy and Plant Breeding, Faculty of Agriculture, Shahid Chamran University, Ahvaz (Ahwaz), Iran*

OLFA CHKIR • *Laboratory of Plant Biotechnology, Faculty of Sciences of Sfax, University of Sfax, Sfax, Tunisia*

BAHAREH DEHSARA • *National Institute of Genetic Engineering and Biotechnology (NIGEB), Tehran, Iran*

NOUREDDINE DRIRA • *Laboratory of Plant Biotechnology, Faculty of Sciences of Sfax–Route Sokra, University of Sfax, Sfax, Tunisia*

MAIADA M. EL-DAWAYATI • *Central Laboratory of Date Palm Research and Development, Agriculture Research Center, Giza, Egypt*

SHERIF EL-SHARABASY • *Central Laboratory of Date Palm Research and Development, Agriculture Research Center, Giza, Egypt*

LOTFI FKI • *Laboratory of Plant Biotechnology, Faculty of Sciences of Sfax, University of Sfax, Sfax, Tunisia*

EZZ EL-DIN G. GADALLA • *Central Laboratory of Date Palm Research and Development, Agricultural Research Center, Giza, Egypt*

JOHN L. GRIFFIS JR. • *Marine and Ecological Sciences, Florida Gulf Coast University, Ft. Myers, FL, USA*

ALI A. HABASHI • *Department of Tissue Culture and Gene Transformation, Agricultural Biotechnology Research Institute of Iran (ABRII), Karaj, Iran*

MONA M. HASSAN • *The Central Laboratory of Date Palm Research and Development, Agricultural Research Center, Giza, Egypt*

LAKHDAR KHELIFI • *Laboratoire des Ressources Génétiques et Biotechnologies, Ecole Nationale Supérieure Agronomique (ENSA, ES1603), Algiers, Algeria*

HUSSAM S.M. KHIERALLAH • *Date Palm Research Unit, College of Agriculture, University of Baghdad, Baghdad, Iraq*

CHIEN-YING KO • *International Cooperation and Development Fund (Taiwan ICDF), Taipei, Taiwan; Taiwan Technical Mission in the Kingdom of Saudi Arabia, Riyadh, Saudi Arabia*

WALID KRIAA • *Laboratory of Plant Biotechnology, Faculty of Sciences of Sfax, University of Sfax, Sfax, Tunisia*

RAJA B. MASMOUDI • *Laboratory of Plant Biotechnology, Faculty of Sciences of Sfax, University of Sfax, Sfax, Tunisia*

HEBAALLAH A. MOHASSEB • *Department of Plant Biotechnology, National Research Centre, Cairo, Egypt*

AMIR MOUSAVI • *National Institute of Genetic Engineering and Biotechnology (NIGEB), Tehran, Iran*

MOUSA MOUSAVI • *Department of Horticulture Science, Faculty of Agriculture, Shahid Chamran University, Ahvaz (Ahwaz), Iran*

POORNANANDA M. NAIK • *Department of Agricultural Biotechnology, College of Agriculture and Food Sciences, King Faisal University, Al-Hassa, Saudi Arabia*

AMENI NASRI • *Laboratory of Plant Biotechnology, Faculty of Sciences of Sfax, University of Sfax, Sfax, Tunisia*

AHMED OTHMANI • *Laboratoire de Culture In Vitro, Centre Régional de Recherches en Agriculture Oasiennes, Degache, Tunisia*

ALAIN RIVAL • *UMR DIADE, Cirad BioS, IRD, Montpellier Cedex 5, France*

MAHMOUD M. SAKER • *Genetic Engineering and Biotechnology Division, Department of Plant Biotechnology, National Research Center, Giza, Egypt*

ABDULLAH J. SAYEGH • *TC Propagation Ltd, Enniscorthy, Wesford, Ireland*

AMEL SELLEMI • *Laboratoire de Culture In Vitro, Centre Régional de Recherches en Agriculture Oasiennes, Degache, Tunisia*

HUSSEIN J. SHAREEF • *Department of Date Palm Tissue Culture, Date Palm Research Center, University of Basrah, Basrah, Iraq*

REHAB SIDKY • *Agriculture Researshe Department, Ministry of Municipality and Environment, Doha, Qatar*

MOHEI EL-DIN M. SOLLIMAN • *Department of Agricultural Biotechnology, College of Agriculture and Food Sciences, King Faisal University, Al-Hassa, Saudi Arabia; Department of Plant Biotechnology, National Research Centre, Cairo, Egypt*

RANIA A. TAHA • *Biotechnology and Micropropagation Lab, Pomology Department, Agricultural and Biological Division, National Research Centre, Giza, Egypt*

KHAYREDDINE TITOUH • *Laboratoire des Ressources Génétiques et Biotechnologies, Ecole Nationale Supérieure Agronomique (ENSA, ES1603), Algiers, Algeria; Division Agriculture de Montagne, Institut National de la Recherche Agronomique d'Algérie, Algiers, Algeria*

EMAN M.M. ZAYED • *Central Laboratory of Date Palm Research and Development, Agriculture Research Center, Giza, Egypt*

ZEINAB E. ZAYED • *Central Laboratory of Date Palm Research and Development, Agriculture Research Center, Giza, Egypt*

# Part I

# Adventitious Organogenesis

# Chapter 1

## Cultivar-Dependent Direct Organogenesis of Date Palm from Shoot Tip Explants

### Larbi Abahmane

### Abstract


A number of public and private laboratories are working on date palm micropropagation to meet the increasing worldwide demand for date palm planting material. A standardized direct organogenesis protocol exists for the production of date palm plantlets to maintain the genetic fidelity of regenerated plants. Organogenesis has the advantage of using low concentrations of plant growth regulators and avoiding the callus phase. In addition, direct regeneration of vegetative buds minimizes the risk of somaclonal variation among plant regenerants. However, in vitro multiplication cycles should be limited in duration by frequent renewal of plant material. This chapter describes a simple and routine organogenesis protocol for date palm multiplication using shoot tip explants.

**Key words** In vitro culture, Micropropagation, Plant acclimatization, Shoot multiplication, True to type, Vegetative buds


## 1  Introduction

Date palm, *Phoenix dactylifera* L., is a perennial, dioecious, and monocotyledonous tree well adapted to arid environments. Typically, date palm propagation is sexually by seed or vegetatively by offshoots. However, both techniques are economically inefficient and fail to meet the demand for large quantities of planting material and the clonal propagation of selected superior genotypes. Plant tissue culture techniques are used to clone a range of important economic palms including coconut, oil palm, and date palm [1–4]. Date palm propagation is possible by somatic embryogenesis, either with embryos produced from embryogenic callus and then germinated to form complete plantlets [3] or through organogenesis where plantlets originate from multiplied clusters of buds without passing through a callus stage. The basis of the organogenesis technique is the exploitation of meristematic tissue potentialities to form new shoots and to utilize lower concentrations of plant growth regulators in the culture media. This technique consists of


Jameel M. Al-Khayri et al. (eds.), *Date Palm Biotechnology Protocols Volume 1: Tissue Culture Applications*,
Methods in Molecular Biology, vol. 1637, DOI 10.1007/978-1-4939-7156-5_1, © Springer Science+Business Media LLC 2017

four main steps: initiation of vegetative buds, shoot multiplication, plantlet elongation, and rooting. The realization of this technique is highly dependent on the success of the first step (initiation) which requires a well-trained laboratory staff [4]. Most problems encountered in the succeeding stages (multiplication, elongation, rooting) may have their origin at the initiation phase. Furthermore, shoots directly initiated from mother tissue without passing through a callus phase yield regenerated plantlets that are true to type [5, 6]. Recent publications highlight different aspects of date palm micropropagation as to culture media and incubation conditions [7–10], type of explants [8, 11, 12], and plant acclimatization [13].

This chapter describes the methods employed to regenerate complete plantlets from shoot tip explants including selection, removal and preparation of offshoots, explant excision and disinfection, bud formation, shoot multiplication, elongation and rooting of regenerated shoots, and plantlet acclimatization.

## 2  Materials

### 2.1  Plant Material and Sterilization

1. Shoot tips of date palm offshoots.
2. Tools needed for offshoot dissection: knife, chain saw.
3. Antioxidant solution: 100 mg/L ascorbic acid and 150 mg/L citric acid.
4. Fungicide solution: 3 g/L benomyl or mancozeb.
5. Sodium hypochlorite solution: 100% commercial bleach.

### 2.2  Culture Medium for Various Culture Stages

1. Basal culture medium: Murashige and Skoog (MS) [14] medium stock solutions (Table 1). The stock solutions and other ingredients are shown in Table 2.
2. Hormonal supplements: Plant growth regulators and other ingredients used for various date palm cultivars are shown in Table 3, in accordance with each culture stage—culture initiation, shoot formation, shoot multiplication and elongation, or rooting.
3. pH adjustment solutions: 0.5N NaOH and 0.5N HCl.
4. Equipment: Stirrer hot plates, pH meter, precision balance, media dispenser, and 0.1–10 ml pipettes, autoclave.
5. Culture medium vessels: Test tubes (150 × 25 mm) or 250 ml flasks, 170 ml baby food jars, or Magenta containers.

### 2.3  Plant Acclimatization

1. Potting mixture: Peat moss and vermiculite at 1:1 (v/v), dispensed in plastic bags 7 cm wide and 13 cm long.
2. Fungicide solution: Benomyl or Pelt 44 at 1 g/L water.

**Table 1**
**Mineral salts used in date palm micropropagation according to the modified Murashige and Skoog [14] formulation**

| Essential elements | Chemical formula | Concentration in stock solution (g/L) | Concentration in the medium (mg/L) |
| --- | --- | --- | --- |
| *Macroelements (20×)* | | | |
| Ammonium nitrate | $NH_4NO_3$ | 33 | 1650 |
| Potassium nitrate | $KNO_3$ | 38 | 1900 |
| Calcium chloride-$2H_2O$ | $CaCl_2 \cdot 2H_2O$ | 8.8 | 440 |
| Magnesium sulfate-$7H_2O$ | $MgSO_4 \cdot 7H_2O$ | 7.4 | 370 |
| Potassium orthophosphate | $KH_2PO_4$ | 3.4 | 170 |
| Sodium phosphate | $NaH_2PO_4 \cdot 2H_2O$ | 3.0 | 150 |
| *Microelements (200×)* | | | |
| Potassium iodide | KI | 0.166 | 0.83 |
| Boric acid | $H_3BO_3$ | 1.24 | 6.2 |
| Manganese sulfate-$1H_2O$ | $MnSO_4 \cdot H_2O$ | 3.38 | 16.9 |
| Zinc sulfate-$7H_2O$ | $ZnSO4 \cdot 7H_2O$ | 1.72 | 8.6 |
| Sodium molybdate-$2H_2O$ | $Na_2MoO4 \cdot 2H_2O$ | 0.050 | 0.25 |
| Cupric sulfate-$5H_2O$ | $CuSO_4 \cdot 5H_2O$ | 0.005 | 0.025 |
| Cobalt chloride-$6H_2O$ | $CoCl_2 \cdot 6H_2O$ | 0.005 | 0.025 |
| *Iron sources (100×)* | | | |
| Ferrous sulfate-$7H_2O$ | $FeSO_4 \cdot 7H_2O$ | 2.78 | 27.8 |
| Ethylenediaminetetraacetic acid disodium | $Na_2EDTA \cdot 2H_2O$ | 3.73 | 37.3 |
| *Vitamins (100×)* | | | |
| *Myo*-inositol | | 10 | 100 |
| Nicotinic acid | | 0.1 | 1 |
| Pyridoxine hydrochloride | | 0.1 | 1 |
| Thiamine hydrochloride | | 0.1 | 1 |
| Biotin | | 0.0025 | 0.025 |

3. Micro-tunnels: Small shelters of 80 cm of height covered by transparent plastic film to maintain a high relative humidity close to saturation.

4. Greenhouse: Equipped with controlled temperature (28 °C) and relative humidity (85%).

**Table 2**
**Murashige and Skoog culture medium preparation**

| Ingredients | Quantities per liter of medium |
| --- | --- |
| Macroelements | 50 ml |
| Microelements | 5 ml |
| Iron source | 10 ml |
| Vitamins | 10 ml |
| Glutamine | 200 mg |
| Sucrose | 30 g |
| Agar | 8 g |
| Growth regulators | Depending on culture stage (*see* Table 3) |
| pH | 5.7 |

## 3   Methods

### 3.1   Medium Preparation

1. Mix stock solutions of basal culture medium as shown in Tables 1 and 2.
2. Add other ingredients according to the culture stage as shown in Table 3.
3. Dispense medium in culture vessels and autoclave culture medium at 121 °C and 1.1 kg/cm$^2$ pressure for 15–25 min according to the culture medium volume per vessel.

### 3.2   Offshoot Preparation

1. Select disease-free offshoots from a well-known elite adult date palm tree.
2. Isolate the offshoot at the point of attachment to the mother tree without damaging its base (*see* **Note 1**).
3. Remove gradually (one by one) the outer leaves and fibrous tissues at their bases until exposure of the shoot tip zone (Fig. 1a).
4. Excise the shoot tip by cutting a circle around the base of the cylindrical shoot tip at a 45° angle. The ultimate size of the excised shoot tip should be about 3–5 cm in width and 6–10 cm in length (Fig. 1b).
5. Soak the excised shoot tip in the antioxidant solution to avoid tissue browning due to the phenolic compounds.

### 3.3   Shoot Tip Disinfection

1. Clean the excised shoot tip with distilled water to remove any organic debris.

**Table 3**

Culture medium strength (MS) and hormonal additives used for different genotypes and culture stages in date palm micropropagation starting with shoot tip explant

| Cultivar | Culture initiation stage | Multiplication stage | Elongation and rooting stage | References |
|---|---|---|---|---|
| Black Bousthami, Bouskri | MS+ (mg/L): NAA(1), IBA(1), NOA(1–5.5), 2-iP(0.1) | – | – | [15] |
| Boufegous, Bouskri, Black Bousthami | MS+ (mg/L): 2-iP(0.1), IBA(0.5), IAA(0.5), NOA (2) | – | – | [16] |
| Moroccan varieties | MS/2+ (mg/L): NAA(1), IAA(1), NOA(1-5.5), 2-iP (0.1–3) | MS+ (mg/L): NOA(2), NAA(1), IAA (1), BA(0.5), 2-iP(1), Kin(1–5) | MS/2+ (mg/L): NAA(1), BAP (0.5), KIN(0.5) , GA3(1–3) | [6] |
| Maktoom | MS/2+ (mg/L): 2-iP(2), BA(1), NAA(1), NOA(1) | MS+ (mg/L): 2-iP(4), BA(2), NAA(1), NOA(1) (Liquid medium) | MS+ (mg/L): GA3(0.5), NAA (0.1) | [17] |
| Sewy | MS/2+ (mg/L): 2-iP(1), Kin (1) NAA(0.5–1) | MS+ (mg/L): Zeatin (1), NOA(0.5) | – | [18] |
| Zagloul | MS+ (mg/L): 2-iP(2), NAA(0.1) | MS+ (mg/L): 2-iP(4), BA(4), NAA(0.5) | – | [19] |
| Zaidi, Hussaini, Asil | MS/2+ (mg/L): IBA(0.1–6), BA(1) | MS/2+ (mg/L): 2-iP(1), TDZ(0.5) | – | [20] |
| Zagloul | MS+ (mg/L): 2ip (2), NAA(1) | MS+ (mg/L): 2-iP(5), Kin(2) | MS+ (mg/L): NAA(1) | [21] |
| Barhee | MS+ (mg/L): NAA(1), NOA(1), BA(1), 2iP(4) | MS+ (mg/L): NAA (1), BAP (1), 2iP (1.5) | MS+ (mg/L): NAA(1) | [22] |
| Mejhool, Mazafati | MS+ (mg/L): NAA(0.5), NOA (0.5), 2-iP(1), BA(1) | – | – | [7] |
| Quntar | – | – | MS+ (mg/L): NAA(1), GA3(0.5) | [8] |

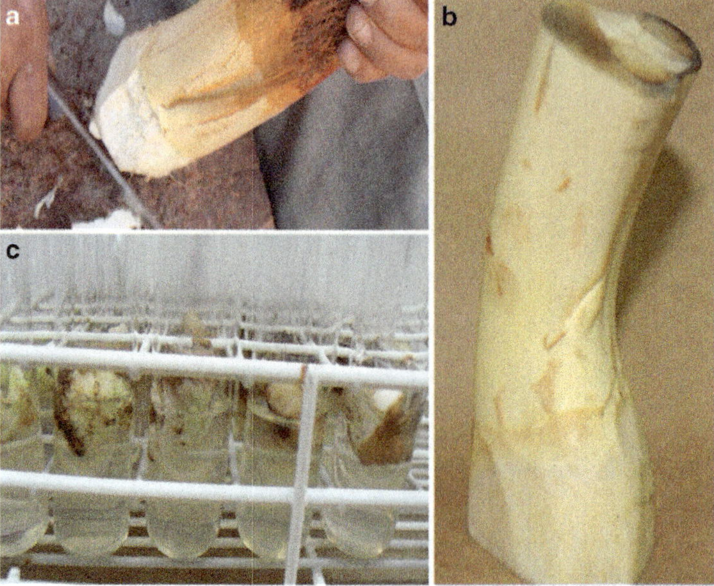

Fig. 1 (a) Offshoot preparation and removal of external leaves. (b) Offshoot shoot tip ready to be disinfected. (c) Explants from shoot tip offshoot at the starting stage

2. Soak in the fungicide solution for 10 min.

3. Rinse three times with sterile distilled water.

4. Soak again in the sodium hypochlorite solution for 20 min.

5. Rinse three times with sterile distilled water under aseptic conditions.

**3.4  Explant Removal**

1. Dissect the sterilized shoot tip using a scalpel and a forceps starting with the gradual removal of the young leaves surrounding the apical dome; cultured explants consist of the proximal ends of the excised leaves.

2. Cut the remaining shoot tip, constituted by very young leaves surrounding the apical dome, into four to six pieces, and transfer them to the culture medium.

3. In general, each offshoot shoot tip yields an average of 15–25 explants.

**3.5  Shoot Formation**

1. Transfer the excised explants, one by one, into starting culture medium (Tables 2 and 3). Make sure that one-half of each explant is imbedded into the culture medium (Fig. 1c).

2. Label each rack with exact references including date, genotype, and culture medium.

3. Incubate in the plant growth room, for 3–6 months, in the dark at 27 °C (*see* **Note 2**).

**Fig. 2** Date palm shoots in the multiplication stage: (**a**) multiple shoots and (**b**) shoot multiplication cultures

4. Check cultures for contaminations after 1 week (*see* **Notes 3–6**).

5. Transfer the explants to fresh culture medium at 1-month intervals. Depending on the genotype and culture medium, shoot formation may require 5–9 months.

***3.6 Shoot Multiplication***

1. Transfer explants showing first bud formation on multiplication media (Tables 2 and 3).

2. Transfer regenerated shoots under a 16-h photoperiod of 14 μmol/m²/s (Fig. 2a, b ).

3. Maintain air temperature in the plant growth room at $27 \pm 1\,°C$ during the illuminated period and at $22 \pm 1\,°C$ during the dark period [23].

4. Split each cluster of buds into two or three small clusters and transfer them to the fresh culture media (*see* **Note 7**).

5. Cut roots at their bases and long leaves to stimulate new bud development.

6. Transfer regenerated shoots to fresh media at 4–5-week intervals (*see* **Note 8**).

***3.7 Shoot Elongation and Rooting***

1. Transfer bud clusters on the shoot elongation medium (Tables 2 and 3).

2. Split bud clusters into smaller ones with two to three buds to enhance shoot elongation (*see* **Note 9**).

3. Keep leaves intact without cutting. Keep roots intact and cut only those exceeding 2–3 cm (Fig. 3a).

4. Separate delicately single shoots having 10 cm in length (Fig. 3b).

5. Transfer single shoots to the rooting medium containing NAA at 0.2–1 mg/L.

6. Transfer to the same fresh medium at 5–6-week intervals.

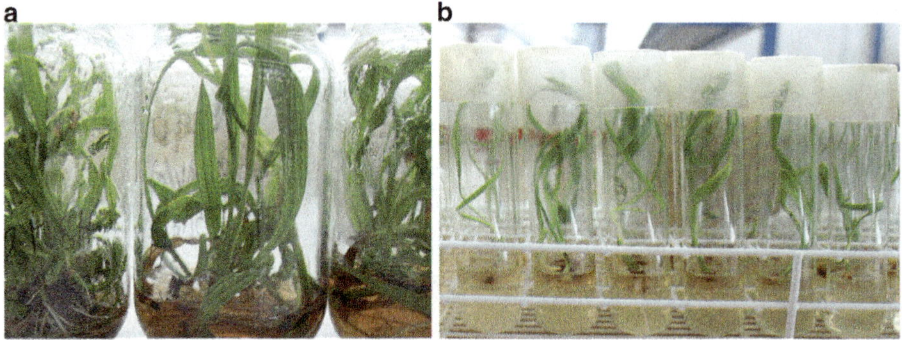

**Fig. 3** (**a**) Shoot elongation. (**b**) Rooting of date palm plantlets

**Fig. 4** (**a**) Well-formed plantlets ready for transfer to the greenhouse. (**b**) Well-acclimatized in vitro plants

7. Incubate cultures in the plant growth room at $27 \pm 1$ °C during the illuminated period and at $22 \pm 1$ °C during the dark period. Maintain a 16-h photoperiod of 40 $\mu mol/m^2/s$ at this step.

***3.8 Plant Acclimatization and Hardening***

1. Select plantlets having required characteristics (2–3 fully opened leaves, 3–4 roots, 10–15 cm long with a well-formed and closed crown) for high survival rate of plants in the greenhouse (Fig. 4a, b) (*see* **Note 10**).

2. Gently remove adhering culture medium from roots by washing in tap water.

3. Dip the roots in the fungicide solution for 5 min.

4. Transfer the plantlets to the potting soil mixture (1:1 peat moss/vermiculite v/v).

5. Incubate plantlets in the greenhouse at 28 °C.

6. Maintain a high relative humidity (RH > 90%) around plantlets by using micro-tunnels (*see* **Note 11**).

7. Control the moisture of substrate by regular watering.

8. Control fungi attacks on plant leaves and crowns by regular spraying of fungicides, 0.5 g/L Pelt 44.

9. Open gradually the micro-tunnels after 4–5 weeks.

10. Maintain plants in the greenhouse for 12 months before transferring to the field. Plantlets have to be regularly watered using a 10% MS solution twice a week.

## 4　Notes

1. While removing offshoots, special care must be given to its connection to the mother tree. In fact, some seedlings can grow up near the trunk and can be confused with true offshoots. Skilled laborers are required to cut and remove offshoots properly without damaging their bases. Suitable offshoots for in vitro culture may have an average weight of 3–6 kg. After removal, clean the offshoots and severely cut all roots and leaves before transportation to the lab. The best time for initiating in vitro cultures from offshoot shoot tips is between the end of date fruit harvesting (November) and the start of the next flowering stage (March) [6].

2. Culturing in the dark during the shoot formation stage enhances bud initiation and prevents oxidation of phenolic compounds that tend to accumulate under light conditions.

3. At the starting stage, avoid contamination, as most of explants are vulnerable to endophytic bacteria. Before explant transfer to fresh media, stringent screening of cultures for contaminants is necessary. Visually check each individual culture under light for any signs of contamination. Contaminated cultures are identifiable by various colors and forms of bacterial colonies on the culture medium or on cultured tissue. Discard all contaminated or suspicious cultures at the beginning of the multiplication process.

4. Date palm tissues contain high levels of caffeoylshikimic acids [24]. These polyphenols accumulate in the culture medium and turn brown over time. Once oxidized by polyphenoloxydases, they form quinones that are highly toxic to cultured tissue. The explant wounding enhances their secretion while preparing explants for in vitro culture or when transferred onto fresh culture media. Several suggestions have been made to reduce polyphenol secretion in the culture medium:

    (a) Presoaking of plant material in antioxidant solution (ascorbic acid, 100 mg/L, and citric acid, 150 mg/L) until transfer onto culture media.

(b) Addition of adsorbents like adenine, glutamine and citrate, activated charcoal, or polyvinylpirrolidone to the culture media [16].

(c) The use of small-sized explants (5 mm length), juvenile tissues (tissues with less lignification), and frequent transfers (3–4-week intervals) onto fresh culture media.

(d) Incubation in the dark during the first 3 months at the starting stage to minimize tissue browning.

5. At the multiplication stage, early rooting decreases the rate of bud multiplication by diverting most tissue nutrients to root formation rather than shoot formation [5]. Studies have shown that root initiation occurs in the presence of high auxin concentrations, 3 mg/L NAA. In contrast, root elongation increases on culture media containing low auxin concentration, 0.1 mg/L. Moreover, time of the year is crucial when explants are cultured for initiating in vitro cultures and seems to have an effect on root formation. Explants cultured, respectively, in July and December (Northern Hemisphere), exhibited rooting percentages of 16 and 40% [25]. Furthermore, low concentrations of mineral nutrients in culture media and incubation of cultures in the dark for a long period also lead to early rooting of buds [5]. All these constraints need attention in date palm tissue culture handling to guarantee a successful establishment of reactive shoots and a good rate of bud multiplication.

6. Date palm tissue cultures are sometimes highly contaminated with endogenous bacteria. Even well-disinfected cultured explants bring with them contaminants in vitro. Numerous studies have confirmed the existence of endogenous bacteria in apparently healthy offshoot tissues. Isolation and identification of these contaminants have shown that they belong to the *Bacillus* genus. Their most important characteristic is endospore formation, which can survive at 80 °C for 30 min and observed after 1 month of culture or even after three to five subcultures. The incidence of this contamination can increase from 20 to 50% of cultures. Antibiotics such as tetracycline (30 µg/ml), streptomycin (10 µg/ml), neomycin (20 µg/ml), and chloramphenicol (30 µg/ml) can control these contaminants [26]. However, in practice, discarding the contaminated cultures at the beginning of the multiplication stage is the best way to control this problem.

7. At the multiplication stage, large containers (370 ml or larger) are preferred over test tubes. In fact, clusters of buds give a high multiplication rate in large containers compared to test tubes. Furthermore, good shoot development occurs when two or three bud clusters are grown in the same container. However, it is very important to check carefully the health of cultures

before transferring each to fresh media to avoid the spread of contaminants.

8. Date palm tissue cultures are susceptible to a vitrification phenomenon. Characterizing this physiological disorder is the development of tissue with lignification deficiency due to an accumulation of water in the cultured tissue [5]. Many factors can enhance this disorder:

   (a) High concentration of plant growth regulators (mainly cytokinins) and some mineral salts (especially ammonium ions) in the culture medium. A study conducted on cv. Aguellid showed that medium containing a high concentration of ammonium nitrate (20,600 meq/L) enhanced shoot growth and causing levels of vitrification to rise from 46 to 53%. In contrast, on culture media containing moderate concentrations of ammonium nitrate (1950 meq/L), vitrification percentages were reduced to 14–19% [27].

   (b) High humidity and accumulation of gases (particularly ethylene) inside culture containers and the use of liquid media [5].

   Measures to reduce vitrification are the following:

   (a) Reduction of hormonal and ammonium concentrations in the media.

   (b) The use of container covers that allow proper release of gases.

   (c) The use of solid instead of liquid media.

   (d) Increasing agar concentration.

9. At the elongation stage, avoid separation of single buds to prevent loss of a large number of buds. Furthermore, in separating a single shoot from the cluster, it is very important to cut in a suitable place to avoid damaging plantlets.

10. Sometimes two plantlets are difficult to separate from each other without losing one of them. They can be transferred together to the greenhouse and easily separated once fully acclimatized.

11. The artificial conditions during in vitro culture result in abnormal plantlet regeneration showing abnormal morphology, anatomy, and physiology. After ex vitro transfer, they are vulnerable to damage and reduced growth from sudden changes in environmental conditions. Growing acclimatizing plants under high relative humidity overcomes withering. The acclimatization phase of in vitro plantlets is crucial in date palm micropropagation.

## References

1. Jayanthi M, Susanthi B, Mohan NM, Mandal PK (2015) In vitro somatic embryogenesis and plantlet regeneration from immature male inflorescence of adult dura and tenera palms of *Elaeis guineensis* (Jacq.) Springer Plus 4:256–262

2. Solís-Ramos LY, Andrade-Torres A, Carbonell LAS, Salín CMO, Serna EC (2012) Somatic embryogenesis in recalcitrant plants. In: Sato KI (ed) Embryogenesis. InTech, Rijeka, Croatia, pp 597–618

3. Al Khayri JM (2007) Date palm (*Phoenix dactylifera* L.) micropropagation. In: Jain SM, Haggman H (eds) Protocols for micropropagation of woody trees and fruits. Springer, Netherlands, pp 509–526

4. Abahmane L (2011) Date palm micropropagation via organogenesis. In: Jain SM, Al-Khayri JM, Johnson D (eds) Date palm biotechnology. Springer, Dordrecht, pp 69–90

5. Al Khateeb AA (2006) Role of cytokinin and auxin on the multiplication stage of date palm (*Phoenix dactylifera* L.) cv. Sukary. Biotechnology 5(3):349–352

6. Beauchesne G, Zaid A, Rhiss A (1989) Meristematic potentialities of bottom of young leaves to rapidly propagate date palm. In: Proceedings of the second symposium on date palm, King Faisal University, Riyadh, 3–6 March 1986

7. Rad MR, Zarghami R, Hassani H, Zakizadeh H (2015) Comparison of vegetative buds formation in two date palm cultivars, Medjool and Mazafati through direct organogenesis. Int J Farm Alli Sci 4(6):549–553

8. Al Mayahi AMW (2015) An efficient protocol for indirect somatic embryogenesis and shoot organogenesis from leaf segment of date palm (*Phoenix dactylifera* L.) cv. Quntar. Afri J Agric Res 10(10):1031–1042

9. Meziani R, Jaiti F, Mazri MA, Anjarne M, Ait Chitt M, El Fadile J, Alem C (2015) Effects of plant growth regulators and light intensity on the micropropagation of date palm (*Phoenix dactylifera* L.) cv. Mejhoul. J Crop Sci Biotech 18(5):325–331

10. Abdulwahed MS (2013) Identification of the effect of different levels of activated charcoal and sucrose on multiplication shoots of date palm (*Phoenix dactylifera* L.) cv. Sufedy *in vitro*. J Hortic Forest 5(9):139–145

11. Fki L, Bouaziza N, Kriaa W, Benjemaa-Masmoudi R, Gargouri-Bouzid R, Rival A, Drira N (2011) Multiple bud cultures of 'Barhee' date palm (*Phoenix dactylifera* L.) and physiological status of regenerated plants. J Plant Physiol 168(14):1694–1700

12. Ibrahim MA, Waheed AM, Al-Taha HA (2013) Plantlet regeneration from root segments of date palm tree (*Phoenix dactylifera* L.) cv. Barhee producing by *in vitro* culture. AAB Bioflux 5(1):45–50

13. Mona MH, Ibrahim MA, Ebrahim MKH, Komor E (2014) Protocol for micropropagated date palm acclimatization: effect of micropropagated plantlet type, soil composition, and acclimatization season. Int J Fruit Sci 14(2):225–233

14. Murashige T, Skoog F (1962) A revised medium for rapid growth and bioassays with tissue cultures. Phys Plant 15:473–497

15. Poulain CA, Rhiss A, Beauchesne G (1979) Multiplication végétative en culture *in vitro* du palmier dattier (*Phoenix dactylifera* L.) C R Acad Agric 11:1151–1154

16. Beauchesne G (1983) Vegetative propagation of date palm (*Phoenix dactylifera* L.) by *in vitro* culture. In: Proceedings of the first symposium on date palm, King Faisal University, pp 698–700

17. Khierallah HSM, Bader SM (2007) Micropropagation of date palm (*Phoenix dactylifera* L.) cv. Mektoom through direct organogenesis. Acta Hort (736):213–224

18. Wanas WH, El Hammady AM, Abo Rawash M, Awad AA (1999) *In vitro* propagation of date palm. 1—Direct organogenesis as affected by cytokinin and auxin levels in the medium. Ann Agric Sci 44(2):19–31

19. Hussain I, Rashid H, Muhammad A, Quraishi A (2001) *In vitro* multiplication of date palm. In: Proceedings of the second International Conference on Date Palm, Al Ain University, pp 432–438

20. Bekheet S (2013) Direct organogenesis of date palm (*Phoenix dactylifera* L.) for propagation of true-to-type plants. Sci Agri 4(3):85–92

21. Bekheet SA, Saker MM (1998) *In vitro* propagation of Egyptian date palm: II—Direct and indirect shoot proliferation from shoot tip explants of *Phoenix dactylifera* L. cv. Zaghlool. In: Proceedings of the first international conference on date palm, Al Ain University, pp 150–157

22. Jazinizadeh E, Zarghami R, Majd A, Iranbakhsh A, Tajaddod G (2015) *In vitro* production of date palm (*Phoenix dactylifera* L.) cv. 'Barhee' plantlets through direct organogenesis. Biol Forum 7(2):566–572

23. Anjarne M, Abahmane L, Bougerfaoui M (2005) Les techniques de micropropagation du palmier dattier (*Phoenix dactylifera* L.): Expérience de l'INRA-Maroc. In: Proceedings of the international symposium on sustainable development of oases systems, INRA-Morocco, pp 86–93

24. Loutfi K, El Hadrami I (2005) *Phoenix dactylifera* date palm. In: Litz RE (ed) Biotechnology of fruit and nut crops. CABI Publishing, Wallingford, pp 144–157

25. Anjarne M, Zaid A (1993) Effets de certains équilibres hormonaux sur l'enracinement précoce des tissus du palmier dattier (*Phoenix dactylifera* L). Al Awamia 82:197–210

26. Leary JV, Nelso N, Tisserat B, Allingham EA (1986) Isolation of pathogenic *Bacillus circulans* from callus cultures and healthy offshoots of date palm (*Phoenix dactylifera* L.) Appl Envir Microbiol 52(5):1173–1176

27. Bougerfaoui M, Zaid A (1993) Effet de la teneur du milieu de culture en ammoniaque sur la vitrification des tissus du palmier dattier cultivés *in vitro*. Al Awamia 82:177–196

# NAA-Induced Direct Organogenesis from Female Immature Inflorescence Explants of Date Palm

Hussam S.M. Khierallah, Saleh M. Bader, and Makki A. Al-Khafaji

## Abstract

Micropropagation has great potential for the multiplication of female and male date palms of commercially grown cultivars by using inflorescences. This approach is simple, convenient, and much faster than the conventional method of using shoot-tip explants. We describe here a stepwise micropropagation procedure using inflorescence explants of Iraqi date palm cultivar Maktoom. Cultured explants were derived from 0.5-cm-long spike segments excised from 8 to 10-cm-long spathes. About 70% formed adventitious buds on Murashige and Skoog (MS) medium supplemented with 2 mg/L naphthalene acetic acid (NAA), 4 mg/L benzylaminopurine (BAP), and 40 g/L sucrose and maintained in the dark for 16 weeks before transferring to normal light conditions. The best multiplication rate was achieved with 3 mg/L 2ip and 2 mg/L; for shoot elongation, the best medium is MS containing 0.5 mg/L BAP, 0.5 mg/L 2ip, and 1 mg/L $GA_3$. Well-developed shoots were cultured for rooting in half MS medium amended with 1 mg/L NAA and 45 g/L sucrose. Plantlets with well-developed roots were successfully hardened in the greenhouse. Inflorescence explants proved to be a promising alternative explant source for micropropagation of date palm cultivars.

**Key words** Adventitious bud formation, Inflorescence, In vitro, Hardening

## 1   Introduction

Traditional propagation of the date palm (*Phoenix dactylifera* L.) is carried out by offshoots, which is an inefficient way to establish new date palm plantations because each tree produces limited numbers of offshoots. Moreover, several genotypes bear offshoots that are difficult to root or produce none at all. Seed-propagated plants produce heterogeneous off-types and require up to 7 years to bear fruit [1]. Thus, there is a great need for rapid and efficient vegetative propagation systems for elite genotypes. For decades, offshoots have been used in date palm tissue culture [2–4]. The potential of inflorescence explants has been tested to develop direct and indirect formation of somatic embryos and organogenesis.

Jameel M. Al-Khayri et al. (eds.), *Date Palm Biotechnology Protocols Volume 1: Tissue Culture Applications*, Methods in Molecular Biology, vol. 1637, DOI 10.1007/978-1-4939-7156-5_2, © Springer Science+Business Media LLC 2017

Inflorescence explants have proved useful in avoiding many constraints that face shoot-tip explants, like high percentage of contamination, heavy browning, and long initiation stage.

Inflorescence-based micropropagation holds great potential for the multiplication of individual recalcitrant female and male date palms and cultivars of commercial interest and is particularly useful when offshoot availability is limited. This type of propagation can be accomplished in a short time with minimal effort as compared to the traditional practice of using shoot-tip explants [5]. Early reports of using floral explants for date palm micropropagation demonstrated the varied potential of female floral initials. First, these initials were able to change from the floral state to vegetative outgrowths by different methods, all of which were correlated to the stage of differentiation at the time of excision. Second, they changed according to the composition of the initial nutrient medium, especially to the growth regulator formula used [6–8]. These floral initials were able to develop directly into complete plantlets. Floral initials also have the potential for induction of adventitious buds or somatic embryos, which in turn develop into complete and separate plantlets [9]. Therefore, Abahmane et al. [10] suggested that a micropropagation technique using tissues of floral spikes at an early stage of growth aided in converting explants from a floral to a vegetative state. Moreover, this early stage can vary by cultivar, climatic conditions, and nutritional status of the parent tree [11]. Kriaa et al. [12] developed a protocol based on the use of mature female flowers, collected at the latest developmental stage before the spathe opens, which is a useful method to avoid tree damage. The major trends of in vitro responding inflorescence explants are direct shoots from initial explants [13, 14], direct and indirect somatic embryogenesis [12, 15, 16], and unfriable callus [17].

The protocol described herein is a stepwise micropropagation procedure using inflorescence explants comprising excision of immature inflorescences from an adult tree at the appropriate time and establishment of the initial explants on a starting medium after surface sterilization of the spathe. Also included are other details on shoot initiation, multiplication, elongation, rooting, hardening, and acclimatization of the Iraqi date cv. Maktoom.

## 2   Materials

### 2.1   *Plant Material and Disinfection*

1. Plant material: Date palm cv. Maktoom floral spathes from adult female tree, 8–10 cm in length (Fig. 1a).

2. Disinfection solution: 0.1% mercuric chloride ($HgCl_2$) solution.

**Fig. 1** In vitro micropropagation stages of date palm using inflorescence explants. (**a**) Floral spathe of 8–10 cm in length used. (**b**) Floral spikes and explant types. (**c**) Adventitious bud formation after 24 weeks of culture. (**d**) Bud multiplication directly from floral explants. (**e**) Elongation. (**f**) Rooting. (**g**) Plantlets acclimatization. (**h**) Plantlets grown under greenhouse conditions

**Table 1**
**Stock solutions of inorganic nutrients, iron source, vitamins, and other additives of modified MS medium [18] used for in vitro culture of inflorescence explants of date palm**

| Components | g/L stock solution | mg/L medium |
|---|---|---|
| *1. Nitrate stock solution* | | |
| $NH_4NO_3$ | 165 | 1650 |
| $KNO_3$ | 190 | 1900 |
| *2. Sulfate stock solution* | | |
| $MgSO_4 \cdot 7H_2O$ | 37 | 370 |
| $MnSO_4 \cdot H_2O$ | 1.69 | 16.9 |
| $ZnSO_4 \cdot 7H_2O$ | 0.86 | 8.6 |
| $CuSO_4 \cdot 5H_2O$ | 0.0025 | 0.025 |
| *3. Halides stock solution* | | |
| $CaCl_2 \cdot 2H_2O$ | 44 | 440 |
| $CoCl_2 \cdot 6H_2O$ | 0.0025 | 0.025 |
| KI | 0.083 | 0.83 |
| *4. B.P.Mo stock solution* | | |
| $H_3BO_3$ | 0.62 | 6.2 |
| $KH_2PO_4$ | 17 | 170 |
| $Na_2MoO_4 \cdot 2H_2O$ | 0.025 | 0.25 |
| *5. Chelated iron stock solution* | | |
| $FeSO_4 \cdot 7H_2O$ | 2.78 | 27.8 |
| $EDTA \cdot Na_2$ | 3.73 | 37.3 |
| *6. Vitamins stock solution* | | |
| Thiamine-HCl | 0.1 | 1 |
| Nicotinic acid | 0.1 | 1 |
| Pyridoxine-HCl | 0.1 | 1 |
| Glycine | 0.2 | 2 |
| Ca-pantothenate | 0.1 | 1 |
| Biotin | 0.01 | 0.1 |
| *7. Other additives add directly to the medium mixture* | | |
| $NaH_2PO_4 \cdot 2H_2O$ | | 170 |
| Adenine hemisulfate | | 40 |
| Glutamine | | 100 |
| Polyvinylpyrrolidone (PVP) | | 2000 |

***2.2 Culture Medium***

1. Basal culture medium: Murashige and Skoog (MS) [18] medium stock solutions (MS stock I, II, III, and IV) (Table 1, *see* **Note 1**).

2. Hormone stock solutions: BAP, 2-isopentenyladenine (2iP), and NAA (1 mg/mL).

3. Adventitious bud formation (ABF) medium: Basal culture medium supplemented with 2 mg/L NAA and 4 mg/L BAP.

4. Bud multiplication (BM) medium: Basal culture medium supplemented with 3 mg/L 2 iP and 2 mg/L BAP.

5. Shoot elongation (SE) medium: Basal culture medium supplemented with 0.5 mg/L BAP, 1.0 mg/L NAA, and 0.05 mg/L $GA_3$.

6. Rooting (RT) medium: Half strength MS salts, 45 g/L sucrose, and NAA 1 mg/L.

7. pH adjustment solutions: 1 N NaOH/1 N HCl.

**2.3 Acclimatization**

1. Potting mixture: Peat moss and perlite at 5:1 (v:v).

2. Fungicide solution: Beltanol 45% and Carbendazim 50% WP, each 1 mL/L.

3. Fertilizer powder: Grow More Fertilizer N-P-K 20-20-20 at 3 g/L.

**2.4 Equipment**

1. Instruments: pH meter, weighing balances, rotary shaker, hot plate, magnetic stirrer, and glass bead sterilizer.

2. Dissection tools: Forceps 30 cm straight, scalpel holder no. 4, and surgical blade nos. 11 and 22.

3. Glassware: 1 L conical flask, 500 mL beaker, 1 mL pipette, and 10 mL graduated cylinder (*see* **Note 2**).

**Table 2**
**Additives supplemented to Murashige and Skoog (MS) medium according to various stages of date palm micropropagation from inflorescence explants**

| Culture stage (medium code) | Composition (mg/L) | | | |
| --- | --- | --- | --- | --- |
| | Basal medium | Additives | Auxin | Cytokinin |
| Adventitious bud formation (ABF) | MS | 40,000 sucrose + 3000 Phytagel MS vitamins + 170 $NaH_2PO_4$ + 100 glutamine + 40 adenine sulfate | 2 NAA | 4 BAP |
| Bud multiplication (BM) | MS | 40,000 sucrose + 7000 agar vit. MS + 170 $NaH_2PO_4$ + 200 glutamine + 40 adenine sulfate | – | 3 2ip 2 BAP |
| Shoot elongation (SE) | MS | 40,000 sucrose + 7000 agar MS vitamins + 170 $NaH_2PO_4$ + 200 glutamine + 80 adenine sulfate | 1 NAA | 0.05 $GA_3$ 0.5 BAP |
| Rooting (RT) | ½ MS | 45,000 sucrose + 7000 agar MS vitamins + 170 $NaH_2PO_4$ + 100 glutamine + 80 adenine sulfate | 1 NAA | – |

## 3  Methods

### 3.1  Medium Preparation

1. Pour 800 mL double-distilled water into 1 L conical flask. Add 10 mL of each stock solution of inorganic nutrients, iron source, and vitamins (Table 1). Dissolve other additives by adding directly into the medium, except agar. Add appropriate amount of plant growth regulators according to the developmental stage (Table 2, *see* **Note 3**).

2. Adjust pH of the medium to 5.7 with 0.1 N NaOH or HCl, before addition of agar. Heat until boiling, dispense 25 mL medium per culture test tube, and cover with polypropylene caps. Sterilize all tubes with media in an autoclave at 121 °C and 1.04 $kg/cm^2$ for 15 min.

3. Culture vessels: Rimless 200 × 25 mm glass test tubes with polypropylene caps, 250 mL Erlenmeyer flask.

### 3.2  Explant Disinfection and Excision

1. Excise floral spathes encasing the immature inflorescence of adult trees of cv. Maktoom collected in early spring (Northern Hemisphere), preferably during the second week of February.

2. Surface-sterilized spathes in 0.1% mercuric chloride ($HgCl_2$) solution containing two drops of Tween-20 for 10 min under aseptic conditions and rinse three times with sterile distilled water.

3. Excise spikes, 3–5-cm-long from 8 to 10-cm-long spathes. Cut the excised spikes into 0.5 cm segments (Fig. 1b) (*see* **Note 4**).

### 3.3  Culture Initiation and Multiplication

1. Culture explants on ABF medium and incubate in the dark at 26 ± 2 °C for 6 weeks (Fig. 1c).

2. Observe formation of adventitious buds resulting from direct organogenesis (approx. Three bud clusters per explant) (*see* **Note 5**).

3. For bud multiplication, transfer initial bud clusters to BM medium and incubate cultures at 16-h photoperiod of cool-white florescent light (40 $\mu mol/m^2/s$) and 27 ± 2 °C (*see* **Note 6**).

### 3.4  Shoot Elongation and Rooting

1. For shoot elongation (SE), transfer the buds, formed from the bud multiplication cultures, to SE medium, and maintain for 6 weeks at 27 ± 2 °C and 16-h photoperiod of cool-white florescent light (40 $\mu mol/m^2/s$).

2. Transfer them to a hormone-free medium for 6 weeks and maintain under the same environmental conditions.

3. Observe shoot elongation reaching a length of about 8.65 cm within this period (Fig. 1e).

4. Isolate in vitro 8-cm-long shoots and culture on RT medium (*see* **Note 7**).

***3.5 Hardening and Acclimatization***

1. Gently rinse the roots of regenerated plantlets with sterile distilled water to remove adhering medium under aseptic conditions. Transfer them to half strength MS medium devoid of sugar and cap with aluminum foil (*see* **Note 8**), and maintain them under high light intensity (185 $\mu$mol/m$^2$/s), provided by cool-white florescent light for 16-h photoperiod, at $27 \pm 2$ °C.

2. Transfer plantlets to the greenhouse, maintained at 30 °C, 40–60% relative humidity, and natural sunlight. Wash the nutrient media adhering to the root system by immersion in sterilized water.

3. Dip plantlets in 1 mL/L systematic fungicide (Beltanol) solution for 1–2 min, and plant them directly in pots ($5 \times 10$ cm) filled with autoclaved peat moss and pearlite (5:1) (v:v). Place planted pots under low transparent plastic tunnels and maintain at 30–35 °C and 90–100% relative humidity.

4. Open the tunnel after 7 days for 10–15 min for ventilation and spray 1 mL/L Carbendazim fungicide. Also, remove dead plants and cut off leaves infected with fungus (Fig. 1g) (*see* **Note 9**).

5. After 2 weeks of planting, open the tunnel by increasing the openings gradually during the subsequent 2 weeks.

6. After 2–3 months of planting, fertilize and irrigate plants as required 2–3 g/L N-P-K fertilizer (20-20-20) gives good results with irrigation and give only if necessary. Transfer successful plants into larger pots $20 \times 25$ cm (Fig. 1h).

7. When plants are 12–18 months old, compound leaves begin to appear. The plants push thick white roots through the pores at the pot bottom within 5–6 months at which time they should again be transplanted into larger pots.

# 4    Notes

1. The most efficient way of preparing MS medium (Table 1) is to prepare stock solutions of inorganic nutrients, iron source, vitamins, and, separately, plant growth hormones. Store them at 4 °C except for the vitamins, which are stored in small batches at $-20$ °C; store stock solutions up to 2–3 months. It is always recommended to prepare fresh plant growth regulator solutions. Any color changes in the stock solutions may be due to precipitation, which can seriously affect the growth

of cultures. Alternatively, use prepared MS medium that is commercially available.

2. Clean all the glassware with liquid detergent and thoroughly wash with tap water. Rinse the glassware with double-distilled water and dry it in a hot air oven at 160 °C for 2 h before use.

3. For preparing stock solutions of BAP, 2iP, and NAA, dissolve 100 mg of each in 3–5 mL 1 N NaOH, and raise the volume to 100 mL by adding double-distilled water. Store in the freezer at −20 °C.

4. The best source of explant is spikes, 3–5 cm long, from Maktoom cv., cut from spathes of 8–10 cm length. Reversion of date palm floral tissue to a vegetative state can take place under tissue culture conditions. Floral initials have the potential for the induction of adventitious buds or somatic embryos, which in turn developed into complete plantlets.

5. The positive effect on shoot formation of ABF medium, characterized by low auxin/cytokinin ratio, indicates that date palm floral explants behave differently from offshoot tissue that requires high concentration of auxin for inducing bud formation and tissue development [3]. Differences between floral and vegetative explants in endogenous hormone contents and their accumulation and sensitivity to exogenous hormones could explain these different requirements.

6. The number of multiplied buds can be increased to 6.6 by raising glutamine concentration up to 200 mg and agitation of the liquid medium (Fig. 1d).

7. The rooting percentage response was 90% for Maktoom cv. (Fig. 1f).

8. Increase, as needed, ventilation between the inside and outside of the tube through gradually punching holes in the aluminum foil, followed by complete removal a few days before transplanting. This gradual approach is beneficial to reduce the relative humidity in the tubes and to increase epicuticular wax development on the leaves.

9. Healthy well-developed date palm plantlets are transplanted into different soil beds in the greenhouse for acclimatization. Many soil mixtures serve as a substrate for date palm acclimatization [14, 19]. Plantlets, selected for this procedure, should have two to three erect leaves, with an average height of 10–20 cm and with well-developed adventitious root systems. These plantlets are transplanted into special pots, 5 cm diameter $\times$ 18 cm in height, containing vermiculite as the soil bed. Placement of small stones at the bottom of pots promotes root growth due to effective drainage of irrigation water.

## References


1. Khierallah HSM, Bader SM, Al-Jboory IJ, Ibrahim KM (2015) Date palm status and perspective in Iraq. In: Al-Khayri JM, Mohan SM, Johnson DV (eds) Date palm genetic resources and utilization. Asia and Europe, vol 2. Springer, Dordrecht, pp 97–152

2. Tisserat B (1979) Propagation of date palm (*Phoenix dactylifera* L.) *in vitro*. J Exp Bot 30:1275–1283

3. Omar MS, Hameed MK, Al-Rawi MS (1992) Micropropagation of date palm (*Phoenix dactylifera* L.). In: Bajaj YPS (ed) Biotechnology in agriculture and forestry. High-tech and micropropagation II, vol 18. Springer-Verlag, Berlin, pp 471–492

4. Khierallah HSM, Bader SM (2007) Micropropagation of date palm (*Phoenix dactylifera* L.) var. Maktoom through direct organogenesis. Acta Hort (736):213–224

5. Abul-Soad AA (2011) Micropropagation of date palm using inflorescence explants. In: Jain SM, Al-Khayri JM, Johnson DV (eds) Date palm biotechnology. Springer, Dordrecht, pp 91–118

6. Drira N (1983) Multiplication vegetative du palmer dattier (*Phoenix dactylifera* L.) par la culture in vitro de bourgeons axillaries et de feuilles que en derivent. CR Acad Sci Paris 296:1077–1082

7. Drira N, Benbadis A (1985) Multiplication végétative du palmer dattier (*Phoenix dactylifera* L.) par révérsion, en culture in vitro, d'ébauches florales de pieds femelles. J Plant Physiol 119:227–235

8. Drira N, Al-Sha'ary A (1993) Analysis of date palm female floral initials potentials by tissue culture. In: Proceedings of the third symposium on date palm, Saudi Arabia. King Faisal University, Al-Hassa, pp 161–170

9. Bhaskaran S, Smith RH (1992) Somatic embryogenesis from shoot tip and immature inflorescence of *Phoenix dactylifera* cv. Barhee. Plant Cell Rep 12:22–25

10. Abahmane L, Bougerfaoui M, Anjarne M (1999) Use of tissue culture techniques for date palm propagation and rehabilitation of palm groves devastated by bayoud disease. In: Proceeding of international symposium on date palm, Assiut University, Assiut, Egypt, 9–11 Nov. pp 385–388

11. Abul-Soad AA, El-Sherbeny NR, Baker SI (2007) Effect of basal salts and sucrose concentrations on morphogenesis in test tubes of female inflorescence of date palm (*Phoenix dactylifera* L.) cv. Zaghloul. Egypt J Agric Res 85 (1B):385–394

12. Kriaa W, Sghaier B, Masmoudi F, Benjemaa R, Drira N (2012) The date palm (*Phoenix dactylifera* L.) micropropagation using completely mature female flowers. C R Biol 335:194–204

13. Khierallah HSM (2007) Micropropagation of two date palm (*Phoenix dactylifera* L.) cultivars using inflorescences and study of the genetic stability using AFLP-PCR markers. PhD dissertation, College of Agriculture, University of Baghdad, Bagdad, Iraq

14. Abul-Soad AA, Mahdi SM (2010) Commercial production of tissue culture date palm (*Phoenix dactylifera* L.) by inflorescence technique. J Genet Eng Biotech 8(2):39–44

15. Abul-Soad AA (2012) Influence of inflorescence explant age and 2,4-D incubation period on somatic embryogenesis of date palm. Emir J Food Agric 24(5):434–443

16. Sidky RA, Eldawyati MM (2012) Proliferation of female inflorescences explants of date palm. Ann Agric Sci 57(2):161–165

17. Zayed EMM, Abdelbar OH (2015) Morphogenesis of immature female inflorescences of date palm *in vitro*. Ann Agric Sci 60(1): 113–120

18. Murashige T, Skoog F (1962) A revised medium for rapid growth and bioassays with tobacco tissue cultures. Physiol Plant 15:473–497

19. Hegazy AE, Kansowa OA, Abul-Soad AA, Nasr MI (2006) Growing behaviors of ex vitro date palm plants after acclimatization. In: Second international conference of genetic engineering and its applications. Sharm El-Sheik City, South Sinai, Egypt, 14–17 November 2006, pp 69–75

# Chapter 3

# Direct Organogenesis from Immature Female Inflorescence of Date Palm by Gradual Reduction of 2,4-D Concentration

Ezz El-Din G. Gadalla

## Abstract

Inflorescences represent an alternative explant source for superior date palm trees, especially those that do not produce offshoots. They provide large numbers of explants free of fungal and bacterial contamination for successful tissue culture initiation. Furthermore, they are characterized by the capacity of plant regeneration within a short time as compared to other explant types. This chapter focuses on the procedures employed for plant regeneration by direct organogenesis using immature female inflorescence explants, including initiation of adventitious buds, differentiation, multiplication, shoot elongation, rooting, and acclimatization. Adding 5 mg/L 2,4-dichlorophenoxyacetic acid (2,4-D) into the initiation medium and gradually reducing it to 1 and then to 0.5 mg/L in the subsequent 2 subcultures, respectively, are determining factors in direct adventitious bud formation from the inflorescence. Bud differentiation is obtained on MS medium containing 0.25 mg/L kinetin (Kin), 0.25 mg/L benzyladenine (BA), 0.25 mg/L abscisic acid (ABA), 0.1 mg/L naphthaleneacetic acid (NAA), and 0.2 g/L activated charcoal (AC). Regenerated shoots exhibit sufficient root formation on MS medium supplemented with 2 mg/L indole butyric acid (IBA) and 1 mg/L NAA and subsequent survival in the greenhouse.

**Key words** Acclimatization, Direct organogenesis, Immature female inflorescence, In vitro, Micropropagation, Tissue culture

## 1   Introduction

Micropropagation of date palm mainly relies on shoot tip explants [1]. However, the use of date palm shoot tips as explants for plant regeneration involves sacrificing one complete offshoot with no guarantee of the desired results. Additionally, offshoots are limited in number, are expensive, and involve substantial manual work to excise the actual explant of the apical dome. Moreover, contamination, browning of the explant, and delayed differentiation of embryonic callus are major limitations in the micropropagation of date palm using offshoot explants [2, 3].

Immature inflorescences are important explant source for elite date palm micropropagation. Inflorescence tissue can be collected with limited damage to the upper parts of the donor plant.

Jameel M. Al-Khayri et al. (eds.), *Date Palm Biotechnology Protocols Volume 1: Tissue Culture Applications,*
Methods in Molecular Biology, vol. 1637, DOI 10.1007/978-1-4939-7156-5_3, © Springer Science+Business Media LLC 2017

Immature inflorescences are protected by sheaths preventing fungal and bacterial infestation and chemical damage associated with sterilization solution [4]. Several researchers have used them as an alternative source of explants in date palm micropropagation [5–10]. They also have the potential for the induction of adventitious buds or somatic embryos, which in turn develop into complete plantlets [11]. Abahmane et al. [12] suggested that the use of floral tissue of floral spikes, at an early stage of growth, changes from a floral to a vegetative state. However, the onset of flowering varies among date palm cultivars and may be influenced by the climatic conditions and nutritional status of the mother tree [13].

A limitation associated with using inflorescence explants is that during the culture-starting stage, roots may form on cultured explants instead of buds. The appearance of roots at this stage inhibits bud formation and leads to culture elimination. At the multiplication stage, early rooting decreases the rate of bud multiplication by diverting most tissue nutrients to root formation rather than shoot formation [14]. Shoot-bud multiplication in date palm depends on the basal formulation of the culture medium, genotype [15], and type and concentration of plant growth regulators [16].

This plant regeneration protocol describes direct induction of adventitious buds from immature inflorescences of female date palm and subsequent growth and plant development until acclimatization.

## 2 Materials

### 2.1 Plant Material and Explant Sterilization

1. Explant source: Immature female inflorescences of date palm Shamiya cv. (dry date cultivar) 8–25 cm in length.

2. Dettol solution: 100 mL/L Dettol in water.

3. Ethanol: 70%.

4. Clorox disinfection solution: 30% Clorox solution (1.6% w/v sodium hypochlorite, NaOCl).

5. Mercuric chloride disinfection solution: 0.1 g/L mercuric chloride.

6. Antioxidant solution: 100 mg/L ascorbic acid and 150 mg/L citric acid.

### 2.2 Culture Medium

1. Basal medium: Murashige and Skoog (MS) salts and vitamins [17] (Table 1).

2. Hormonal stock solutions (1 mg/mL): 2,4-dichlorophenoxy acetic acid (2,4-D), indole butyric acid (IBA), indole acetic acid (IAA), naphthalene acetic acid (NAA), gibberellic acid (GA$_3$), 2-isopentenyladenine (2iP), benzyladenine (BA), 6-furfurylaminopurine (kinetin), and abscisic acid (ABA).

**Table 1**
**Composition of basal medium of Murashige and Skoog [17]**

| Constituent | Concentration (mg/L) |
| --- | --- |
| *Macronutrients* | |
| $NH_4NO_3$ | 1650 |
| $KNO_3$ | 1900 |
| $CaCl_2 \cdot 2H_2O$ | 440 |
| $MgSO_4 \cdot 7H_2O$ | 370 |
| $KH_2PO_4$ | 170 |
| *Micronutrients* | |
| $MnSO_4 \cdot 4H_2O$ | 22.30 |
| $ZnSO_4 \cdot 4H_2O$ | 8.60 |
| $H_3BO_3$ | 6.20 |
| KI | 0.83 |
| $NaMoO_4 \cdot 2H_2O$ | 0.25 |
| $CuSO_4 \cdot 5H_2O$ | 0.025 |
| $CoCl_2 \cdot 6H_2O$ | 0.025 |
| *Iron* | |
| $Na_2EDTA$ | 37.25 |
| $FeSO_4 \cdot 7H_2O$ | 27.85 |
| *Vitamins* | |
| Nicotinic acid | 0.5 |
| Pyridoxine-HCl | 0.5 |
| Thiamine-HCl | 5 |
| Myo-inositol | 100 |
| Biotin | 0.5 |
| *Amino acid* | |
| Glycine | 2 |
| glutamine | 200 |
| *Sodium and potassium* | |
| $NaH_2PO_4$ | 170 |
| $KH_2PO_4$ | 120 |
| *Sucrose* | 30,000 |
| *Agar* | 7000 |

3. Culture medium for various culture stages: Murashige and Skoog medium (Table 1) containing hormones and activated charcoal (AC) as specified in Table 2, including initiation medium (MI1), modified initiation medium (MI2), modified initiation medium (MI3), differentiation medium (DM), multiplication medium (MM) (*see* **Note 1**), elongation medium (EM) (*see* **Note 2**), and rooting medium (RM) (*see* **Note 3**).

4. pH adjustment solutions: 0.1 and 1 N each of KOH and HCl.

***2.3 Acclimatization of Plantlets***

1. Fungicide solution: 2 g/L Topsin M 70 in water.

2. Plastic pots: 5 cm diameter × 18 cm height.

**Table 2**
**Different culture stages and their corresponding growth regulators and activated charcoal supplemented to the MS medium (Table 1)**

| Culture stage | Stage duration (weeks) | Plant growth regulators (mg/L) | | | | | | | | | AC (g/L) |
|---|---|---|---|---|---|---|---|---|---|---|---|
| | | 2,4-D | 2iP | NAA | IAA | KI | BA | IBA | GA$_3$ | ABA | |
| Initiation medium 1 (MI1) | 4 weeks | 5 | 3 | | | | | | | | 1.5 |
| Modified initiation medium 2 (MI2) | 8 weeks | 1 | 3 | 1 | 2 | | | | | | 1 |
| Modified initiation medium 3 (MI3) | 4 weeks | 0.5 | 3 | 0.5 | | | | | | | 0.5 |
| Differentiation medium (DM) | 3 weeks | | | 0.1 | | 0.25 | 0.25 | | | 0.25 | 0.2 |
| Multiplication medium (MM) | 9–12 months (4-week interval) | | 0.1 | 0.05 | 0.2 | 0.5 | 0.5 | | | | 0.2 |
| Elongation medium (EM) | 3 weeks | | | | | | 1 | 1 | 1 | | 0.2 |
| Rooting medium (RM) | 8 week (4-week interval) | | | 1 | | | | 2 | | | |

3. Potting soil mix: peat moss and sand at 2:1 (v/v).

4. Greenhouse.

***2.4  Equipment***

1. Glassware and culture vessels: beakers (500, 1000 mL), graduated cylinders (500, 1000 mL), glass culture jars (200, 300 mL), and culture tubes (15, 25 cm).

2. Surgical tools: forceps and scalpels.

3. Instruments: sterilizer, laminar airflow hood, growth chamber, precision balance, magnetic stirrer, microwave oven, pH meter, and autoclave.

# 3  Methods

***3.1  Medium Preparation***

1. Prepare 1 mg/mL individual stock solutions of each plant growth regulator. Dissolve 2,4-D, IBA, IAA, NAA, GA$_3$, 2iP, BA, and kinetin individually in 95% ethanol or 1 N sodium hydroxide (NaOH) and ABA using 1 N Hydrochloric acid (HCl), and then make up the required volume by adding double-distilled water. Store stock solutions at 4 °C until use.

2. To prepare 1 L of medium, weigh 4.4 g of commercially available powdered Murashige and Skoog (MS) medium (Table 1), and dissolve it in 700 mL distilled water.

3. Add hormonal concentrations and activated charcoal according to the culture stage, as shown in Table 2, and complete the volume to 800 mL with distilled water.

4. Adjust the pH of the medium to 5.7 using NaOH and HCl solutions.

5. Weigh 7 g/L agar, add to 200 mL distilled water, and heat until dissolved.

6. Add the melted agar (200 mL) to the 800 mL culture medium and mix well.

7. Distribute the culture medium into 200 mL culture jars (35 mL/jar), and cap with polypropylene closures.

8. Autoclave the culture medium jars for 20 min at 121 °C and 1.1 kg/cm$^2$.

9. Store the culture medium jars at room temperature in the dark until use, for up to 1 week.

*3.2 Explant Preparation*

1. Excise immature female inflorescence, encased in the spathe, before emergence of the spathe, from adult date palm in late January to early February (Fig. 1a, b; *see* **Note 4**).

2. Immediately refrigerate the excised spathes at 5 °C until use, for up to 3 days (*see* **Note 5**).

3. Wash the spathe with tap water for 10 min and soak in Dettol solution for 5 min.

4. Sterilize spathes with Clorox solution for 5 min, and then rinse them three times with sterilized water.

5. Remove the outer protective sheath, and cut each spikelet into 1–3 cm long pieces which carry many florets (female flower initials).

6. Dip spikelets in the antioxidant solution for 5 min (*see* **Note 6**).

7. Briefly dip spikelets into 0.1 g/L mercuric chloride solution for 2 s immediately before culturing (*see* **Note 7**).

*3.3 Initiation of Adventitious Buds*

1. Culture sterilized 1 cm spikelet segments on initiation medium (MI1, Table 2), and incubate cultures at 26 ± 2 °C in the dark for 4 weeks.

2. Transfer the explants to modified initiation medium (MI2, Table 2), incubate cultures at 26 ± 2 °C for 8 weeks in the dark, and subculture at 4-week intervals (Fig. 1c, d).

**Fig. 1** Stages of direct organogenesis from date palm immature inflorescence explants: (**a, b**) immature inflorescence explants, (**c, d**) induction of adventitious buds on the spikelet explants, (**e**) formation of direct shoot from the flower, (**f**) formation of direct shoot from the flower, (**g**) multiplication stage, (**h**) rooting of plantlets, (**i**) acclimatization of plantlets

3. Transfer the explants on modified initiation medium (MI3, Table 2), and incubate at 26 ± 2 °C in the dark for 4 weeks (*see* **Notes 8** and **9**).

### 3.4 Differentiation

1. Transfer the explants to the differentiation medium (DM, Table 2).

2. Incubate cultures at 26 °C ± 2 °C in the dark.

3. Observe bud development from the inflorescence explants, 3 weeks after transfer on the differentiation medium (Fig. 1e, f).

### 3.5 Multiplication

1. Subculture cluster containing 3–4 direct buds onto multiplication medium (MM, Table 2) (Fig. 1g).

2. Incubate cultures at 16-h photoperiod (40 µmol/m²/s) and 26 ± 2 °C.

3. Repeat subculturing several times at 4-week intervals to obtain stock cultures of shoots.

**3.6  Elongation Stage**

1. Culture the resultant individual shoots on MS elongation medium (EM, Table 2).

2. Incubate cultures at 16-h photoperiod (50 $\mu$mol/m$^2$/s) and $26 \pm 2$ °C for 3 weeks.

**3.7  Rooting Formation Stage**

1. Culture an individual healthy shoot of about 10 cm in length with 2–3 leaves on rooting media (RM) Table 2 (Fig. 1h) for 8 weeks (4-week interval).

2. Incubate cultures at 16-h photoperiod (80 $\mu$mol/m$^2$/s) and $26 \pm 2$ °C.

**3.8  Acclimatization**

1. Select the healthy plantlets, about 10 cm in length with 2–3 leaves, and wash the roots under running tap water to remove the adhering solidified culture medium.

2. Dip plantlet into the fungicide solution for 3 min.

3. Transfer the plantlets into plastic pots (Torpedo) trays, filled with potting mixture and place the trays in the greenhouse under plastic-sheet tunnel. Maintain under 16-h photoperiod (135 $\mu$mol/m$^2$/s) and 85–90% relative humidity at 27–30 °C for 2 weeks (*see* **Note 10**).

4. Irrigate plantlets with water after 2 weeks, and fertilize with 10% MS salt solution after additional 4 weeks.

5. Reduce relative humidity gradually to 45–65% by partially removing the plastic sheets over a period of 8 weeks and then to completely expose to the ambient conditions.

6. Spray plantlets regularly with fungicide solution (2 g/L Topsin M 70) once every 2 weeks during the first 8 weeks of the acclimatization process (*see* **Note 11**).

7. Transplant the plantlets to larger pots (60–80 mm diameter) after 6 months following the start of the acclimatization process. Place the plantlets in a shaded area of the nursery for further growth (Fig. 1i).

---

## 4  Notes

1. Concentrations of AC higher than 0.2 g/L decrease the bud multiplication rate.

2. Add GA$_3$ to the culture medium by filter sterilization to avoid breakdown by heating during autoclaving.

3. Using a combination of auxins (1 mg/L NAA and 2 mg/L IBA) produces a good root system.

4. Use the best responsive spathe size (length of 12 cm); however, 8–25 cm length can be used.

5. Keeping inflorescences at room temperature more than 3 days causes excessive browning and contamination.

6. The explant should be kept in the antioxidant solution during handling to avoid tissue browning due to oxidation of phenolic compounds.

7. Dipping spikelets in 0.1 g/L mercuric chloride solution for 2 s and immediately culturing, without rinsing in water, is effective for reducing contamination and rarely affects explant response.

8. Gradual reduction of the 2,4-D concentration from 5 to 0.5 mg/L of the initiation culture medium results in three types of response: direct shoot formation as well as direct and indirect somatic embryogenesis.

9. Avoid extending the culture duration on modified initiation medium (MI3) more than 3–4 weeks, such condition leads to root formation which inhibits bud formation.

10. Maintain 85–90% relative humidity and 27–30 °C during the first 7–10 days of transplanting to avoid leaf wilting. After 4 weeks, relative humidity reduce gradually to 45–65% by partially removing the plastic sheets, then plantlets are exposed to ambient environment.

11. Spraying plantlets with fungicide solution (Topsin M 70 at 2 g/L) increases survival rate by protecting them from fungal infestation and subsequent death.

## References


1. Khattab MM, Ibrahim IA, Gadalla EG (2003) In vitro propagation of Egyptian dry date palm. 1—Effect of explant, time of culture on browning, callus formation and type of media on somatic embryogenesis. Bull Fac Agric Cairo Univ 54(4):555–568

2. Sharma DR, Kumari R, Chowdury JB (1980) In vitro culture of female date palm (*Phoenix dactylifera* L.) tissues. Euphytica 29:169–174

3. Bhaskaran SH, Smith RH (1992) Somatic embryogenesis from shoot tip and immature inflorescence of *Phoenix dactylifera* L. cv. Barhee. Plant Cell Rep 12:22–25

4. Sidky RA, El-Dawyati MM (2012) Proliferation of female inflorescences explants of date palm. Ann Agric Sci 57(2):161–165

5. Loutfi K (1999) Organogenèse et embryogenèse somatique à partir des tissus floraux du palmier dattier (*Phoenix dactylifera* L.) cultivés in vitro. Aspects histologiques et caryologiedes *vitro* plants. Thèse doctorat Es-Sciences, Université Cadi Ayyad Marrakech

6. El-Korchi B (2007) Large scale in vitro propagation of a rare and unique male date palm (*Phoenix dactylifera* L.) using inflorescence technique. Acta Hort (736):243–254

7. Loutfi K, Chlyah H (1998) Vegetative multiplication of date palm from the in vitro cultured inflorescence: effect of some growth regulator combinations and organogenetic potential of various cultivars. Agronomy 18:573–580

8. Masmoudi-Allouche F, Meziou B, Kriaâ W, Gargouri-Bouzid R, Drira N (2010) In vitro flowering induction in date palm (*Phoenix dactylifera* L.). J Plant Growth Regul 29: 35–43

9. Stino RG, El-Kosary S, Hassan MM, Kinawy AA (2015) Direct embryogenesis from inflorescences culture of Sewy date palm (*Phoenix dactylifera* L.). J Biol Chem Environ Sci 10 (1):173–186

10. Gadalla EG, Hassan MM, Al-Sharabasy SF (2015) Effect of growth regulators on somatic embryogenesis of date palm inflorescence cv. Sewi. In: 2nd Minia international conference on agriculture and irrigation in the Nile Basin countries, 23–25 March, pp 503–515

11. Drira N, Al-Shaary A (1993) Analysis of date palm female floral initials potentials by tissue culture. In: Proceedings of the third symposium on date palm, King Faisal University, Al-Hassa, Saudi Arabia, pp 161–170

12. Abahmane L, Bougerfaoui M, Anjarne M (1999) Use of tissue culture techniques for date palm propagation and rehabilitation of palm groves devastated by bayoud disease. In: Proceedings of the international symposium on date palm, Assiut University, Assiut, Egypt, 9–11 Nov, pp 385–388

13. Abul-Soad AA (2003) Biotechnological studies of date palm: micropropagation of inflorescence, molecular biology, and secondary metabolites. PhD dissertation, Pomology Department, Faculty of Agriculture, Cairo University

14. Al Khateeb AA (2008) The problems facing the use of tissue culture technique in date palm (*Phoenix dactylifera* L.). Sci J King Faisal Univ 9:85–104

15. Mazri MA, Meziani R (2015) Micropropagation of date palm: a review. In Vitro Cell Dev Biol Plant 4:160–164

16. Abahmane L (2011) Date palm micropropagation via organogenesis. In: Jain SM, Al-Khayri JM, Johnson DV (eds) Date palm biotechnology. Springer, Dordrecht, pp 69–90

17. Murashige T, Skoog FA (1962) A revised medium for rapid growth and bioassays with tobacco tissue cultures. Physiol Plant 15:473–479

# Chapter 4

# Optimized Direct Organogenesis from Shoot Tip Explants of Date Palm

## Rehab Sidky

## Abstract


In vitro propagation is an available alternative to produce uniform and good-quality planting material to establish large-scale date palm cultivation in a short time. This study was carried out to achieve organogenesis and multiplication directly from shoot tips without callus formation, thus avoiding any possibility of undesirable genetic variability among the regenerated plants. The shoot tips explants are cultured on Murashige and Skoog (MS) medium supplemented with 1 mg/L naphthaleneacetic acid (NAA), 1 mg/L naphthoxyacetic acid (NOA), 2.5 mg/L benzyladenine (BA), and 2.5 mg/L isopentenyladenine (2iP). Numerous adventitious buds appeared from the shoot tip explants in darkness after six subcultures at 4-week intervals. Vegetative buds pass through three stages: initiation bud formation, vegetative bud differentiation, and shoot bud proliferation. Shoots are transferred onto medium containing low concentrations of growth regulators for shoot multiplication. The organogenesis protocol described herein consists of six steps: initiation of meristematic buds, multiplication, elongation, rooting, pre-acclimatization, and finally plant acclimatization.

**Key words** In vitro, Direct organogenesis, Micropropagation, Tissue culture


## 1 Introduction

Date palm, *Phoenix dactylifera* L., is one of the most important fruit trees in the Middle East and Saharan and sub-Saharan regions of Africa. In some areas, it is the only tree that provides food, shelter, and fuel to the communities. Dates are not only a staple food but also are an important export cash crop [1]. Date palms produce shoots from axillary shoot meristems and inflorescences from floral meristems [2]. The use of offshoots is the most conventional vegetative technique for date palm propagation [3]. This method permits the preservation of true-to-type features of multiplied genotypes. However, the average number of offshoots per palm is very limited over the life span of the tree and is restricted to the juvenile stage [4]. Plant tissue culture techniques have been used to clone a wide range of date palm cultivars worldwide.


Jameel M. Al-Khayri et al. (eds.), *Date Palm Biotechnology Protocols Volume 1: Tissue Culture Applications*,
Methods in Molecular Biology, vol. 1637, DOI 10.1007/978-1-4939-7156-5_4, © Springer Science+Business Media LLC 2017

Using these techniques, date palm can be micropropagated by organogenesis in which plantlets are produced from multiplied buds without passing through the callus stage [5, 6]. The organogenesis technique is based for exploitation of the meristematic potentialities of shoot tip explants to form new shoots. The plant growth regulators supplemented to the culture media are used at minimal concentrations. Since vegetative buds come directly from the mother plant tissue, plantlets produced are identical to the mother tree. However, the success of this technique is highly dependent on the success of the first multiplication step (initiation) which requires well-trained staff. Furthermore, factors such as the explant source, time of culture, number of subcultures, plant growth regulator, genotype, and media composition are capable of inducing in vitro variability [7].

This chapter describes a method to propagate date palm from shoot tips directly without callus formation including shoot tip sterilization, explant isolation, and culture medium compositions of all stages of plant regeneration.

## 2 Materials

### 2.1 Plant Materials and Sterilization

1. Date palm, cv. Siwy, offshoots provided shoot tips as explant sources.
2. Antioxidant solution: 100 mg/L ascorbic acid and 150 mg/L citric acid.
3. Fungicide solution: Ridomil 1 g/L (*see* **Note 1**).
4. Disinfection solution: Clorox 50 and 20%.

### 2.2 Culture Medium

1. Basal culture medium: Murashige and Skoog (MS) [8] medium (Table 1).
2. Plant growth regulators and other additives: Specified in Table 2 for various culture stages (initiation, multiplication, elongation, rooting, and pre-acclimatization).
3. Culture vessels: Test tubes 25 × 150 mm, jars 150 and 250 ml.

### 2.3 Acclimatization

1. Potting mixture: Peat moss, vermiculite, and perlite at 1:1:1.
2. Pots: 5 × 18 cm.
3. Plastic tunnel: Plastic white opaque, relative humidity (90–95%), and a constant temperature of 25–26 °C.
4. Greenhouse: Relative humidity of (80–90%) and a constant temperature of $25 \pm 1$ °C.
5. Fertilizer solution: 2–3 g/L N-P-K (17-17-17).
6. Fungicide solution: 1 g/L Topsin M 70.

**Table 1**
**Composition of basal culture medium of Murashige and Skoog [8]**

| Constituent | Concentration (mg/L) |
| --- | --- |
| *Macronutrients* | |
| $NH_4NO_3$ | 1650 |
| $KNO_3$ | 1900 |
| $CaCl_2 \cdot 2H_2O$ | 440 |
| $MgSO_4 \cdot 7H_2O$ | 370 |
| $KH_2PO_4$ | 170 |
| *Micronutrients* | |
| $MnSO_4 \cdot 4H_2O$ | 22.30 |
| $ZnSO_4 \cdot 4H_2O$ | 8.60 |
| $H_3BO_3$ | 6.20 |
| KI | 0.83 |
| $NaMoO_4 \cdot 2H_2O$ | 0.25 |
| $CuSO_4 \cdot 5H_2O$ | 0.025 |
| $CoCl_2 \cdot 6H_2O$ | 0.025 |
| *Iron* | |
| $Na_2EDTA$ | 37.25 |
| $FeSO_4 \cdot 7H_2O$ | 27.85 |
| *Vitamins* | |
| Nicotinic acid | 0.5 |
| Pyridoxine-HCl | 0.5 |
| Thiamine-HCl | 0.1 |
| Myo-inositol | 100 |
| *Amino acid* | |
| Glycine | 2 |

# 3  Methods

## 3.1  Preparation of Offshoots

1. Choose 3–5-year-old date palm offshoots preferably during November to March (*see* **Note 2**).

2. Select disease-free offshoots (*see* **Note 3**).

3. Collect offshoots from celebrated farms, preferably offshoots of 7–10 kg weight.

4. Remove the bases of leaves with a sharp knife until around 5.5 cm in diameter to obtain offshoot hearts. Cut a few more leaves to 5 cm that remain to protect soft tissues from sterilization agent.

5. Transfer shoot tips to a chilled antioxidant solution [9] to avoid tissue browning.

**Table 2**
**Medium additives supplemented to the MS culture medium used for the different stages of date palm organogenesis protocol**

| Medium | Composition (mg/L) | Plant growth regulators (mg/L) |
| --- | --- | --- |
| 1. Initiation | MS containing $NaH_2PO_4 \cdot 2H_2O$ 170, myo-inositol 100, adenine sulfate 40, thiamine HCl 0.4, sucrose 3%, polyvinylpyrrolidone (PVP) 0.2%, agar 0.7%, pH 5.7 | 1 NAA, 1 NOA, 2.5 BA, 2.5 2iP |
| 2. Multiplication | 1. MS containing $NaH_2PO_4 \cdot 2H_2O$ 170, myo-inositol 100 adenine sulfate 40, thiamine HCl 0.4, sucrose 50% and solidified with agar 0.7%<br>2. ½MS containing 10 silver nitrate, $NaH_2PO_4 \cdot 2H_2O$, 170 myo-inositol 100 adenine sulfate 40, thiamine HCl 0.4, sucrose 5%, agar 0.7%, pH 5.7 | 0.5 BA, 0.5 2iP |
| 3. Elongation | ½MS containing myo-inositol 100, $KH_2PO_4$ 170, thiamine HCl 0.4, AC 100, sucrose 5%, agar 0.7%, pH 5.7 | 0.5 $GA_3$, 1 NAA |
| 4. Rooting | ½MS containing myo-inositol 100, $KH_2PO_4$ 170, thiamine HCl 0.4, AC 1000, sucrose 4%, agar 0.7%, pH 5.7 | 0.1 NAA, 4 paclobutrazol |
| 5. Pre-acclimatization | ½MS containing PEG MW8000, sucrose 3%, pH 5.2 | 0.1 NAA |

### 3.2 Sterilization

1. Soak the excised shoot tips (Fig. 1a) in a fungicide solution for 20 min.

2. Soak shoot tips directly in 50% Clorox disinfection solution (sodium hypochlorite) for 20 min (*see* **Note 4**).

3. Remove one layer of outer leaves using scalpel and forceps (Fig. 1b).

4. Soak again shoot tips in a commercial 20% Clorox solution (sodium hypochlorite) for 10 min.

5. Rinse three times with sterile distilled water.

### 3.3 Culture of Shoot Tips

1. Place isolated shoot tips in sterilized Petri dishes.

2. Remove the young leaves surrounding the apical dome gently. The explants consist of the bottom of the excised leaves.

3. Carefully separate young leaves closely surrounding the apical dome.

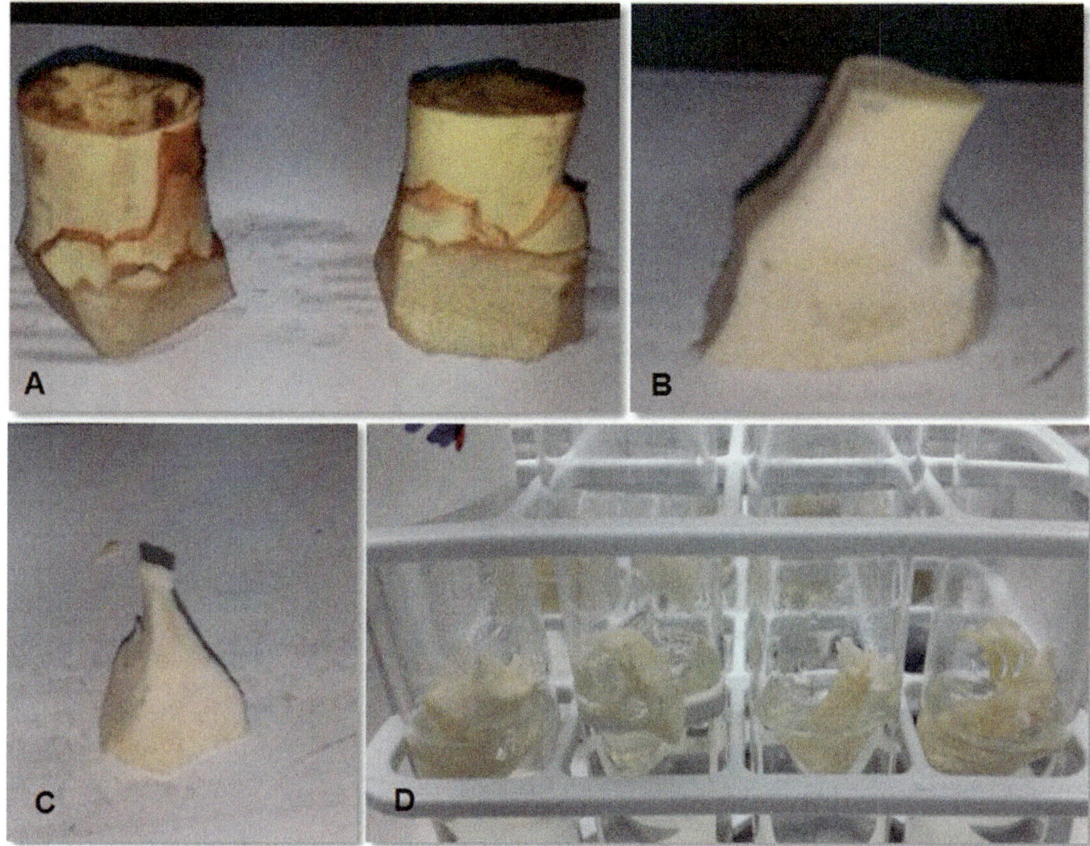

**Fig. 1** Plant materials (offshoots) used as a source of explants. (**a**) Shoot tip from offshoot ready to be disinfected, (**b**) explant after remove one layer of leave, (**c**) apical meristems ready to cut into 4–6 pieces, (**d**) apical meristems in initiation media

4. Cut the entire shoot tip into 4–6 pieces longitudinally, and then transfer to the culture medium for initiation stage.

5. Axillary buds found among the young leaves are also suitable as explants.

6. Culture each explant (apical meristems) in test tubes (Fig. 1c, d).

**3.4 Initiation Stage**

1. Culture explants in the initiation medium (*see* **Note 5**).

2. Subculture explants on the same media at 4-week intervals for 6 months until the initiation of vegetative buds (Fig. 2a).

3. Incubate cultures in the darkness, at $27 \pm 1$ °C during the initiation stage.

**Fig. 2** Development stages of vegetative bud on multiplication stage of date palm cv. Siwy (**a**) initiation bud formation, (**b**) vegetative bud differentiation, (**c**) shoot bud proliferation

***3.5 Shoot Multiplication***

Vegetative buds pass through three stages: initiation of bud formation (Fig. 2a), vegetative bud differentiation (Fig. 2b), and shoot bud proliferation (Fig. 2c).

1. Divide the vegetative buds (Fig. 2b) into small clumps, each containing three buds for multiplication (*see* **Note 6**).

2. Subculture them on multiplication medium (full strength) twice in test tubes at 4-week intervals; *see* Table 2. Incubate under light (25 μmol/m$^2$/s) for more vegetative buds differentiation.

3. After two subcultures transfer the differentiated buds in the test tubes filled with multiplication medium (Table 2), and subculture for 12 weeks at 4-week intervals for proliferation of shoots. Incubate under light 40 μmol/m$^2$/s.

4. Transfer proliferation shoots to MS half-strength medium into 250 ml-jars supplemented with 10 mg/L silver nitrate under light 40 μmol/m$^2$/s (*see* **Note 7**).

5. All cultures are maintained in a plant growth room at 25 ± 2 °C for all stages with 16-h photoperiod.

**Fig. 3** Development stages of vegetative bud on multiplication stage of date palm cv. Siwy (**a**) shoot bud multiplication, (**b**) shoot elongation, (**c**) plantlets formed root

### 3.6 Shoot Elongation and Rooting

1. Separate individual shoots from the multiplied shoot clusters (Fig. 3a) (*see* **Note 8**).

2. Transfer individual shoots directly to the elongation medium, MS half-strength medium; *see* Table 2 (Fig. 3b).

3. Use healthy shoots to initiate a good root system after shoot elongation.

4. Transfer shoots to the rooting media [10], half-strength MS basal medium (with supplements indicated in Table 2) (*see* **Note 9**).

5. On this media, thickness of shoots is increased, accelerating root formation and promoting secondary root formation for 12 weeks, at 6-week intervals under light 40 $\mu mol/m^2/s$.

6. Transfer plantlets,10 cm in length with 2–3 leaves and well-developed roots from standard jars 250 ml to small tubes (25 × 150 mm) (Fig. 3c).

### 3.7 Pre-acclimatization and Acclimatization

1. Transfer plantlets from standard small tubes (25 × 150 mm) diameter to larger tubes (28 × 250 mm) (*see* **Note 10**).

2. Culture rooted plantlets (Fig. 3c) in pre-acclimatization medium [10], half-strength MS liquid medium (Table 2), and after 6 weeks transfer plantlets to the greenhouse.

3. Remove individually healthy plantlets with specifications: 10–15 cm in length, 10 mm shoot base in diameter from test tubes.

4. Gently remove agar sticking to the roots in running tap water; soak plants in fungicide for 15 min.

**Fig. 4** Acclimatization of tissue culture-derived plantlets: (**a**) plantlets under plastic tunnel in greenhouse, (**b**) plantlets after the first leaf appeared

5. Transfer plants in pots (5 × 18 cm) filled with soil containing a mixture of peat moss, vermiculite, and perlite at 1:1:1 [10], and grow them in the greenhouse under a large plastic tunnel (40 × 20 cm) until beginning the first leaf appear (Fig. 4a) by maintaining high relative humidity (90–95%).

6. Remove plastic tunnel after the first leaf appeared to reduce humidity, and allow the plants to adapt to the large tunnel house (9 × 3 m) (Fig. 4b).

## 4  Notes

1. The most efficient way to sterilize shoot tips is to soak in a fungicide solution.

2. The best growth and bud regeneration, and lowest rate or tissue browning, occur when cultures are established during November to March.

3. Select offshoots from a well-known elite adult date palm.

4. Most effective way of sterilization of shoot tips is by soaking directly in a Clorox solution without rinsing shoot tips with distilled water.

5. The success of vegetative bud formation is highly dependent on the success of first-step initiation.

6. Bud initiation is controlled by several factors that may act in concert with composition of culture media, cutting of explants and incubation conditions.

7. Silver nitrate has proved to be a very potent inhibitor of ethylene action and is widely used in plant tissue culture; ethylene

suppresses shoot organogenesis in vitro. Silver nitrate enhanced shoot-bud multiplication of date palm cv. Zagloul presented as shoot bud number and shoot bud length [11].

8. The composition of elongation media is important for the development plants, $GA_3$ with NAA plays important role, GA3 gave the significantly the longest shoots and roots [12]. In the elongation stage, MS medium with 0.5 mg/L GA3 and 0.1 mg/L NAA enhanced plantlet length [13].

9. Leaf width or the diameter of elongated shoots tends to broaden during the elongation stage. When these shoots are exposed to exceptional conditions like overheating or growth deficiency, they tend to twist and become thin and delicate. The larger the leaf width of the elongated shoots, the more growth and developments are observed. Using $GA_3$ with NAA is more effective to promote shoot length and leaf width [14].

10. The pre-acclimatization of plantlets is carried in the laboratory to reduce the ex vitro stress in the greenhouse for enhancing the plantlet survival rate.

## References

1. El Hadrami A, Al-Khayri JM (2012) Socioeconomic and traditional importance of date palm. Emir J Food Agri 24:371–385

2. Sudhersan C, Abo El-Nil M, Hussain J (2001) Hapaxanthic axillary shoots in date palm plants grown in vitro and in vivo. Palms 45:84–89

3. Aleid, SM, Al-Khayri JM, Al-Bahrany AM (2015) Date palm status and perspective in Saudi Arabia. In: Al-Khayri JM, Jain SM, Johnson DV (eds) Date palm genetic resources and utilization. Asia and Europe, vol 2. Springer, Dordrecht, pp 49–95

4. Tisserat B (1983) Tissue culture of date palm—a new method to propagate an ancient crop and a short discussion of the California date industry. Principes 27:105–117

5. Al Khateeb AA (2006) Role of cytokinin and auxin on the multiplication stage of date palm (*Phoenix dactylifera* L.) cv. Sukry. Biotechnology 5(3):349–352

6. Abahmane L (2011) Date palm micropropagation via organogenesis. In: Jain SM, Al-Khayri JM, Johnson DV (eds) Date palm biotechnology. Springer, Dordrecht, pp 69–90

7. Yu J, Holland JB, McMullen MD, Buckler ES (2008) Genetic design and statistical power of nested association mapping in maize. Genetics 178:539–551

8. Murashige T, Skoog F (1962) A revised medium for rapid growth and bioassays with tissues culture. Phys Plant 15:473–497

9. Al-Khayri JM, Al-Bahrany AM (2004) Genotype-dependent in vitro response of date palm (*Phoenix dactylifera* L.) cultivars to silver nitrate. Sci Hortic 99:153–162

10. Sidky RA, Zaid ZE, El-Bana A (2009) Optimized protocol for in vitro rooting of date palm (*Phoenix dactylifera* L.). Egypt J Agric Res 87: 277–288

11. Bekheet SA (2013) Direct organogenesis of date palm (*Phoenix dactylifera* L.) for propagation of true-to-type plants. Sci Agric 4:85–92

12. Rasmia SS, Zeinab EZ, Sidky RA (2011) Effect of Ammonium Nitrate and $GA_3$ on growth and development of date palm plantlets in vitro and acclimatization stage. Res J Agri Biol Sci 7:17–22

13. Khierallah SM, Bader SM (2007) Micropropagation of date palm (*Phoenix dactylifera* L.) var. Maktoom through direct organogenesis. Acta Hort (736):213–224

14. Abul-Soad AA, Zaid ZE, Sidky RA (2006) Improved method for the micropropagation of date palm (*Phoenix dactylifera* L.) through elongation and rooting stages. Bull Fac Agric Cairo Univ 57:791–801

# Direct Organogenesis and Indirect Somatic Embryogenesis by In Vitro Reversion of Mature Female Floral Buds to a Vegetative State

**Eman M.M. Zayed**

## Abstract



This protocol describes in vitro plant regeneration from mature female inflorescence explants of date palm (*Phoenix dactylifera* L.) by reversion of floral state (reproductive phase) to the vegetative state. The mature female inflorescence (fully developed) is cultured on MS induction medium containing 10 mg/L 2,4-dichlorophenoxyacetic acid (2,4-D), 3 mg/L 2-isopentenyladenine (2iP), and 2 mg/L paclobutrazol (PBZ) or 2 mg/L abscisic acid (ABA). The basal part of the petals has meristematic cells, which can be induced to initiate callus or direct shoot formation depending on the plant growth regulator amendments. Callus forms on the induction medium supplemented with PBZ after 12 weeks, whereas it differentiates into somatic embryos on a medium containing 0.1 mg/L naphthaleneacetic acid (NAA). Direct shoots are regenerated on the induction medium amended with ABA after 24 weeks. Procedures for plant regeneration from mature female inflorescence explants are described, and histological changes which occur during the reversion process are presented.

**Key words** Abscisic acid (ABA), Callus, Direct organogenesis, In vitro, Mature female inflorescence, Paclobutrazol (PBZ), Reversion floral bud



## 1  Introduction

Micropropagation of date palm (*Phoenix dactylifera* L.) is hampered by the limited number off offshoots needed for multiplication. Alternatively, immature and mature inflorescence tissue represents an abundant and successful source of explants for date palm micropropagation [1–7]. However, excision of immature female inflorescence may damage the growing tip of the mother tree due to intrusive isolation of the spathe. Alternatively, mature inflorescences are isolated easily without harming the vegetative tissue of the mother plant, to produce ideal mature inflorescence explants.

The mature female inflorescence is a novel source for in vitro propagation of species of the Arecaceae family [4]. Its large flower



Jameel M. Al-Khayri et al. (eds.), *Date Palm Biotechnology Protocols Volume 1: Tissue Culture Applications,*
Methods in Molecular Biology, vol. 1637, DOI 10.1007/978-1-4939-7156-5_5, © Springer Science+Business Media LLC 2017

spikes that reach 30–40 cm in length are an inexpensive source of explants. Moreover, the mature female inflorescence remains healthy longer due to the surrounding protective spathe. Mature flower tissues have zones of meristematic cells at the base of the sepals and petals that have the capability to stimulate vegetative growth [5, 7].

Histological analyses of female flowers of date palm at the time of culture showed the presence of three sepals, three petals, three carpels, six staminodes, and three separate ovules, each ovule connected to the base of the ovary. This confirms that these flowers are at the final developmental stage [5, 7, 8]. Successive transverse and longitudinal section analysis of differentiating mature floral explants revealed that callus was initiated from the basal part of the flowers, especially at the sepals and petals [5, 7]. In addition, histological analyses showed that shoot primordia were initiated from the basal part of the petals, while carpels and stamens degenerated [1, 7].

In vitro reversion of floral phase to the vegetative state is a complicated process, which requires controlling different factors [6, 7]. Exogenous application of plant growth retardants at appropriate concentrations, and occurrence of competent cells in the explants are required to stimulate induction of organ differentiation [9, 10]. Paclobutrazol (PBZ) can stimulate embryogenesis since it inhibits gibberellin biosynthesis and leads to the ability to alter cytokinin levels in a new endogenous hormone balance [11, 12]. The addition of abscisic acid (inhibitor) to media may change the commitment of cells from cell division to differentiation and influence morphogenesis in a number of plants by modifying the effects of other hormones, notably cytokinins and gibberellins, as well as auxins [9, 13].

In this protocol, mature female inflorescence explants of date palm are used for plant regeneration through the reversion process. This chapter also provides histological evidence to confirm callus and organ initiation occurrence from the somatic cells (sepals and petals) away from the sexual organs (carpels).

## 2  Materials

### 2.1  Plant Material and Disinfectants

1. Mature female inflorescences (spathes) of date palm Siwy cv. (semidry) 40–50 cm in length, green in color.

2. Disinfection solutions: 70% ethanol solution and 0.01% (w/v) mercuric chloride ($HgCl_2$) solution.

3. Antioxidant solution: 100 mg/L ascorbic acid and 150 mg/L citric acid.

4. Pesticide solutions: 3 g/L Vitavax fungicide; 5 g/L Benlate fungicide; 3 mL/L Coragen insecticide.

***2.2 Culture Media Composition***

1. Basal culture medium: Murashige and Skoog salts and vitamins (MS) [14] (Table 1) containing plant growth regulators according to culture stages shown in Table 2.

2. pH adjustment solutions: 0.1 and 1 M KOH and 0.1 and 1 M HCl.

3. Callus induction medium (CI): MS medium (Table 1) containing 10 mg/L 2,4-dichlorophenoxyacetic acid (2,4-D), 3 mg/L 2-isopentenyladenine (2iP), and 2 mg/L paclobutrazol (PBZ).

4. Shoot induction medium (SI): MS medium (Table 1) containing 10 mg/L 2,4-dichlorophenoxyacetic acid (2,4-D), 3 mg/L 2-isopentenyladenine (2iP), and 2 mg/L abscisic acid (ABA).

5. Somatic embryo development (ED): MS medium (Table 1) containing 0.1 mg/L 1-naphthaleneacetic acid (NAA) [15].

6. Plant regeneration (PR): MS medium (Table 1) amended with 0.1 mg/L NAA and 0.05 mg/L benzyladenine (BA) [16].

7. Pre-acclimatization (PA): MS liquid medium (¼ MS) containing 0.1 mg/L NAA, 10 g/L sucrose, pH 5.2 [4].

***2.3 Histological Analysis***

*2.3.1 Reagents and Solutions*

1. Samples: Mature female flowers at consecutive differentiation developmental stages.

2. Fixation: FAA solution: formalin, acetic acid, ethyl alcohol (5:5:90 by volume).

3. Dehydration: Gradient concentration of ethanol alcohol series: 50, 70, 85, 95%, and absolute (anhydrous) alcohol.

4. Cleaning: Mixtures of xylene and absolute alcohol: (1:3), (1:1), and (3:1), and pure xylene.

5. Embedding in paraffin: Paraffin wax melting point 54–56 °C.

6. Sectioning: Microtome.

7. Staining: Double stain of 1% Safranin "O" and 0.01% Fast green FCF.

8. Slide mounting: Haupt's adhesive consisting of 1 g gelatin, 2 g phenol crystal, and 100 mL warm distilled water.

*2.3.2 Equipment for Histological Analysis*

1. Tools and supplies: Sharp razor blades, filter paper, fine brush, pencil, origami dish or suitable mold, needle, wooden or (metal, plastic) chucks and clean black sheets.

2. Glassware: Glass vials (10–20 mL), Coplin jars, forceps, slide box, slides, and long cover glass (24 × 45 mm).

3. Instruments: Vacuum pump, desiccators, oven, differential heated and embedding hot plate, rotary microtome with sharp blade, and microscope fitted with a camera.

**Table 1**
**Chemical composition of modified Murashige and Skoog (MS) medium [14]**

| MS (mg/L) | |
| --- | --- |
| *Macronutrients* | |
| $KNO_3$ | 1900 |
| $NH_4NO_3$ | 1650 |
| $MgSO_4 \cdot 7H_2O$ | 370 |
| $KH_2PO_4$ | 170 |
| $NaH_2PO_4 \cdot H_2O$ | 170 |
| $CaCl_2 \cdot 2H_2O$ | 440 |
| *Micronutrients* | |
| $H_3BO_3$ | 6.2 |
| $MnSO_4 \cdot 2H_2O$ | 22.3 |
| $ZnSO_4 \cdot 7H_2O$ | 8.6 |
| $Na_2MoO_4 \cdot 2H_2O$ | 0.25 |
| $CuSO_4 \cdot 5H_2O$ | 0.025 |
| $CoCl_2 \cdot 6H_2O$ | 0.025 |
| KI | 0.83 |
| *Iron source* | |
| $FeSO_4 \cdot 7H_2O$ | 27.8 |
| $Na_2EDTA \cdot 2H_2O$ | 37.3 |
| *Vitamins and organic supplements* | |
| *myo*-Inositol | 100 |
| Glutamine | 200 |
| Nicotinic acid | 0.5 |
| Pyridoxine·HCl | 0.5 |
| Thiamine·HCl | 0.1 |
| Glycine | 2.0 |
| Adenine sulfate | 40 |
| Ascorbic acid | 100 |
| Citric acid | 150 |
| *Carbon source* | |
| Sucrose | 30,000 |

**Table 2**
**Hormonal and activated charcoal supplements to the culture media for developmental stages of mature female flowers of date palm**

| Media additives | Culture stages | | | | |
| --- | --- | --- | --- | --- | --- |
| | Callus induction (CI) | Shoot induction (SI) | Embryo development (ED) | Plant regeneration (PR) | Pre-acclimatization (PA) |
| 2,4-Dichlorophenoxyacetic acid (2,4-D) | 10 mg/L | 10 mg/L | – | – | – |
| 2-Isopentenyladenine (2iP) | 3 mg/L | 3 mg/L | – | – | – |
| Paclobutrazol (PBZ) | 2 mg/L | – | – | – | – |
| Abscisic acid (ABA) | – | 2 mg/L | – | – | – |
| Naphthaleneacetic acid (NAA) | – | – | 0.1 mg/L | 0.1 mg/L | 0.1 mg/L |
| Benzyladenine (BA) | – | – | – | 0.05 mg/L | – |
| Activated charcoal | 1 g/L | 1 g/L | – | – | – |

## 3  Methods

### 3.1  Medium Preparation

1. Prepare MS medium (Table 1) in double-distilled water and supplement with 30 g/L sucrose (*see* **Note 1**).

2. Prepare hormone stock solutions: Dissolve 2,4-D (1 mg/mL), PBZ (1 mg/mL), and NAA (1 mg/mL) in a few drops of absolute ethanol; 2iP (1 mg/mL) in a few drops of 1 M HCl; ABA (1 mg/mL) and BA (1 mg/mL) in 1 M KOH, and then make up the required volume by adding distilled water.

3. Mix the components of the culture medium, adjust pH to 5.7 for all culture media except the pre-acclimatization medium to pH 5.2, using the solution 0.1 and 1 M KOH and 0.1 and 1 M HCl, and then add 1 g/L activated charcoal together with 6 g/L agar (*see* **Note 2**).

4. Dispense medium into 200 mL culture jars (40 mL per jar) cover with polypropylene caps, and autoclave at 121 °C and pressure of 1.1 kg/cm$^2$ for 20 min.

### 3.2  Explant Preparation and Sterilization

1. Collect mature female inflorescences (spathes) at the last developmental stage before the spathe splits open but after total emergence from adult mother trees of date palm at the end of spring, around March. Spathes excised from the last visible part by using a hatchet without removing or harming the leaves (*see* **Notes 3** and **4**).

2. Put the collected spathes enclosed within the hard protective sheath in paper bags and refrigerate at 4 °C for 1–4 days until use.

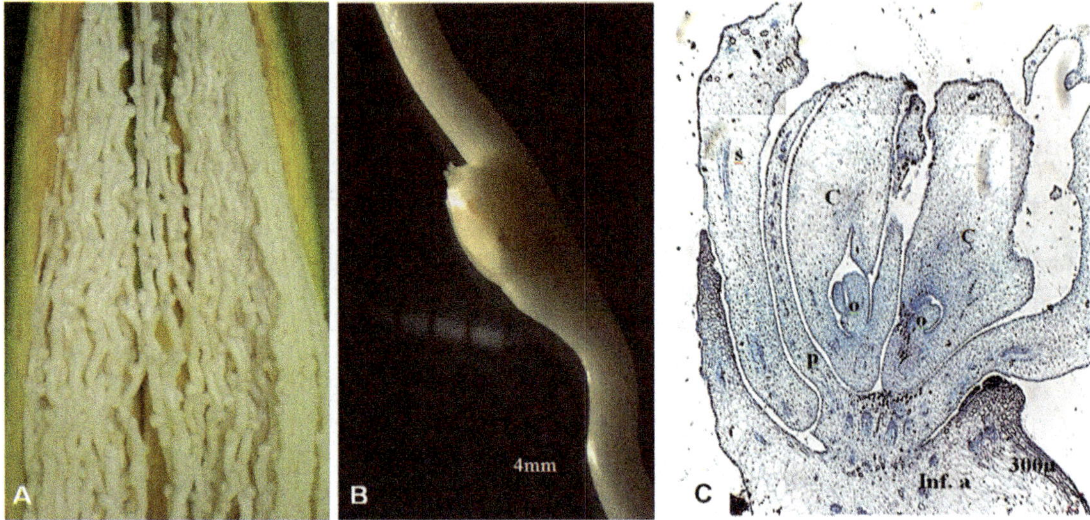


**Fig. 1** Morphology and histology of mature female inflorescence. (**a**) Morphology of the mature inflorescence, (**b**) mature flower *explant* on the axis of female inflorescence of date palm. Note that both the flower and the inflorescence axis have a white color, (**c**) longitudinal section of the same flower emerged with inflorescence axis (*inf. a*) showing the presence of sepals (*s*), petals (*p*), and two separated carpels (*c*) with ovules (*o*)

3. Scrape spathe edges by knife and clean with a piece of cotton two to three times.

4. Wash spathe well by brushing with soap and water, and then rinse under running water for 1 h to remove dust.

5. Transfer spathe into a laminar flow chamber and immerse in 70% ethanol for 1–2 min and wash with sterile distilled water (*see* **Note 5**).

6. Surface sterilize spathe by soaking in the 0.01% (w/v) mercuric chloride solution for 1 h, and then rinse with sterile distilled water three times to remove traces of mercuric chloride.

7. Place spathe horizontally on the surface of the laminar hood, and remove part of the external protective sheath in rectangle-shape cut by using a sterilized scalpel blade (Fig. 1a).

8. Soak inflorescence explants in antioxidants solution for 2 min before cutting them to prevent tissue browning.

9. Cut spikelets, 2–3 cm long bearing 2–4 mature flowers (Fig. 1b, c) (*see* **Note 6**).

**3.3 Induction of Callus from Mature Flowers**

1. Culture explants horizontally (1–3 segments per jar) (*see* Note 7) on the CI medium (Table 2).

2. Keep cultures in plant growth room at 25 ± 2 °C in the dark (*see* **Note 8**). Transfer culture explants at 6–8 week intervals onto fresh CI medium, and maintain under the same incubation conditions (*see* **Notes 9** and **10**).

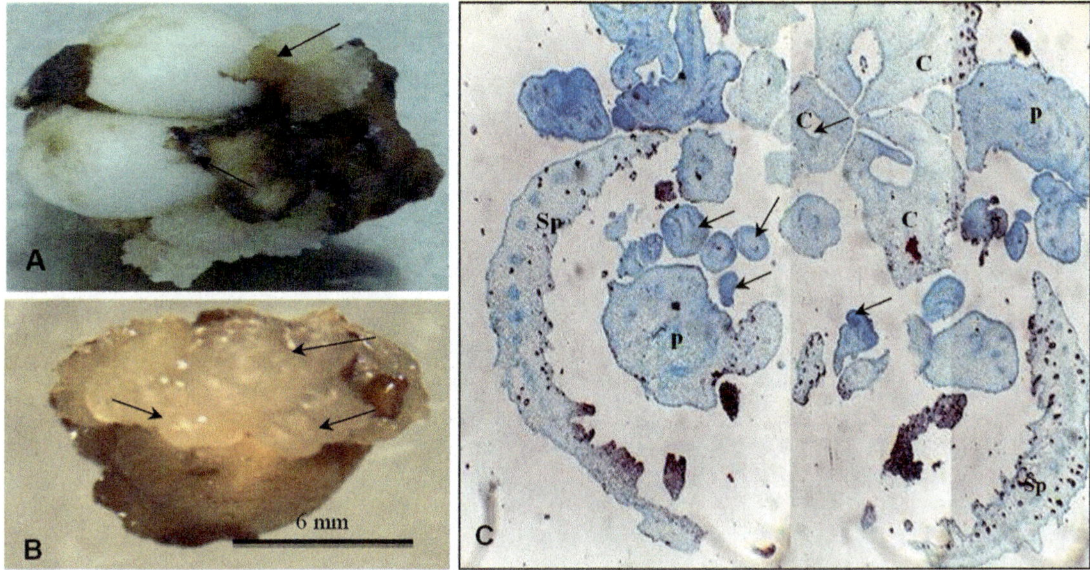

**Fig. 2** Initiation and developmental stages of the callus formation. (**a**) Mature female flower after 4–6 weeks of culturing on PBZ showing the flower swelling and callus development at the basal part between petals and carpels, (**b**) embryogenic callus masses has a translucent appearance with irregular arising from the floral perianth, (**c**) transverse section of the same flower in the *arrows* indicate the callus masses near the sepals and petals

3. Mature flowers start swelling initially and detach easily from inflorescence axis. Callus initiates at the basal part of the flower after 4–6 weeks of culture on the CI medium (Fig. 2a) (*see* **Note 11**), subsequently embryogenic callus is produced from the basal part (sepals and petals) of the flower at the end of the 16th week of culture (Fig. 2b, c) (*see* **Notes 12 and 13**).

4. Transfer embryogenic callus onto ED medium (Table 2), and keep in the dark at 25 ± 2 °C for 8 weeks and subculture at 4-week intervals.

5. Transfer germinated somatic embryos to PR medium (Table 2), and incubate at 27 ± 2 °C and 16-h photoperiod, 20 μmol/m$^2$/s of cool-white fluorescent lamps for further growth.

***3.4 Induction of Direct Shoot Formation from Mature Flowers***

1. Mature flowers swelling consequently were followed by shoot initiation from the basal part of flowers after 12 weeks of culture on S1 medium (Fig. 3a, b; Table 2) (*see* **Notes 14 and 15**).

2. Developing shoot buds form visible adventitious shoots after 24 weeks. Transfer the cultured explants to fresh SI medium at 6-week intervals.

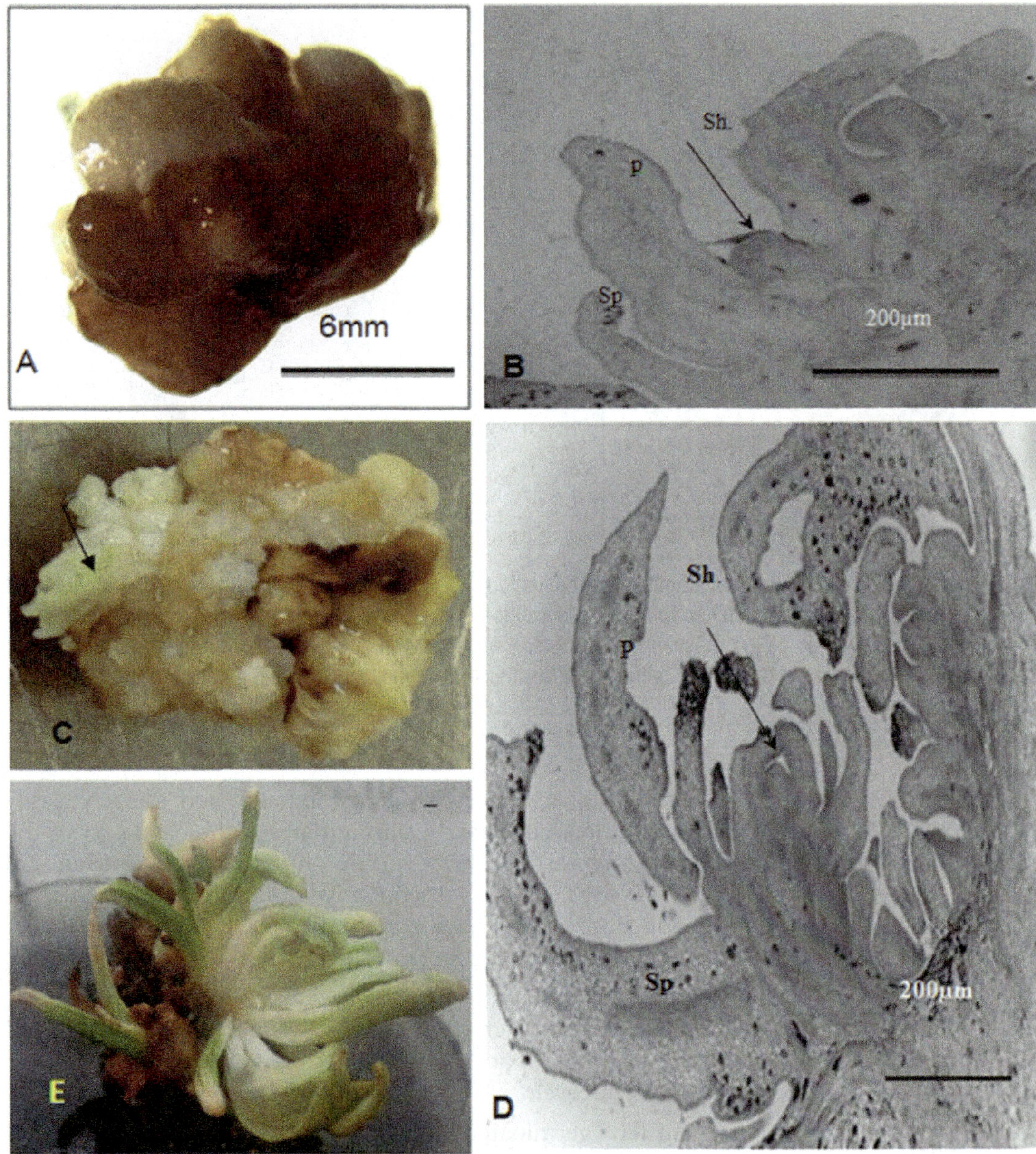



**Fig. 3** Initiation and formation of direct shoot. (**a**) Direct shoot initiation within the flower without the form of callus structures, (**b**) longitudinal section of mature pistillate flower showing in vitro induced cell divisions occurred in the basal part of the petal lead to form the first shoot initiation after 12 weeks of culturing on ABA, (**c**) new leaves arising from the surface of the mature flower after 24 weeks of culturing, (**d**) shoot meristem consisting of a mass of meristematic cells differentiated into meristematic apex surrounded by leaf primordial, (**e**) shoot formation

3. Incubate the cultures under 16-h photoperiod, 40 $\mu mol/m^2/s$ (cool-white fluorescent lamps) at $27 \pm 2$ °C (Fig. 3c) (*see* **Notes 16** and **17**).

4. Transfer adventitious shoots to PR medium (Table 2) and incubate at $27 \pm 2$ °C and 16-h photoperiod, 20 $\mu mol/m^2/s$ of cool-white fluorescent lamps for further growth (Fig. 3e).

### 3.5  Somatic Embryo Germination and Plant Formation

1. Culture mature somatic embryos and shoots on PR medium (Table 2) and maintain at $27 \pm 2$ °C under 16-h photoperiod of 40 $\mu mol/m^2/s$ provided by cool-white fluorescent lamps.

2. The mature somatic embryos are maintained for 12–18 months during which they are subcultured at 2-month intervals. Somatic embryos start to germinate 6–8 months after culturing, and complete plant formation is obtained within 3–4 months weeks after germination.

### 3.6  Pre-acclimatization and Acclimatization

1. Culture plantlets on PA medium (Table 2).

2. Incubate cultures for 6 weeks at $27 \pm 2$ °C under 16-h photoperiod of light intensity 80 $\mu mol/m^2/s$ provided by cool-white fluorescent lamps.

3. Transfer plantlets into the greenhouse, and immerse in 0.5% (w/v) Benlate fungicide solution for 1–2 min. Plant them in plastic pots $5 \times 18$ cm (Torpedo) containing a mixture of peat moss, vermiculite, and sand (1:1:1, v/v/v).

4. Keep pots in the greenhouse under natural daylight and high relative humidity 90% by covering with white polyethylene. After 1 week, punch holes in the cover, and remove completely after 1 month.

### 3.7  Histological Analysis

1. Collect the samples of mature female flowers at the time of culture and during the culture at different stages of callus and shoot development.

2. Place samples immediately in FAA solution for 24 h. During the fixation, the samples are placed under vacuum (15 min for three to four times) to remove the air inside the samples. In between these three times, add new amounts of FAA solution.

3. Dehydrate the samples to remove water from tissues in an ethanol alcohol series for 10–20 min in each ethanol grade.

4. Clear dehydrated samples by immersed in xylene-ethanol baths (3:1, 1:1, 1:3) (1 h per step).

5. Infiltrate the tissues with the paraffin wax to replace the clearing agent within the tissue and then the tissue embedded in

paraffin wax (the samples should be completely covered by molten wax).

6. Cut serial transverse and longitudinal sections (8 μm in thickness) by using microtome.

7. Stain the sections with Safranin-Fast green combination (*see* **Note 18**). Following staining, tissue sections affixed with Canada balsam to slides are typically covered by glass. These coverings protect tissue specimens from the environment and facilitate microscopic examination and micrography [17]. Then analyze under a light microscope to observe the morphological changes.

## 4  Notes

1. Sucrose added to the culture media must not exceed 30 g/L due to inflorescence containing a high level of total soluble sugars and its fractions [6, 7].

2. Add activated charcoal to induction medium to avoid tissue browning and absorb the excessive quantity of plant growth regulators.

3. Separate mature female inflorescences without any damage to the crown of tree to maintain the subsequent growth of mother trees.

4. Treat the wounded region of mother tree with 3 g/L Vitavax fungicide and 3 mL/L Coragen pesticide to protect it from infection by diseases and pests after spathes excision.

5. Sterilize the laminar flow surface by 70% ethanol before use.

6. Longitudinal section of mature female flower of date palm revealing the presence of three free short imbricate sepals, three largely imbricate petals, three separate carpels that are plump and swollen, and three separated ovules; one ovule is initiated in the base of the carpel. The tips of carpels and the stigma are exposed at the apex of the mature flower [7].

7. Keep explants in contact with culture media to maximize compounds of media uptake.

8. To induce the morphogenesis potential, and avoid oxidation of phenolic under the light.

9. When phenolic compounds are released into the culture medium, transfer cultures into fresh media to avoid toxic effect on explants.

10. Regularly monitor the growth of cultures and discard contaminated cultures and explants with poor growth.

11. Cross section shows that callus initiated from different zones of the mature female flowers: the surface of the perianth segments, the basal part (abscission zone), and from the sepals and petals.

12. Flowers that form callus enlarge drastically in size while that form shoots increase slightly in size.

13. The highest callus formation is produced from mature flower explants when cultured on 2 mg/L PBZ [7].

14. Direct shoots initiate from the basal part of the flower are mainly from meristematic cells. These shoot meristems consist of a mass of meristematic cells that later differentiated and developed, subsequently forming adventitious shoots.

15. Mature female flowers explants initiated shoots directly forms sepals and petals, while carpels turned brown and degenerated.

16. In sexual organs, maturation of monocots seems to occur in a basipetal direction (the distal parts of floral organs mature faster than the basal part) [8]. However, the histological analysis shows some meristematic cells (most reactive) at the basal part responsible for differentiation of callus or organs [7]. Meristem identity is an important aspect of plant development which is stable except during events such as switch-over from vegetative to reproductive growth or less frequently in reversion from floral to vegetative growth has been observed in response to inherent plant signals or environmental cues [18].

17. Application of exogenous specific hormones essential for stimulating differentiation of meristematic cells at the basal part of floral tissues. Subsequently, these cells initiate callus or shoots.

18. Staining of histological sections is done to increase the contrast of tissue through color that reveals structural details and allows observation of specific features.

19. An overview of this protocol for in vitro plant regeneration of date palm from mature female inflorescence through indirect somatic embryogenesis and direct organogenesis is presented in Fig. 4.

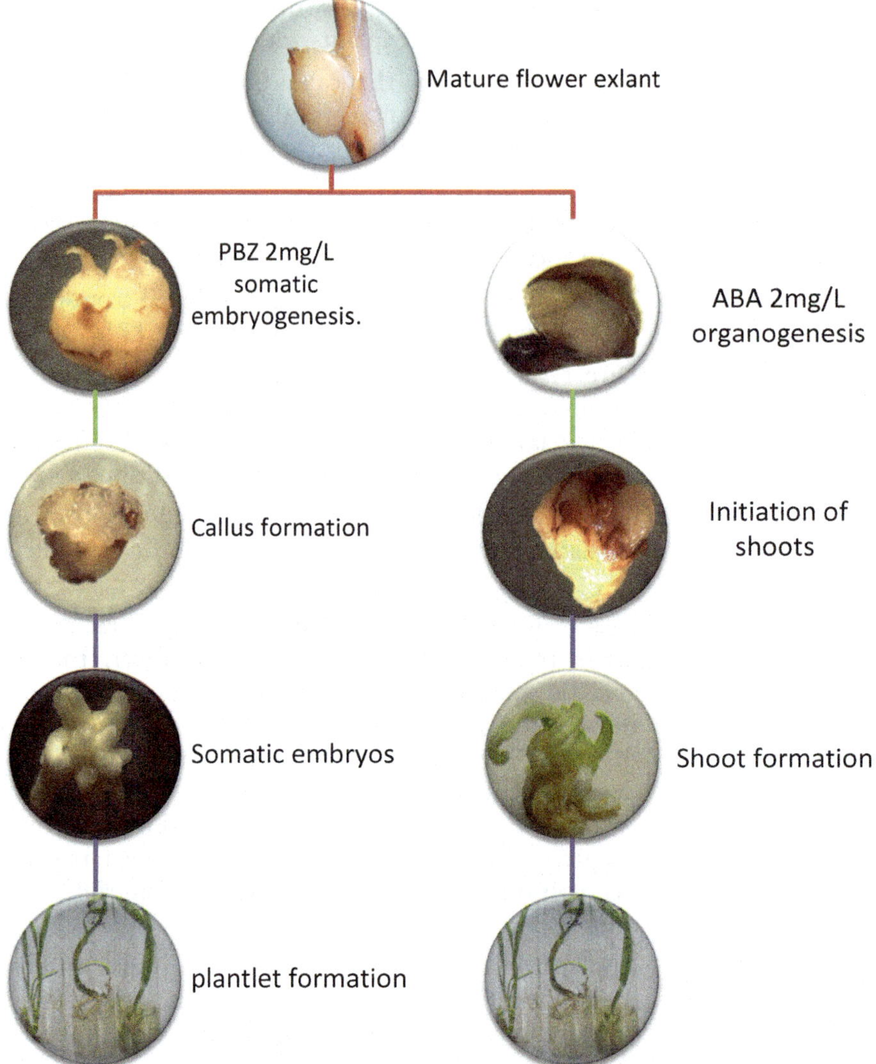


**Fig. 4** General overview of the protocol for plant regeneration of date palm from mature female inflorescence through indirect somatic embryogenesis and direct organogenesis (*see* **Note 19**)

## References


1. Loutfi K, Chlyah H (1998) Vegetative multiplication of date palms from in vitro cultured inflorescences: effect of some growth regulator combinations and organogenetic potential of various cultivars. Agronomie 18:573–580

2. Abahman L (2010) Date palm (*Phoenix dactylifera* L.) micropropagation from inflorescence tissues by using somatic embryogenesis technique. Acta Hortic 882:827–832

3. Abul-Soad AA (2011) Micropropagation of date palm using inflorescence explants. In: Jain SM, Al-Khayri JM, Johnson DV (eds) Date palm biotechnology. Springer, Dordrecht, pp 91–118

4. Zayed EMM (2011) Propagation of *Phoenix dactylifera* L. *Chamaerops humilis* L. and *Hyophorbe verschaffeltii* L. palms by using tissue culture technique. Ph.D. Thesis, Department of Horticulture, Faculty of Agriculture Cairo University, Egypt, p 167

5. Kriaa W, Sghaier B, Masmoudi F, Benjemaa R, Drira N (2012) The date palm (*Phoenix*

*dactylifera* L.) micropropagation using completely mature female flowers. C R Biol 335:194–204

6. Zayed EMM, Abdelbar OH (2015) Morphogenesis of immature female inflorescences of date palm in vitro. Ann Agric Sci 60 (1):113–120

7. Zayed EMM, Zein El-Din AFM, Manaf HH, Abdelbar OH (2016) Floral reversion of mature inflorescences of date palm in vitro. Ann Agric Sci 61(1):125–133

8. De Mason D, Stolte W, Tisserat B (1982) Floral development in *Phoenix dactylifera*. Can J Bot 60(8):1437–1446

9. George EF, Hall MA, De Klerk GJ (eds) (2008) Plant growth regulators III. In: Plant propagation by tissue culture. Springer, Dordrecht, pp 227–283

10. Rademacher W (1991) Biochemical effects of plant growth retardants. In: Gausman HW (ed) Plant biochemical regulators. Marcel Dekker Inc., New York, pp 169–200

11. Marshall JG, Rutledge RG, Blumwald E, Dumbroff EB (2000) Reduction in turgid water volume in jack pine, white spruce and black spruce in response to drought and paclobutrazol. Tree Physiol 20:701–707

12. Suboti A, Jevremovi S, Trifunovi M, Petri M, Miloševi S, Grubiši D (2009) The influence of gibberellic acid and paclobutrazol on induction of somatic embryogenesis in wild type and hairy root cultures of *Centaurium erythraea* Gillib. Afr J Biotech 8(14):3223–3228

13. Sethi U, Basu A, Mukherjee SG (1990) Role of inhibitors in the induction of differentiation in callus cultures of *Brassica*, *Datura* and *Nicotiana*. Plant Cell Rep 8:598–600

14. Murashige T, Skoog F (1962) A revised medium for rapid growth and bioassays with tobacco tissue cultures. Physiol Plant 15:473–497

15. Mater AA (1986) In vitro propagation of *Phoenix dactylifera* L. Date Palm J 4(2):137–152

16. Omar MS (1988) In vitro response of various date palm explants. Date Palm J 6(2):371–388

17. Johansen DA (1940) Plant microtechnique. McGraw-Hill Book Co., New York, pp 126–156

18. Huala E, Sussex IM (1993) Determination and cell interaction in reproductive meristems. Plant Cell 5:1157–1165

# Part II

# Somatic Embryogenesis

# Chapter 6

# Enhanced Indirect Somatic Embryogenesis of Date Palm Using Low Levels of Seawater

Rania A. Taha

## Abstract

Date palm tolerates salinity, drought, and high temperatures. Arid and semiarid zones, especially the Middle East region, need a huge number of date palms for cultivation. To meet this demand, tissue culture techniques have great potential for mass production of plantlets, especially using the indirect embryogenesis technique; any improvement of these techniques is a worthy objective. Low levels of salinity can enhance growth and development of tolerant plants. A low level of seawater, a natural source of salinity, reduces the time required for micropropagation processes of date palm cv. Malkaby when added to MS medium. Medium containing seawater at 500 ppm total dissolved solid (TDS) (12.2 mL/L) improves callus proliferation, whereas 1500 ppm (36.59 mL/L) enhances plant regeneration including multiplication of secondary embryos, embryo germination, and rooting.

**Key words** Callogenesis, Date palm, Micropropagation, Salinity, Seawater, Somatic embryogenesis

## 1 Introduction

Date palm is a promising plant for combating adverse cultivation conditions including desertification, lack of water resources, high temperature, drought, and salinity [1]. It grows very well under adverse conditions especially in arid and semiarid regions and produces highly valued date fruits. Date palm fruit is rich in carbohydrates and nutrients. It contains sugars in mostly inverted form, dietary fiber, high levels of essential amino acids and minerals [2], as well as powerful and beneficial ingredients [3].

Tissue culture techniques have become an ideal method to produce large numbers of date palm plantlets [4]. Although somatic embryogenesis is widely used commercially for propagating palms, this technique takes a long time (2–3 years) to produce acclimatized plants.

Various approaches are followed to enhance callogenesis, embryogenesis, maturation, and germination of date palm somatic embryos. Supplementing the culture medium with polyethylene

Jameel M. Al-Khayri et al. (eds.), *Date Palm Biotechnology Protocols Volume 1: Tissue Culture Applications*, Methods in Molecular Biology, vol. 1637, DOI 10.1007/978-1-4939-7156-5_6, © Springer Science+Business Media LLC 2017

glycol, growth retardants, and sorbitol was reported to increase embryo multiplication, germination, rooting, and plantlet survival during acclimatization [5, 6]. Biotin, thiamine, coconut water [7, 8], medium modification [9], and some techniques like partial desiccation also improve somatic embryogenesis of date palms [10].

A simple and rapid approach to achieve enhanced response of all culture stages of date palm embryogenesis is required in order to reduce time and cost for efficient commercial production. Low levels of salinity can enhance growth and development of in vitro date palm callus culture [11]. Moreover, the number of somatic embryos of cv. Zagloul was enhanced in response to low concentration (25 mM) of sodium chloride (NaCl). However, higher concentrations of NaCl reduced callus growth and the number of somatic embryos [11, 12]. In addition, utilization of seawater, in all in vitro culture stages, proved to be highly effective for enhancing somatic embryogenesis of cv. Malkaby [13]. This chapter focuses on the method of using seawater to improve callus proliferation, somatic embryogenesis, germination, rooting, and acclimatization in date palm micropropagation.

## 2  Materials

### 2.1  Plant Materials and Sterilization

1. Date palm offshoots of Malkaby cv., weighing 5–7 kg.
2. Fungicide solution: Benlate, 1 g/L.
3. Antibiotic solution: Penicillin, 500 mg/L.
4. Disinfectant solution: 50% commercial bleach (2.6 g/L w/v sodium hypochlorite) containing 0.1% v/v Tween 20.
5. Antioxidant solution: α-tocopherol, 400 mg/L.

### 2.2  Basal Culture Medium

1. Murashige and Skoog (MS) basal medium [14] (Table 1).
2. pH adjustment solutions: KOH and HCl solutions at 0.1, 0.5, and 1 N each.
3. Hormones stock solutions: 2,4-dichlorophenoxyacetic acid (2,4-D), isopentenyl adenine (2iP), naphthaleneacetic acid (NAA) and 6-benzylaminopurine (BAP), abscisic acid (ABA) each at 1 mg/mL.

### 2.3  Medium Additives Used for Various Culture Stages

1. Callus induction (M1): MS inorganic salts (Table 1), 100 mg/L glutamine, 10 mg/L 2,4-D, 3 mg/L 2-iP, 1.5 g/L activated charcoal (AC), 30 g/L sucrose, and solidified with 7 g/L agar.
2. Callus proliferation (M2): ¾ MS inorganic salts (Table 1), 100 mg/L glutamine, 10 mg/L 2,4-D, 3 mg/L 2-iP, seawater at 500 ppm TDS (36.59 mL seawater/L medium), 1.5 g/L AC, 30 g/L sucrose and solidified with 7 g/L agar (*see* **Note 1**).

**Table 1**
**Components of Murashige and Skoog (MS) medium [14]**

| Components | Final concentration (mg/L) |
| --- | --- |
| *Macro salts* | |
| Ammonium nitrate ($NH_4NO_3$) | 1650 |
| Calcium chloride ($CaCl_2 \cdot 2H_2O$) | 440 |
| Magnesium sulfate ($MgSO_4 \cdot 7H_2O$) | 370 |
| Potassium phosphate ($KH_2PO_4$) | 170 |
| Potassium nitrate ($KNO_3$) | 1900 |
| *Minor salts* | |
| Boric acid ($H_3BO_3$) | 6.2 |
| Cobalt chloride ($CoCl_2 \cdot 6H_2O$) | 0.025 |
| Cupric sulfate ($CuSO_4 \cdot 5H_2O$) | 0.025 |
| Ferrous sulfate ($FeSO_4 \cdot 7H_2O$) | 27.8 |
| $Na_2EDTA \cdot 2H_2O$ | 37.2 |
| Manganese sulfate ($MnSO_4 \cdot 4H_2O$) | 22.3 |
| Potassium iodide (KI) | 0.83 |
| Sodium molybdate ($Na_2MoO_4 \cdot 2H_2O$) | 0.25 |
| Zinc sulfate ($ZnSO_4 \cdot 7H_2O$) | 8.6 |
| *Vitamins* | |
| *Myo*-inositol | 100 |
| Niacin | 0.5 |
| Pyridoxine HCl | 0.5 |
| Thiamine HCl | 0.1 |
| Glycine | 2 |

3. Embryo cultures (M3): ½ MS inorganic salts (Table 1) supplemented with 0.1 mg/L NAA, 0.05 mg/L BA, seawater at 1500 ppm TDS (36.59 mL/L), 1 g/L AC, 30 g/L sucrose, and 7 g/L agar (*see* **Note 1**).

4. Rooting stage (M4): ½ MS solid medium (Table 1), 0.1 mg/L NAA and seawater at 1500 ppm TDS (36.59 mL/L), 1.5 g/L AC, 40 g/L sucrose, and 7 g/L agar.

**2.4  Equipment**

1. Tools: Scalpels, forceps, hand or electric saw, pruning shears, magnetic bars.

2. Glassware: Beakers assortment (1000, 2000, and 3000 mL).

3. Culture vessels: Glass jars of 250 mL, glass test tubes (25 × 150 mm and 25 × 250 mm), with polypropylene closures.

4. Instruments: pH meter, microwave oven, autoclave, laminar airflow hood, magnetic stirrer, and growth chamber.

5. Acclimatization supplies: Plastic pots (5 × 18 and 12 × 18 cm), potting mixture (peat moss and perlite at 2:1 v/v ratio), and polyethylene bags and sheets.

6. Facilities: Culture room, greenhouse, and shade house.

# 3 Methods

### 3.1 Preparation of Stock Solutions and Culture Media

1. Prepare NAA, 2,4-D, and ABA stock solutions by dissolving 0.1 g of each growth hormone in a few drops of KOH 0.1 N, and add distilled water to make up the volume to 100 mL. Store at −20 °C.

2. Prepare 2iP and BA stock solutions by dissolving 0.1 g of each cytokinin in a few drops of HCl 0.1 N and add distilled water to make up the volume to 100 mL. Store at −20 °C.

3. Prepare MS basal medium in distilled water according to Table 1.

4. This protocol has four main culture stages: callus induction (M1); callus proliferation and embryogenic callus (M2); embryo proliferation, germination, and elongation (M3); and rooting (M4). *See* Table 2 for preparing the suitable medium for each culture stage.

**Table 2**
**Ingredients of culture medium for various culture stages of date palm somatic embryogenesis**

| Ingredients | Callus induction, M1 | Callus proliferation, M2 | Embryo culture, M3 | Rooting, M4 |
|---|---|---|---|---|
| MS | Full | 3/4 | 1/2 | 1/2 |
| 2,4-D (mg/L) | 10.0 | – | – | – |
| 2 iP (mg/L) | 3 | 6 | – | – |
| NAA (mg/L) | – | 10 | 0.1 | 0.1 |
| BA (mg/L) | – | – | 0.05 | – |
| Seawater (mL/L) | – | 12.20 (500 ppm TDS) | 36.59 (1500 ppm TDS) | 36.59 (1500 ppm TDS) |
| Glutamine (mg/L) | 100 | 100 | – | – |
| Sucrose (g/L) | 30 | 30 | 30 | 40 |
| AC (g/L) | 1.5 | 1.5 | 1 | 1.5 |
| Agar (g/L) | 7 | 7 | 7 | 7 |

5. Adjust the pH of the media to 5.8 with HCl and KOH solutions.

6. Add agar 0.7% (w/v) for the media. Heat the media in a microwave oven for 10–15 min.

7. Dispense the media into suitable containers; 15 mL into $25 \times 150$ mm glass test tubes for callus initiation, 25 mL into $25 \times 250$ mm glass test tubes for rooting medium, 30 mL into 250 mL glass jars for callus proliferation and embryo cultures.

8. Sterilize the media by autoclaving at 121 °C for 20 min at 1.1 kg/cm$^2$ pressure.

9. Store the autoclaved media at room temperature until use for a minimum of 7 days.

**3.2 Surface Sterilization of Plant Source Material**

1. Separate offshoots, weighing 5–7 kg, from adult date palms, bring them to the lab, and remove outer leaves gradually.

2. Make several cuts at the base and leaves with hand or electric saw and pruning shears until the white primordial leaves appear with base size $10 \times 3$ cm approximately.

3. Insert explants in systemic fungicidal solution for 15 min then in an antibiotic solution for 30 min.

4. Place explants in an antioxidant solution and incubate in the refrigerator for 24 h at 5 °C (*see* **Note 2**).

5. Sterilize explants by immersion in 50% commercial bleach for 20 min with 2–3 drops of Tween 20 then wash three times with sterilized distilled water under laminar airflow hood (*see* **Note 3**).

6. Store in sterile distilled water until culture initiation.

**3.3 Environmental Conditions for Culture**

1. Incubate in vitro cultures in an incubator or plant growth room at $25 \pm 2$ °C, under dark conditions; for callus cultures and fluorescent tubes (30 μmol/m$^2$/s), with a 16-h photoperiod; for embryo cultures.

2. At the acclimatization stage, incubate the regenerated in vitro plants in a growth room at $26 \pm 2$ °C, with a 16-h photoperiod and 50 μmol/m$^2$/s light intensity.

**3.4 Callus Induction**

1. Sterilize and flame all instruments and laminar airflow hood before use.

2. Under aseptic conditions, cut shoot tips longitudinally into several parts.

3. Slightly insert segments into 15 mL callus induction medium (M1) in a test tube (Table 2). A minimum of 100 explants per genotype is recommended for initiation (Fig. 1a).

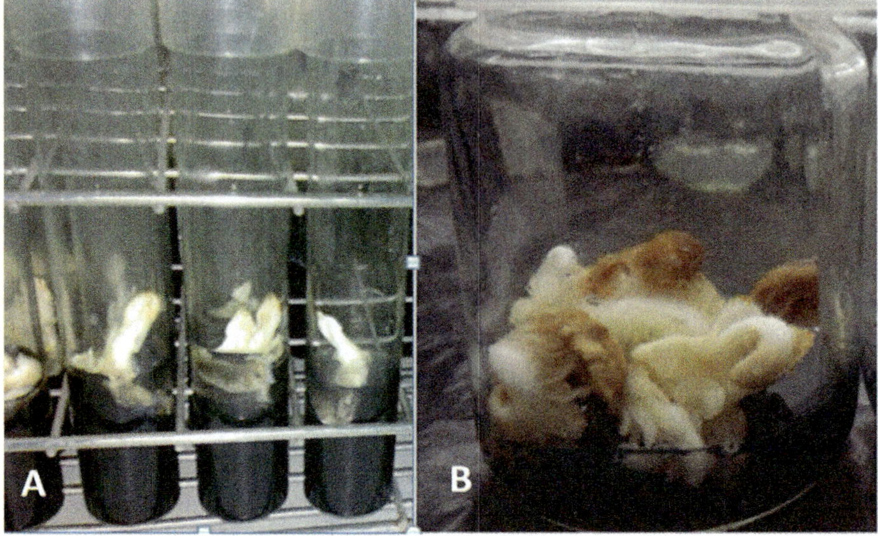

**Fig. 1** Culture initiation stage of date palm cv. Malkaby: (**a**) shoot tip explants cultured in the initiation medium and (**b**) explant swelling after 6 weeks

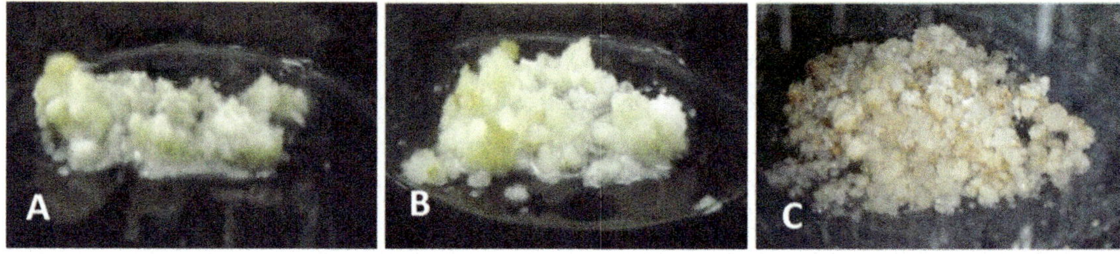

**Fig. 2** Effect of seawater on proliferation of date palm callus cultures of cv. Malkaby: (**a**) control, (**b**) at 500 ppm, and (**c**) embryogenic callus at 500 ppm

4. Seal the tube with transparent parafilm, and incubate them in the dark in a plant growth chamber or incubator.

5. After 6 weeks, transfer swollen explants (Fig. 1b) to the fresh culture medium with the same components then repeat it gradually every 6 weeks until callus appearance (*see* **Note 4**).

*3.5 Callus Proliferation*

1. Select explants producing prolific callus on the induction medium.

2. Transfer prolific callus to M2 medium (Table 2), and maintain in plant growth room or incubator in complete darkness.

3. Subculture regularly at 6-week intervals to the same fresh medium for increasing mass production of callus (Fig. 2a, b; *see* **Note 5**) until embryogenic callus appearance.

*3.6 Embryo Culture*

1. Transfer embryogenic callus cultures (Fig. 2c) produced at the previous stage, into embryo proliferation medium (M3).

**Fig. 3** Effect of seawater on shoot elongation of germinated embryos of Malkaby cv.: (**a**) control and (**b**) medium containing 1500 ppm TDS seawater

2. Excise embryo clusters and transfer to the same medium for continued embryo proliferation.

3. Maintain somatic embryo cultures in plant growth room, with light conditions ($30\ \mu mol/m^2/s$, for 16-h/day), and regularly subculture at 3-week intervals to the same fresh medium to increase mass production of somatic embryos and germination (Fig. 3a, b).

*3.7  Rooting*

1. Excise germinated somatic embryos (3–4 cm long) produced in the previous stage, and transfer to rooting medium (M4) (Fig. 4a).

2. After 4 weeks, reculture rooted plantlets into the same medium supplemented with 0.1 mg/L ABA, void of agar, and charcoal for 4 weeks (Fig. 4b; *see* **Note 6**).

*3.8  Acclimatization*

1. Sterilize soil mixture (peat moss and perlite; 2:1 by volume), irrigate with water, and fill pots.

2. Select well-rooted plantlets with 10–12 cm in length, 2–3 leaves, and thick base.

3. Wash with tab water to remove agar residue.

4. Cut leaves to approximately one-half their length.

5. Immerse plantlets in fungicide solution then plant in the soil pots.

**Fig. 4** Effect of seawater on rooting of date palm cv. Malkaby at 1500 ppm. Notice the thickness of the leaflets (**a**), the thickness and number of rootlets (**b**)

6. Immediately, cover plantlets with plastic bags or polyethylene sheets in the greenhouse to maintain relative humidity at 80%, under $27 \pm 2\ °C$, approximately (*see* **Note 7**).

7. Irrigate with ½ MS solution weekly or as needed.

8. Gradually, reduce humidity by making cuts in plastic bags or gradual removal of polyethylene sheets (*see* **Note 7**).

9. After 24 weeks of ex vitro growth, transfer surviving plants to larger pots ($12 \times 18$ cm in length) filled with the same previous planting medium, and maintain in the shaded greenhouse (*see* **Notes 8–10**).

## 4  Notes

1. The component of both cations and anions of seawater are presented in Fig. 5. Seawater used had a pH of 8.02 and total dissolved solid (TDS) at 41,000 ppm. To calculate the amount which should be added, use the following equation:

   Volume (mL) = 500 ppm × 1000 mL/41,000 ppm = 12.195 mL.
   Also 1500 ppm × 1000 mL/41,000 ppm = 36.585 mL

2. Phenolic oxidation is a serious problem for date palm micropropagation. Tocopherol is an antioxidant used to prevent browning and promote growth response [15]. However, citric and ascorbic acids can be used as an alternative to this antioxidant.

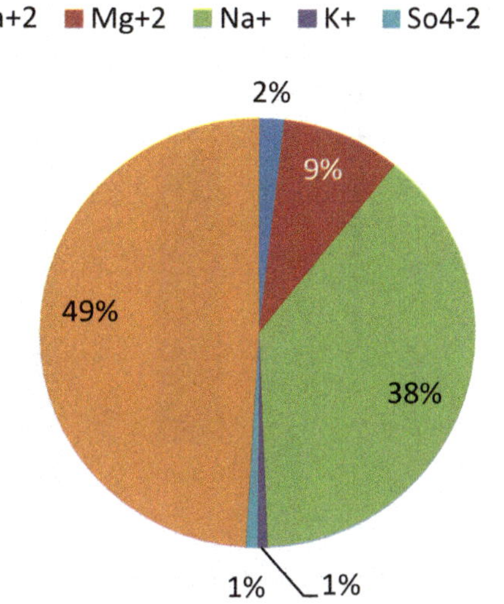


**Fig. 5** The ionic composition of seawater, expressed as percentage of total soluble salts, used in the culture medium of date palm cv. Malkaby

3. Contamination with fungi and bacteria is a major problem in date palm in vitro culture. Sterilization method in this protocol gave minimum 80% success.

4. The explant successful response is assessed by percent of swollen explants, swelling degree, and formation of callus.

5. Prolonged callus proliferation stage could enhance somaclonal variation. Seawater addition can multiply callus mass production and that could lead to avoiding the need for more subcultures [16]. It can stimulate growth and developments of plants which tolerate salinity like date palm and jojoba [13, 17].

6. Rooting stage needs at least two steps, root induction and then hairy root proliferation. Addition of ABA or ancymidol to the second rooting medium could stimulate root induction, hairy roots, and base thickness of date palm plantlets [5].

7. The process of acclimatization of plants in the greenhouse needs particular care to avoid dryness and contamination. Therefore, spraying with fungicides and discarding infected plants may be needed, gradually. Moreover, reducing humidity can be done by making small cuts and bigger cuts and then removing plastic bags, gradually at 1-week intervals.

8. Surviving plants show a new leaflet proving good root growth.

9. Plants grown ex vitro in a shaded greenhouse are maintained at 18–27 °C.

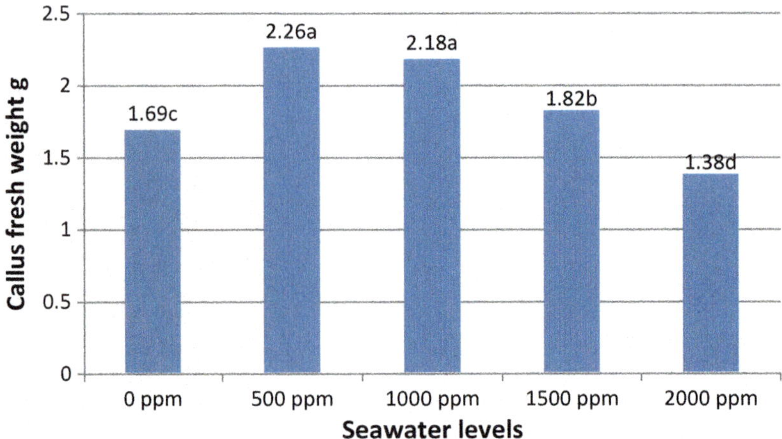

**Fig. 6** Effect of low levels of seawater on callus fresh weight (g). This figure was constructed based on data published in [13]

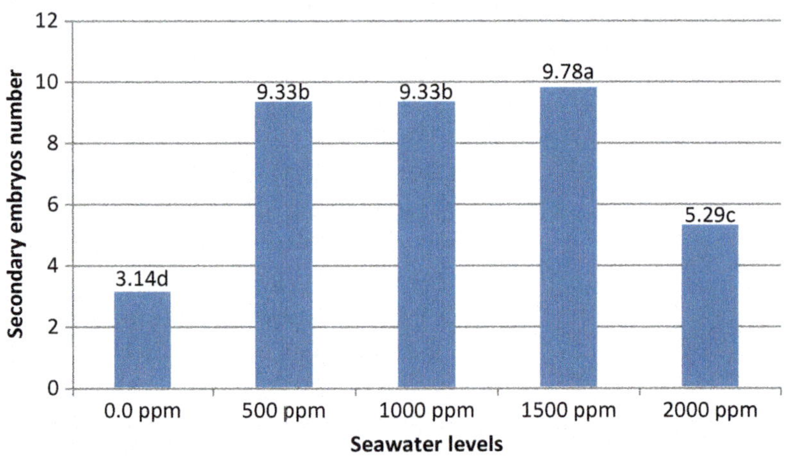

**Fig. 7** Effect of low levels of seawater on embryo numbers. This figure was constructed based on data published in [13]

10. Seawater concentrations used in this protocol are suitable for Malkaby cultivar; the optimum concentration for other date palm cultivars may vary. Figures 6, 7, 8, 9, 10 show the effect of various concentrations of seawater on date palm cv. Malkaby. Figure 11 shows the effect of various levels of seawater on some chemical analysis of in vitro leaflets of cv. Malkaby.

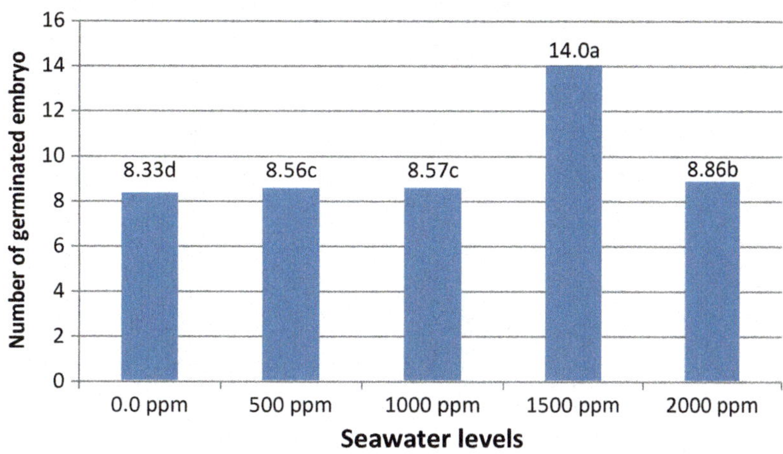

Fig. 8 Effect of low levels of seawater on germinated embryo numbers. This figure was constructed based on data published in [13]

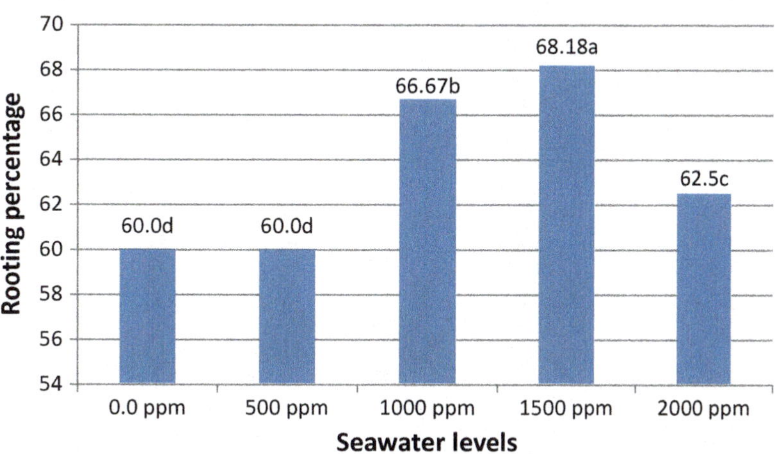

Fig. 9 Effect of low levels of seawater on percentage of rooted plantlets. This figure was constructed based on data published in [13]

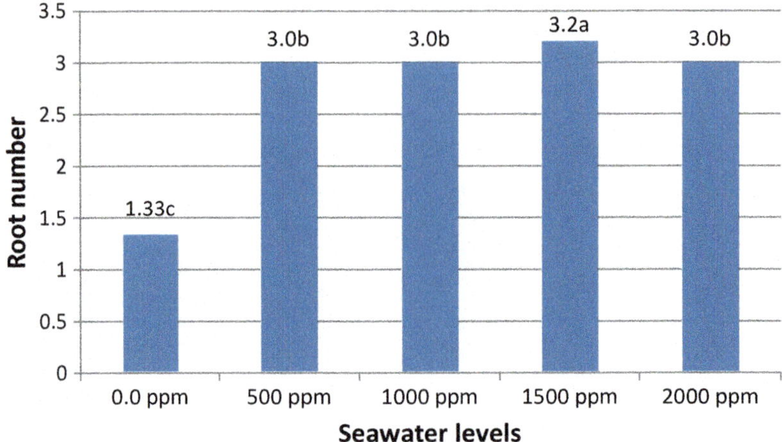

Fig. 10 Effect of low levels of seawater on root numbers per plantlet. This figure was constructed based on data published in [13]

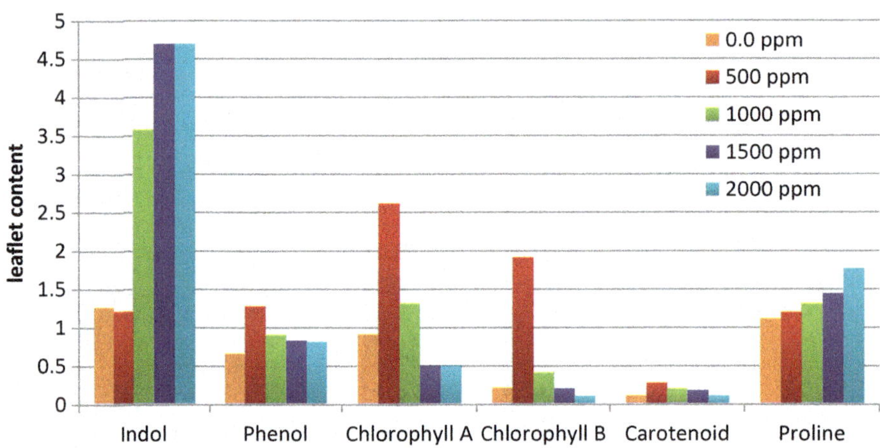

Fig. 11 Effect of low levels of seawater on biochemical analysis of in vitro date palm leaflets. This figure was constructed based on data published in [13]

## References


1. Al-Khalifah NS, Shanavaskhan AE (2012) Micropropagation of date palms. In: Asia-Pacific Consortium on Agricultural Biotechnology and Association of Agricultural Research Institutions in the Near East and North Africa, p. 54

2. Siddiq M, Greiby I (2014) Overview of date fruit production, postharvest handling, processing, and nutrition. In: Siddiq M, Aleid SM, Kader AA (eds) Dates: postharvest science, processing technology and health benefits. John Wiley and Sons Ltd., Chichester. doi:10.1002/9781118292419.ch1

3. El-Far AH, Shaheen HM, Abdel-Daim MM, Al Jaouni SK, Mousa SA (2016) Date palm (*Phoenix dactylifera*): protection and remedy food. Curr Trends Nutraceuticle 1(2):9

4. Jain SM (2007) Recent advances in date palm tissue culture and mutagenesis. Acta Hortic 736:205–211

5. Ibrahim AI, Hassan MM, Taha RA (2011) Morphological studies on date palm micropropagation as a response to growth retardants application. In: Proceedings of the 3rd international conference of genetics & it's

applications, Sharm El-Sheikh, Egypt, 5–8 Oct. 2011, pp 291–304

6. Hassan MM, Taha RA (2012) Callogenesis, somatic embryogenesis and regeneration of date palm (*Phoenix dactylifera* L.) cultivars affected by carbohydrate sources. Int J Agric Res 7:231–242

7. Al-Khayri JM (2001) Optimization of biotin and thiamine requirements for somatic embryogenesis of date palm (*Phoenix dactylifera* L.) In Vitro Cell Dev Biol Plant 37:453–456

8. Al-Khayri JM (2010) Somatic embryogenesis of date palm (*Phoenix dactylifera* L.) improved by coconut water. Biotechnology 9:477–484

9. Ibrahim K, Alromaihi KB, Elmeer KMS (2009) Influence of different media on in vitro roots and leaves of date palm somatic embryos cvs. Kapakap and Tharlaj. Am Eurasian J Agric Environ Sci 6:100–103

10. Ibrahim IA, Hassan MM, Taha RA (2012) Partial desiccation improves plant regeneration of date palm in vitro cultures. Wudpecker J Agric Res 1:208–214

11. Al-Khayri JM (2002) Growth, proline accumulation and ion content in sodium chloride stressed callus of date palm. In Vitro Cell Dev Biol Plant 38:79–82

12. Ibraheem YM, Pinker I, Böhme M (2012) The effect of sodium chloride-stress on 'Zaghloul' date palm somatic embryogenesis. Acta Hortic 961:367–373

13. Taha RA, Hassan MM (2014) Effect of low levels of salinity on development of date palm embryogenic cultures. Asian J Agric Sci 6:69–74

14. Murashige T, Skoog F (1962) A revised medium for rapid growth and bioassays with tobacco tissue cultures. Physiol Plant 15:473–497

15. Mustafa NS, Taha RA, Hassan SAM, Zaid NSM, Mustafa EA (2013) Overcoming phenolic accumulation of date palm in vitro culture using Tochopherol and cold pretreatment. Middle-East J Sci Res 15:344–350

16. Taha RA (2014) Effect of growth regulators and salinity levels on in vitro cultures of jojoba plants. World Appl Sci J 31:751–758

17. Fayek MA, Shaban EA, Zayed NS, El-Obeidy AA, Taha RA (2010) Effect of salt stress on chemical and physiological contents of jojoba (*Simmondsia chinensis* (link) Schneider) using in vitro culture. World J Agric Sci 6:446–450

# Chapter 7

# Enhanced Indirect Somatic Embryogenesis from Shoot-Tip Explants of Date Palm by Gradual Reductions of 2,4-D Concentration

## Zeinab E. Zayed

## Abstract

Shoot-tip explants obtained from offshoots of adult date palms are an excellent source for callus induction and subsequent somatic embryogenesis. In this protocol, the shoot-tip explants are transferred sequentially to a series of media containing gradually reduced concentrations of plant growth hormones: (a) 10 mg/L 2,4-dichlorophenoxy acetic acid (2,4-D) and 3 mg/L 2-isopentenyl adenine (2iP), (b) 7 mg/L 2,4-D and 1 mg/L 2iP, (c) 5 mg/L 2,4-D and 1 mg/L 2iP, and (d) 3 mg/L 2,4-D and 1 mg/L 2iP. Embryogenic callus differentiates into somatic embryos upon transfer to MS medium containing 0.5 mg/L abscisic acid (ABA) and 0.1 mg/L naphthalene acetic acid (NAA). Well-matured somatic embryos germinate on a medium containing 0.1 mg/L NAA. Repeated, multiple, and secondary somatic embryos are induced to produce normal well-developed somatic embryos upon transfer to MS medium containing 0.1 mg/L NAA and 0.05 mg/L benzyladenine (BA). This protocol is potentially applicable for commercial micropropagation of date palm.

**Key words** Acclimatization, Callus, In vitro, Micropropagation, Somatic embryo germination, Somatic embryo maturation, Rooting

## 1 Introduction

The traditional method of date palm (*Phoenix dactylifera* L.) propagation is by offshoots. However, this method is restricted by the limited number of available offshoots and reduced field survival rates. Tissue culture techniques offer an effective means of multiplication for large-scale commercial production. Date palm can be regenerated through direct somatic embryogenesis [1] and, most commonly, through indirect somatic embryogenesis which is intermediated with a callus phase [2].

Murashige and Skoog (MS) basal salt medium [3] is the choice for date palm tissue culture. The addition of 2,4-D is commonly used for callogenesis in date palm which has been reported at a wide range (1–100 mg/L) [4]. Occasionally this auxin is combined with

Jameel M. Al-Khayri et al. (eds.), *Date Palm Biotechnology Protocols Volume 1: Tissue Culture Applications*,
Methods in Molecular Biology, vol. 1637, DOI 10.1007/978-1-4939-7156-5_7, © Springer Science+Business Media LLC 2017

a cytokinin like 2-isopentyladenine (2iP) [5]. Abscisic acid (ABA) and polyethylene glycol (PEG) have been shown to play an important role in the synchronization [6] and maturation of date palm somatic embryos [7].

Embryogenic callus produces a large number of somatic embryos repetitively after continuous callus growth on MS medium containing 0.1 mg/L naphthalene acetic acid (NAA) [4]. After repeated subcultures, embryogenic nodular callus leads to the formation of proembryos originating on the callus surface [8]. During the differentiation process of date palm, different shapes of somatic embryos are observed. They can be categorized as follows: (a) individual embryos, with a normal appearance and growth; (b) repeated, multiple, and secondary somatic embryos which are ideal for date palm commercial micropropagation, as they produce a massive number of somatic embryos; and (c) occasional malformed structures that are observed that normally fail to produce plantlets [9].

This protocol describes the procedures for date palm in vitro plant regeneration based on indirect somatic embryogenesis starting from shoot-tip explants. It describes requirements for various culture stages including callus production, callus differentiation, development of somatic embryos, rooting, and acclimatization of plantlets. This protocol may be applicable for large-scale commercial production of date palm plantlets.

## 2    Materials

### 2.1    Plant Material and Disinfectants

1. Young offshoots 2–4 years old, 5–7 kg in weight, and 50–80 cm in length of date palm Siwy cv. (Fig. 1a–c, *see* **Notes 1–3**).

2. Antioxidant solution: 100 mg/L ascorbic acid and 150 mg/L citric (*see* **Note 4**).

3. Clorox solution: 20% (v/v) commercial bleach (Clorox), 5.25% w/v sodium hypochlorite plus two drops Tween 20 for 100 mL solution.

4. Mercuric chloride solution: 0.2 g/L mercuric chloride ($HgCl_2$).

### 2.2    Culture Media

1. Basal culture medium: Murashige and Skoog (MS) [3] medium stock solutions (MS stock I, II, III, and IV) (Table 1).

2. Callus induction (Medium I, Table 2): Murashige and Skoog (MS) [3] basal nutrient medium (Table 1) containing 30 g/L sucrose, 7 g/L agar-agar, 1.5 g/L activated charcoal (AC) (*see* **Note 5**), 45 mg/L adenine-sulfate, 200 mg/L glutamine, 170 mg/L $NaH_2PO_4$, and 500 mg/L casein hydrolysate

**Fig. 1** Date palm offshoots provide a source of explants: **(a)** mother date palm characterized by strong growth, and free of injuries, fungi, and insects with good yield, **(b)** offshoot attached to the mother plant of date palm, **(c)** offshoot after separation from the mother plant

(*see* **Note 6**). In addition to 3–10 mg/L 2,4-D and 1–3 mg/L 2iP for callus induction medium I a, b, c, and d.

3. Maturation of somatic embryos (Medium II, Table 2): MS basal nutrient medium (Table 1), 40 g/L sucrose (*see* **Note 7**), 7 g/L agar, 30% coconut milk, 45 mg/L adenine sulfate, 170 mg/L $NaH_2PO_4.2H_2O$, 200 mg/L $KH_2PO_4$, 200 mg/L glutamine, 0.5 mg/L ABA (*see* **Note 8**), and 0.1 mg/L NAA.

4. Germination and conversion of somatic embryos to plantlets (Medium III, Table 2): ¾ MS basal nutrient medium (Table 1), 40 g/L sucrose, 7 g/L agar, 30% coconut water, 170 mg/L $NaH_2PO_4.2H_2O$, 200 mg/L $KH_2PO_4$, 200 mg/L glutamine, 0.1 mg/L biotin, 0.2 mg/L calcium pantothenate, 0.1 mg/L NAA, and 0.05 mg/L benzyl adenine (BA).

5. Nutrient medium component for rooting stage (Medium IV, Table 2): ½ MS basal nutrient medium (Table 1), 50 g/L

**Table 1**
**Chemical composition of MS basal nutrient medium used for date palm in vitro during indirect somatic embryogenesis protocol [3]**

| Medium composition | Stock concentration (mg/L) | Final concentration in culture medium (mg/L) |
| --- | --- | --- |
| *Stock I: Major inorganic nutrients (20× stock) use 50 mL to prepare 1 L medium* | | |
| $NH_4NO_3$ | 33,000 | 1650 |
| $KNO_3$ | 38,000 | 1900 |
| $CaCl_2 \cdot 2H_2O$ | 8800 | 440 |
| $MgSO_4 \cdot 2H_2O$ | 7400 | 370 |
| $KH_2PO_4$ | 3400 | 170 |
| $NaH_2PO_4.H_2O$ | 3400 | 170 |
| *Stock II: Minor inorganic nutrients (200× stock) use 5 mL to prepare 1 L medium* | | |
| KI | 166 | 0.83 |
| $H_3BO_3$ | 1240 | 6.2 |
| $MnSO_4 \cdot 2H_2O$ | 4460 | 22.3 |
| $ZnSO_4 \cdot 7H_2O$ | 1720 | 8.6 |
| $Na_2.MoO_4 \cdot 2H_2O$ | 50 | 0.25 |
| $CuSO_4 \cdot 5H_2O$ | 5 | 0.025 |
| $CoCl_2 \cdot 6H_2O$ | 5 | 0.025 |
| *Stock III:* Iron source *(200× stock) use 5 mL to prepare 1 L medium* | | |
| $FeSO_4 \cdot 7H_2O$ | 5560 | 27.8 |
| $Na_2EDTA \cdot 2H_2O$ | 7460 | 37.3 |
| *Stock IV: Vitamins (200× stock) use 5 mL to prepare 1 L medium* | | |
| *myo*-Inositol | 25,000 | 125 |
| Nicotinic acid | 200 | 1 |
| Pyridoxine·HCl | 200 | 1 |
| Thiamine·HCl | 200 | 1 |
| Glycine | 400 | 2 |

sucrose, 7 g/L agar, 0.1 mg/L biotin, 0.2 mg/L calcium pantothenate, 1 mg/L NAA, 1 mg/L indolebutyric acid (IBA), 1 g/L activated charcoal (AC), and 0.4 mg/L paclobutrazol (PBZ) (*see* **Note 9**).

6. Pre-acclimatization (Medium V, Table 2): ¼ MS basal liquid nutrient medium (Table 1) and 10 g/L sucrose, 6 g/L polyethylene glycol 8000(PEG) (*see* **Note 10**).

**Table 2**
**Hormonal, activated charcoal, agar, and sucrose supplements to the MS culture medium (Table 1) according to the culture stages of date palm indirect somatic embryogenesis**

| | Culture phase | | | | |
| Media additives | Callus induction Medium I (a, b, c, d) | Maturation Medium II | Germination Medium III | Rooting Medium IV | Pre-acclimatization Medium V |
|---|---|---|---|---|---|
| 2,4-D | (a) 10 mg/L<br>(b) 7 mg/L<br>(c) 5 mg/L<br>(d) 3 mg/L | – | – | – | – |
| 2iP | (a) 3 mg/L<br>(b) 1 mg/L<br>(c) 1 mg/L<br>(d) 1 mg/L | – | – | – | – |
| NAA | – | 0.1 mg/L | 0.1 mg/L | 1 mg/L | – |
| ABA | – | 0.5 mg/L | – | – | – |
| BA | – | | 0.05 mg/L | – | – |
| IBA | – | – | – | 1 mg/L | – |
| PBZ | – | – | – | 0.4 mg/L | – |
| AC | 1.5 g/L | – | – | 1 g/L | – |
| Agar | 7 g/L | 7 g/L | 7 g/L | 7 g/L | – |
| Sucrose | 30 g/L | 40 g/L | 40 g/L | 50 g/l | 10 g/L |

7. Acclimatization: ¼ MS inorganic salts solution; fungicide solution (Benlate 0.1% w/v).

*2.3 Equipment*

1. Glassware and plastic ware: Glass culture jars (150 and 250 mL), test tubes (2.5 × 25 cm), Erlenmeyer flask, pipettes (0.1–10 mL), beakers, Petri dishes, aluminum foil, and covering using polypropylene.

2. Surgical tools: Axe, saw, forcipes, and scalpel.

3. Acclimatizing: Plastic pots (18 cm height and 5 cm diameter), potting soil mixture (peat moss, vermiculite and sand mixed at 1:1:1 v/v/v), polyethylene bags, and greenhouse.

# 3   Methods

*3.1 Prepare Nutrient Media*

1. Mix (MS) salts solutions with other components for each prepared nutrient medium.

2. Adjust pH of all culture media to 5.8 by adding few drops of HCl or NaOH solution prior to the addition of agar to culture nutrient medium.

3. Add agar, make up the desired volume of media by distilled water, and heat the solution until the agar is fully dissolved.

4. Dispense medium into small jar (150 mL) at 35 mL/jar for shoot-tip establishment, callus induction, and germination media while 45 mL/jar for rooting medium. Test tubes (2.5 × 25 cm) containing 20 mL culture medium are used for pre-acclimatization stage.

5. Cap the jars immediately with polypropylene, and autoclave for 15 min at 121 °C and 1.1 kg/cm$^2$.

### 3.2 Prepare and Sterilize Explants

1. Remove from offshoots the white soft leaves until the shoot tip appears. The apical meristem plus a few primordial leaves (shoot tip) used as explants material (Fig. 2a).

2. Dip the shoot-tip explants in antioxidant solution for 15 min.

3. Surface sterilize explants (shoot tip) under aseptic conditions, and divide into two steps:

    (a) Immerse the explants in Clorox solution for 5 min and then rinse twice with sterile distilled water.

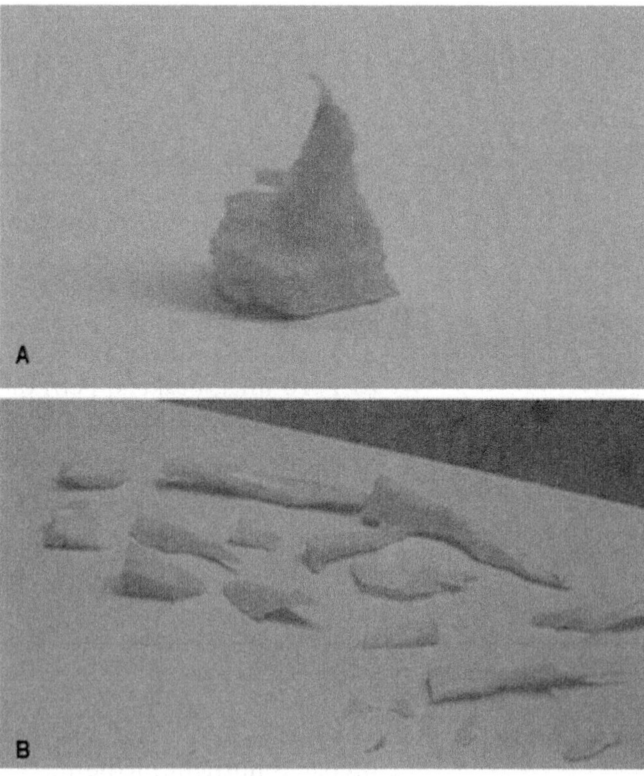

**Fig. 2** Shoot-tip explant preparation: **(a)** shoot-tip explants, **(b)** sections of explants, cut longitudinally into 4–8 sections

(b) Transfer the explants to Petri dishes and remove two most exterior primordial leaves, surface sterilize in 0.2 g/L mercuric chloride ($HgCl_2$) for 1 h, and rinse the explants three times with sterile distilled water.

4. Remove all primordial leaves except the two pairs surrounding the shoot-tip explants.

**3.3 Culture of Explants and Callus Induction**

1. Cut the surface disinfected shoot tip longitudinally into 4–8 sections and inoculate explants individually on callus induction medium (Medium I a; Table 2) (Fig. 2b; *see* **Note 11**).

2. Transfer the shoot-tip explants sequentially to a series of media containing gradually reduced concentrations of 2,4-D in addition to 2iP (Medium I b, c, d; Table 2) from the first to the fifth subculture. Incubate culture jars in total darkness at $25 \pm 2\,°C$ (*see* **Note 12**).

3. Check culture jars to be sure that explants are free of any fungal or bacterial contamination, and subculture contamination-free cultures at 4-week intervals.

4. At the end of the third subculture, compact callus is formed and then develops into friable callus at the end of the fifth subculture (Fig. 3a) (*see* **Note 13**).

5. Friable embryogenic callus is harvested which contains free nodules of proembryos (Fig. 3b).

**3.4 Maturation of Somatic Embryos**

1. Culture the embryogenic callus and culture on somatic embryo maturation medium (Medium II; Table 2; Fig. 3c).

2. Incubate the cultures in a growth room at $24 \pm 1\,°C$ and 16-h photoperiod ($45\ \mu mol/m^2/s$) provided by cool-white fluorescent lamps.

**3.5 Germination and Conversion of Somatic Embryos to Plantlets**

1. Five different forms of somatic embryos are observed: individual (normal), repeated, multiple, secondary embryos, and any other form of somatic embryo (malformed structures) (Fig. 4a–e; *see* **Notes 14–16**).

2. Transfer individual embryos (normal embryo) to the germination medium (Medium III; Table 2).

3. Repeated, multiple, and secondary somatic embryos are transferred to a fresh germination medium (Medium III) for at least four subcultures.

4. Incubate cultures at $27 \pm 2\,°C$ and 16-h photoperiod provided by cool-white fluorescent light ($100\ \mu mol/m^2/s$).

5. After the somatic embryos convert to plantlets, transfer to the rooting stage (Fig. 5a, b).

**Fig. 3** Callus multiplication and somatic embryogenesis: **(a)** callus compact appearance on the surface of explants, **(b)** friable calli consist of free nodules of proembryos (embryonic callus), **(c)** differentiation of somatic embryos

**3.6  Rooting Stage**

1. Collect the plantlets derived from somatic embryogenesis after separating the shoots. The plantlets 5 cm in length with two expanded leaves and an adequate root system (two to three roots) are transferred to rooting medium (Medium IV) for two subcultures at 4-week intervals (Fig. 5c; Table 2).

2. Transfer the plantlets to test tubes (2.5 × 25 cm) containing nutrient culture medium for pre-acclimatization (Medium V) (Table 2). All tubes are covered with aluminum foil caps and incubated at 16-h photoperiod of cool-white fluorescent light (200 $\mu$mol/m$^2$/s) and 27 $\pm$ 2 °C for 1 month as a pre-acclimatization stage.

Fig. 4 (a) Normal somatic embryo with one root and folded tip (individual embryo). (b) Repeated embryo, all embryos originated from the same point; sometimes this point originates embryogenic callus developed within the same culture into secondary embryos. (c) Multiple embryo; notice no embryogenic callus around the attached point and a synchronization development of shoots. (d) Secondary embryo on the top of primary embryo coat: *1.* root of primary embryo, *2.* callus derived from the coat base, *3.* secondary embryos, *4.* shoot of primary embryo. (e) Abnormal somatic embryos (malformed embryos)

**3.7  Acclimatization Stage**

1. Rinse the plantlets thoroughly with tap water, and then immerse in fungicide solution (Benlate, 0.1% w/v) for 5 min.

**Fig. 5** Date palm tissue culture final stages: **(a)** cluster of shoots, **(b)** elongation of shoots, **(c)** rooting, **(d)** acclimatization

2. Transfer the plantlets into plastic pots containing soil mix, and cover with polyethylene bags to maintain high relative humidity (90–95%).

3. Reduce the relative humidity gradually after 1 week from planting, through holes punched in the polyethylene bag, and remove the bags completely after 4 weeks (Fig. 5d, *see* **Note 17**).

4. Maintained the plants in the greenhouse under natural light at 27 ± 2 °C and 50–60% relative humidity, and irrigate with ¼ MS inorganic salts, once a week and spray with fungicide as needed.

## 4  Notes

1. The offshoots are taken from the mother date palms and transferred to the laboratory on the same day to prevent dryness and to reduce secretion of phenolic compounds that can cause inhibition of tissue response.

2. The propagation process starts with the selection of healthy offshoots from mother date palm trees. The selected mother plants should exhibit strong growth with good yield and free of physical injuries, fungal diseases, and insects (Fig. 1a–c).

3. The offshoots are preferably separated from the mother plant date palm in October and November (autumn) or March and April (spring).

4. The explants are immersed for 15 min in antioxidant solution to avoid browning. Browning in palm tissue culture is toxic and can cause the death of explants.

5. Browning can be decreased by adding activated charcoal to the nutrient medium, using moderate pH 5–5.6, and maintaining cultures in the dark at 23 °C.

6. The addition of casein hydrolysate to callus medium accelerates callus production. Casein hydrolysate is a mixture of 18 amino acids, source of calcium, phosphate, several microelements, and vitamins [10, 11].

7. Increasing sucrose to 40 g/L improves maturation of somatic embryo and then reduces synchronous process which is noticed during somatic embryos development.

8. Adding 0.5 mg/L ABA for 4 weeks to the maturation medium (Medium II) promotes embryo maturation and supports the accumulation of storage proteins, lipids, and starch. ABA suppresses the formation of aberrant embryo structures and prevents precocious germination of somatic embryos [12].

9. The main target of adding 0.4 mg/L PBZ to the rooting medium (Medium IV) is to increase the trunk thickness of plantlets, accelerate root formation and promote root branching (secondary roots) for successful transplanting.

10. The addition of PEG 8000 (6 g/L) to pre-acclimatization medium (Medium V) increases wax deposition and decreases water loss when plantlets are transferred ex vitro.

11. The sterilization process of explants involves two disinfectants to assure eliminating fungus and bacterial organisms in order to maximize tissue survival.

12. Gradually decreasing the concentration of 2,4-D from 10 to 3 mg/L encourages growth of callus and accelerates the development to embryonic callus (proembryos).

13. Shoot tips initiate white yellowish callus with a granular appearance in a medium containing 3 mg/L 2,4-D and 1 mg/L 2iP. The nodules are somatic proembryos located on the surface of the callus.

14. The individual normal somatic embryo germinates directly into the shoot and root to obtain hardy plantlets. Repeated embryos are clusters of 3–4 embryos arising repetitively, which are usually of normal morphology. Multiple embryos are secondary embryos which consist 3–4 embryos and can occur on the base of the original embryo (Fig. 4a–c).

15. The secondary embryos are accessory embryos growing on original embryos. The original somatic embryo can generate new somatic embryos (Fig. 4d).

16. Malformed structures fail to complete their growth and development process to form somatic embryos (Fig. 4e).

17. The plantlets are covered with polyethylene bags to prevent sudden exposure to ambient atmosphere and subsequent plant dehydration [13].

## Acknowledgment

The author expresses sincere thanks and appreciation to agronomists Ahmed Hassan El-Tanboly and Mahmoud Ali Hamza for their vital help to achieve this work.

## References

1. Sudhersan C, Abo El-Nil M, Al-Baiz A (1993) Occurrence of direct somatic embryogenesis on the sword leaf of in vitro plantlets of *Phoenix dactylifera* L., cultivar Barhee. Curr Sci 65:887–888

2. Al-Khayri JM (2007) Date palm *Phoenix dactylifera* L. micropropagation. In: Jain SM, Haggman H (eds) Protocols for micropropagation of woody trees and fruits. Springer, The Netherlands, pp 509–526

3. Murashige T, Skoog F (1962) A revised medium for rapid growth and bioassays with tobacco tissue culture. Physiol Plant 15:473–479

4. Mater AA (1986) In vitro propagation of (*Phoenix dactylifera* L.) Date Palm J 4(2):137–152

5. Ammirato PV, Steward FC (1971) Some effects of the environment on development of embryos from cultured free cells. Bot Gaz 132:149–158

6. Al-Khayri JM, Al-Bahrany AM (2012) Effect of abscisic acid and polyethylene glycol on the synchronization of somatic embryo development in date palm (*Phoenix dactylifera* L.) Biotechnol 11(6):318–325

7. Zaid ZE, Abo-El-Soaud AA, Sidky RA (2008) Effect of plant growth regulators and polyethylene glycol on maturation of date palm somatic embryos. Egypt J Biotechnol 30:47–56

8. Fki L, Masmoudi R, Drira N, Rival A (2003) An optimized protocol for plant regeneration from embryogenic suspension cultures of date palm (*Phoenix dactylifera* L.) cv. Deglet nour. Plant Cell Rep 21:517–524

9. Zaid ZE, Gomaa AH, Ibramim IA (2004) In vitro growth and development of different somatic embryos shapes of date palm genotypes. In: The second international conference on date palm, El-Arish, North Sinai, Egypt, 6–8 Oct. 2004, pp 139–156

10. Khierallah HSM, Hussein NH (2013) The role of coconut water and casein hydrolysate on somatic embryogenesis of date palm and genetic stability detection using RAPD marker. Res Biotechnol 3(4):20–28

11. Hosny S, Hammad G, El Sharbasy S, Zayed Z (2016) Effect of coconut milk, casein hydrolysate and yeast extract on the proliferation of *in vitro* Barhi date palm (*Phoenix dactylifera* L.) J Hort Sci Ornament Plant 8(1):46–54

12. Zaid ZE, Gomaa AH, Ibramim IA (2004) Somatic embryogenesis production of date palm (Phoenix dactylifera L.) by application of abscisic acid (ABA). In: International conference engineering & application, Sharm El-Sheikh, South Sinai, Eygpt, 8–11 Apr. 2004, pp 467–481

13. Zaid A, De Wet PF (2002) Date palm propagation. In: Zaid A (ed) Date palm cultivation. Rev. Ed. Plant production and protection paper 156. Food and Agriculture Organization United Nations, Rome, pp 73–105

# Chapter 8

# Indirect Somatic Embryogenesis from Mature Inflorescence Explants of Date Palm

## Ali M. Al-Ali, Chien-Ying Ko, Sultan A. Al-Sulaiman, Sami O. Al-Otaibi, Abd Ulmoneem H. Al-Khamees, and Megahed H. Ammar

## Abstract


Due to the limitations associated with shoot tip explants in the micropropagation of date palm, inflorescence explants are an ideal alternative. This chapter focuses on the protocol for the induction of callus from inflorescence tissue, establishment for proliferation of somatic embryos, germination, elongation, rooting, and acclimatization. Female inflorescences, 30–40 cm in length, cv. Shaishee, were used for culture initiation. After disinfection, the outer inflorescence cover (spathe) is cut open, and the spikelet explants, 1 cm long, are cultured on modified Murashige and Skoog (MS) medium containing 100 mg/L 2,4-D, 3 mg/L kinetin, and 3 mg/L 2ip and incubated at $25 \pm 2\,^\circ\mathrm{C}$ in the dark. Callus obtained after 6–8 months of culturing is transferred to the culture medium to induce somatic embryogenesis and plant regeneration. Well-developed regenerated shoots are cultured on MS medium containing 0.2 mg/L NAA for root induction and plantlets acclimatized in the greenhouse before transfer to the field.

**Key words** Callus, Inflorescent, Micropropagation, Somatic embryo


## 1 Introduction

Date palm (*Phoenix dactylifera* L.) is one of the most important economic crops for fresh fruit and food processing and is used as a landscape tree in the Middle East and North Africa [1]. The number of date palms in the Kingdom of Saudi Arabia is estimated at 21 million trees; about 400 different cultivars are spread over the diverse agricultural areas of the Kingdom [2]. The conventional method of date palm production is carried out using vegetatively propagated offshoots for field cultivation, which has the limitation that each palm produces a low number of offshoots [3]. This is due to offshoots being produced only during the vegetative development phase of the palm and on genotypic dependence. Micropropagation of date palm from meristems [4, 5] and immature inflorescences [6–8] for initiation has been reported and may be applied efficiently to regenerate a large number of disease-free


Jameel M. Al-Khayri et al. (eds.), *Date Palm Biotechnology Protocols Volume 1: Tissue Culture Applications*,
Methods in Molecular Biology, vol. 1637, DOI 10.1007/978-1-4939-7156-5_8, © Springer Science+Business Media LLC 2017

plants [9–14]. However, available offshoots, which provide meristem tissue for explants, are produced in limited numbers in some rare cultivars. Inflorescences, as a source of explants materials, can overcome this limitation. Thus, establishing a simple initiation protocol to increase the survival of explants is very important for commercial production. The purpose of this chapter is to provide a reproducible and reliable protocol to establish efficient date palm micropropagation by using inflorescences.

## 2   Materials

### 2.1   Plant Material and Explant Preparation

1. Date palm female inflorescence of cultivar Shaishee (*see* **Note 1**).
2. Disinfection solution: 70% ethanol.

### 2.2   Culture Media

1. Basal medium: Murashige and Skoog (MS) [15] medium stock solutions (Table 1). Store stock solutions in the refrigerator at 4 °C. Use chemicals of analytical grade.
2. Solution for pH adjustment: 1 N NaOH and 1 N HCl.
3. Hormones and other additives are listed in Table 2 according to the culture stages.
4. Callus induction medium (I): MS medium containing 200 mg/L glutamine, 2.5 mg/L Ca-pantothenate, 170 mg/L $NaH_2PO_4$, 100 mg/L 2,4-D, 3 mg/L kinetin, 3 mg/L 2ip, 75 mg/L citric acid, 75 mg/L ascorbic acid, 1.5 g/L activated charcoal, 0.8 g/L phytagel, and 4 g/L agar (Table 2).
5. Somatic embryo induction medium (II): MS medium containing 200 mg/L glutamine, 2.5 mg/L Ca-pantothenate, 170 mg/L $NaH_2PO_4$, 2.5 mg/L NAA, 3 mg/L kinetin, 3 mg/L 2ip, 75 mg/L citric acid, 75 mg/L ascorbic acid, 1.5 g/L activated charcoal, 0.8 g/L phytagel, and 4 g/L agar.
6. Somatic embryo proliferation and germination medium (III): MS medium containing 200 mg/L glutamine, 2.5 mg/L Ca-pantothenate, 170 mg/L $NaH_2PO_4$, 0.8 g/L phytagel, and 4 g/L agar.
7. Shoot elongation (IV): MS medium containing 200 mg/L glutamine, 2.5 mg/L Ca-pantothenate, 170 mg/L $NaH_2PO_4$, 1.0 mg/L NAA, 0.5 mg/L 2ip, 1.5 g/L activated charcoal, 0.8 g/L phytagel, and 4 g/L agar.
8. Rooting (V): MS medium containing 200 mg/L glutamine, 2.5 mg/L Ca-pantothenate, 170 mg/L $NaH_2PO_4$, 0.2 mg/L NAA, 0.8 g/L phytagel, and 4 g/L agar.

**Table 1**
**Chemical composition of the modified MS medium [15]**

| Chemical constituents | Concentration (mg/L) | Volume per liter (mL) |
|---|---|---|
| *Major inorganic nutrients* | | |
| $KNO_3$ | 19,000 | 100 |
| $NH_4NO_3$ | 16,500 | |
| $CaCl_2 \cdot 2H_2O$ | 4400 | |
| $MgSO_4 \cdot 2H_2O$ | 3700 | |
| $KH_2PO_4$ | 1700 | |
| $NaH_2PO_4.H_2O$ | 1955 | |
| *Minor inorganic nutrients* | | |
| KI | 166 | 5 |
| $H_3BO_3$ | 1340 | |
| $MnSO_4 \cdot 2H_2O$ | 3377 | |
| $ZnSO_4 \cdot 7H_2O$ | 1860 | |
| $Na_2.MoO_4 \cdot 2H_2O$ | 50 | |
| $CuSO_4 \cdot 5H_2O$ | 5 | |
| $CoCl_2 \cdot 6H_2O$ | 5 | |
| *Iron source* | | |
| Fe Na EDTA | 7340 | 5 |
| *Organic supplements* | | |
| *Myo*-inositol | 25,000 | 5 |
| *Vitamins and others* | | |
| L-glutamine | 40,000 | 5 |
| Nicotinic acid | 1000 | |
| Pyridoxine·HCl | 500 | |
| Thiamine·HCl | 1000 | |
| Glycine | 600 | |
| Calcium pantothenate | 500 | |
| *Carbon source* | | |
| Sucrose | 30 g/L | |

**2.3 Plant Acclimatization**

1. Peat moss.
2. Sand.
3. Perlite.
4. Yellow containers, plastic film, and 7.5 cm plastic pots.
5. Fungicide solution: 0.6 g/L Carbomar, a systemic fungicide.
6. Plant foods (Peters professional, 20-20-20).
7. Sodium hypochlorite.

**2.4  Equipment**

1. Culture tubes, 15 and 20 cm.
2. GA7 Magenta box.
3. Micropipettes, 1000 μL.
4. pH meter.
5. Media dispenser.
6. Saran wrap plastic cling film.
7. Forceps, scissors, and scalpel.
8. Flame source.

# 3   Methods

**3.1  Medium Preparation**

1. Mix all stock solution and bring to the final volume of 1 L with distilled water.
2. Add other ingredients as specified in Table 2 according to culture stage.
3. Adjust the pH of the medium to 5.7 with 1 N HCl and 1 N NaOH.

**Table 2**
**Hormonal, activated charcoal, antioxidants, phytagel, and agar supplements to the culture medium used for inflorescence of date palm callus induction, proliferation, and germination of somatic embryos, elongation, and rooting**

| Media additives | Culture phase (medium code) | | | | |
| --- | --- | --- | --- | --- | --- |
| | Culture initiation (I) | Embryogenic callus (II) | Somatic embryogenesis (III) | Shoot elongation (IV) | Rooting (V) |
| 2,4-D | 100 mg/L | – | – | – | – |
| 2iP | 3 mg/L | 3 mg/L | – | 0.5 mg/L | – |
| NAA | – | 2.5 mg/L | – | 1 mg/L | 0.2 mg/L |
| Kinetin | 3 mg/L | 3 mg/L | – | – | – |
| A C | 1.5 g/L | 1.5 g/L | – | 1.5 g/L | – |
| Ascorbic acid | 75 mg/L | 75 mg/L | – | – | – |
| Citric acid | 75 mg/L | 75 mg/L | – | – | – |
| Phytagel | 0.8 g/L | 0.8 g/L | 0.8 g/L | 0.8 g/L | 0.8 g/L |
| Agar | 4 g/L | 4 g/L | 4 g/L | 4 g/L | 4 g/L |

Abbreviations: 2,4-dichlorophenoxyacetic acid (2,4-D), 2-isopentenyladenine (2iP), naphthaleneacetic acid (NAA), Activated charcoal (AC)

4. Add 4 g/L agar and 0.8 g/L phytagel. Heat until agar is dissolved.

5. Dispense medium in culture vessels and autoclave the medium for 20 min at 121 °C.

### 3.2  Culture Initiation

1. Isolate female inflorescence of Shaishee cv., 30–40 cm in length (Fig. 1a, b).

**Fig. 1** The regeneration and somatic embryo formation processes of the inflorescence tissue of date palm: (**a**) plant material, (**b**) complete inflorescence (30–40 cm), (**c**) inflorescence explants (1 cm), (**d**) explants in medium, (**e**) callus formation, (**f**) callus proliferation, (**g**) white callus, (**h**) somatic embryo proliferation, (**i**) plantlet regeneration, (**j**) plantlet elongation, (**k**) medium used for establishment of proper rooting system

2. Sterilize the exterior surface of inflorescence spathe by wiping twice with 70% ethanol inside the laminar flow hood (*see* **Note 2**).

3. Cut the spathe open to expose the inflorescence and isolate 1-cm-long spikelets to use as explants (Fig. 1c, d).

4. Culture the explants on the initiation medium (Table 2, I; *see* **Note 3**).

5. Incubate cultures at $25 \pm 2$ °C in the dark.

6. Transfer explants to a fresh culture medium at a 4-week interval for 6–8 months until callus initiation (Fig. 1e, f; *see* **Note 4**).

### *3.3 Callus Induce and Somatic Embryo Formation*

1. Maintain callus on the medium containing NAA (2.5 mg/L by replacing 2,4-D from the callus medium) (Table 2, II).

2. Incubate cultures at $25 \pm 2$ °C in the dark.

3. Somatic embryos will be obtained after culturing for 3–4 months by subculturing at 4-week intervals (Fig. 1g).

### *3.4 Proliferation of Somatic Embryos*

1. Separate somatic embryos with a scalpel and culture on a modified MS medium (Table 2, III).

2. Incubate cultures at $25 \pm 2$ °C in the dark.

3. A large number of somatic embryos will be produced after culturing for 1–2 months; subculture on the fresh culture medium at 4-week intervals (Fig. 1h).

### *3.5 Germination of Somatic Embryos*

1. Culture somatic embryos on a modified of MS salt (Table 2, III), subculture on the fresh culture medium at each 4-week interval. Somatic embryos would germinate and develop in plantlets with both roots and shoots (Fig. 1i).

2. Different developmental stages of somatic embryos are observed including globular embryo, early cotyledon, mature, beginning of germination, growing of shoots, and whole plantlet (Fig. 2).

3. Incubate cultures at $25 \pm 2$ °C and 12-h photoperiod with light source provided by florescent lamps (GE lighting, Daylight, F40D-EX) of 30 $\mu mol/m^2/s$ photon flux.

### *3.6 Plantlet Elongation and Rooting*

1. Elongate regenerated shoots onto MS medium (Table 2, IV) for 4–6 weeks in order to get well-developed shoots (Fig. 1j), at $25 \pm 2$ °C, 12-h photoperiod light source provided by florescent lamps (GE lighting, Daylight, F40D-EX) of 30 $\mu mol/m^2/s$ photon flux.

2. Culture well-developed shoots onto MS medium supplemented with 0.2 mg/L NAA and 30 g/L sucrose for rooting after 4–6 weeks of culturing (Fig. 1k; Table 2, V), at $25 \pm 2$ °C, 12-h photoperiod, light source provided by florescent lamps (GE lighting, Daylight, F40D-EX) of 40 $\mu mol/m^2/s$.

**Fig. 2** Various different stages of the somatic embryo development: (**a**) globular embryo, (**b**) early cotyledon, (**c**) mature, (**d**) beginning of germination, (**e**) growth of plantlet, (**f**) whole plantlet

### 3.7 Plant Acclimatization

1. Select regenerated plantlets 13–15 cm in length, with proper rooting system (*see* **Note 5**).

2. Rinse plantlets gently under a slow stream of water to remove residual agar medium sticking to roots.

3. Spray plantlets with a 0.6 g/L fungicide (Carbomar, systemic fungicide) to prevent fungal infection (*see* **Note 6**).

4. Transfer the plantlets in 7.5 cm plastic pots (*see* **Note 7**) filled with peat moss, sand, and perlite mixture (1:1:1 ratio) (Fig. 3a, b), and maintain them in a container covered with plastic sheet.

5. Keep container in the culture room at 25 ± 2 °C under 12-h photoperiod with light source provided by fluorescent lamps (GE lighting, daylight, F40D-EX) of 40 $\mu$mol/m$^2$/s photon flux for acclimation (Fig. 3c).

6. During acclimation period, open container every week to check for fungal infection, and immediately remove infected material.

7. Irrigate plantlets every 14 days (winter—November to April) and 7 days (summer—May to October) with 0.3 g/L N-P-K fertilizer (Peters, 20-20-20) dissolved in distilled water. Well-formed plantlets are normally obtained after 2–3 months (Fig. 3d).

8. Transfer plantlets to 70% shaded greenhouse further acclimation and uniform growth (Fig. 3e, f) in preparation for future plantation in the field.

**Fig. 3** Plant acclimatization system of date palm by inflorescence culture: (**a**) establishment of proper rooting system, (**b**) transplanting, (**c**) container used for establishment of acclimatization system, (**d**) hardening process, (**e, f**) uniform production of date palm established from inflorescence tissue culture plants

## 4   Notes

1. The spathe proper size for culturing inflorescence is 30–40 cm long and 9–11 cm wide. Make sure that inflorescence spathe is intact and has not split open. As the spathe elongates to over 45 cm, it splits open which may cause severe contamination upon culturing. In smaller spathes, 25 cm long, the explant tissue is too fragile and most explants die.

2. Make certain that all tissue-handling tools and culture vessels are properly sterilized, and ensure that all work is under a fully sterilized environment.

3. The tube cap and GA7 container are capped, and the cap is wrapped with plastic cling film.

4. Regular subculturing of cultures is crucial to avoid nutrient depletion from the culture medium and to ensure normal growth and development.

5. Use plants having at least three new roots per plant for the acclimatization stage. This is the most important stage in the protocol for the recovery of tissue culture plants. Failure to control the environment can cause low survival.

6. Prepare fungicide solution in the distilled water and store at 4 °C; it should be used within 24 h.

7. The 7.5 cm plastic pots are surface-sterilized in 1% sodium hypochlorite for 30 min and rinsed with sterile distilled water two times.

## References


1. Chao CT, Krueger RR (2007) The date palm (*Phoenix dactylifera* L.): overview of biology, uses, and cultivation. HortScience 42 (5):1077–1082

2. Khaled MF (2006) The famous date varieties in the Kingdom of Saudi Arabia. Ministry of Agriculture Kingdom of Saudi Arabia, Riyadh

3. Zaid A, de Wet PF (2002) Date palm propagation. In: Zaid A (ed) Date palm cultivation. FAO plant production and protection paper 156. Rome. pp 73–105

4. Tisserat B (1984) Propagation of date palm by shoot tip cultures. HortScience 19:230–231

5. Zaid A, Tisserat BH (1983) *In vitro* shoot tip differentiation in *Phoenix dactylifera* L. Date Palm J 2:163–182

6. Abahmane L (2010) Date Palm (*Phoenix dactylifera* L.) micropropagation from inflorescence tissues by using somatic embryogenesis technique. Acta Hort 882:827–832

7. Abahmane L (2013) Recent achievements in date palm (*Phoenix dactylifera* L.) micropropagation from inflorescence tissues. Emir J Food Agric 25(11):863–874

8. Abul-Soad AA (2012) Influence of inflorescence explant age and 2,4-D incubation period on somatic embryogenesis of date palm. Emir J Food Agric 24(5):434–443

9. Al-Khayri JM (2010) Somatic embryogenesis of date palm (*Phoenix dactylifera* L.) improved by coconut water. Biotechnology 9:477–484

10. Al-Khayri JM (2007) Date palm *Phoenix dactylifera* L. micropropagation. In: Jain SM, Haggman H (eds) Protocols for micropropagation of woody trees and fruits. Springer, Berlin, pp 509–526

11. Al-Khayri JM (2001) Optimization of biotin and thiamine requirements for somatic embryogenesis of date palm (*Phoenix dactylifera* L.) In Vitro Cell Dev Biol Plant 37:453–456

12. Fki L, Masmoudi R, Drira N, Rival A (2003) An optimized protocol for plant regeneration from embryogenic suspension cultures of date palm, *Phoenix dactylifera* L., cv. Deglet Nour. Plant Cell Rep 21:517–524

13. Tisserat B (1981) Production of free-living date palms through tissue culture. Date Palm J 1(1):43–54

14. Veramendi J, Navarro L (1996) Influence of physical conditions of nutrient medium and sucrose on somatic embryogenesis of date palm. Plant Cell Tissue Organ Cult 45:159–164

15. Murashige T, Skoog F (1962) A revised medium for rapid growth and bioassays with tobacco tissue culture. Physiol Plant 15:473–497

# Indirect Somatic Embryogenesis of Date Palm Using Juvenile Leaf Explants and Low 2,4-D Concentration

## Lotfi Fki, Walid Kriaa, Ameni Nasri, Emna Baklouti, Olfa Chkir, Raja B. Masmoudi, Alain Rival, and Noureddine Drira

## Abstract

This chapter describes an efficient protocol for large-scale micropropagation of date palm. Somatic embryo-derived plants are regenerated from highly proliferating suspension cultures. Friable embryogenic callus is initiated from juvenile leaves using slightly modified Murashige and Skoog (MS) medium supplemented with 0.1 mg/L 2,4-dichlorophenoxyacetic acid (2,4-D). Suspension cultures consisting of proembryonic masses are established from highly competent callus for somatic embryogenesis using half-strength MS medium enriched with 0.1 mg/L 2,4-D and 300 mg/L activated charcoal. The productivity of cultures increased 20-fold when embryogenic cell suspensions were used instead of standard protocols on solidified media. The overall production of somatic embryos mostly exceeds 10,000 units per liter per month. Partial desiccation of mature somatic embryos, corresponding to a decrease in water content from 90 down to 75%, significantly improved germination rates.

**Key words** Acclimatization, Callus, Cell suspension, Somatic embryos, In vitro plants

## 1 Introduction

Conventionally, date palm (*Phoenix dactylifera* L.) is propagated from offshoots, which are limited in quantity, only 10–15 produced over the life of a date palm. Seedling date palms show variability in field performance because of their genetic heterogeneity. Since 1970, extensive efforts have been undertaken in order to mass propagate date palm through in vitro tissue culture [1–8]. In order to establish aseptic cultures, various explants have been used, including zygotic embryos, shoot tips, leaves or inflorescences [1], and date palm plant regeneration through adventitious organogenesis [2]. This approach is known for being slow but risky in generating somaclonal variation. The simple development of lateral buds has also been explored for the successful regeneration

Jameel M. Al-Khayri et al. (eds.), *Date Palm Biotechnology Protocols Volume 1: Tissue Culture Applications*,
Methods in Molecular Biology, vol. 1637, DOI 10.1007/978-1-4939-7156-5_9, © Springer Science+Business Media LLC 2017

of whole date palm plantlets [3]. However, somatic embryogenesis has higher potential for mass propagation [1, 3, 4]. There are two morphogenetic pathways ensuring the production of somatic embryos. The first pathway is direct somatic embryogenesis, which is yet to be fully developed for large-scale date palm plant regeneration. The second pathway is an indirect method which is based on the induction of embryogenic callus. Plant regeneration via somatic embryogenesis occurs in five steps: initiation of embryogenic cultures, proliferation of embryogenic cultures, prematuration of somatic embryos, maturation of somatic embryos, and germination of somatic embryos. Initiation and proliferation occur on MS medium supplemented with 2,4-dichlorophenoxyacetic acid (2,4-D), which induces differentiation of localized meristematic cells. These cells develop into mature somatic embryos after transfer to a medium containing either a low concentration or devoid of auxin. Mature somatic embryos germinate readily and have functional roots and shoot apices. This chapter describes an efficient method for date palm plant regeneration based on the establishment of embryogenic suspension cultures by using a low 2,4-D concentration.

## 2     Materials

### 2.1   Plant Material and Disinfection

1. Juvenile leaves surrounding the apical dome in the shoot tip of any date palm cultivar. These leaves should be less than 0.5 cm in length (*see* **Note 1**).

2. Disinfectant solution: 0.1 g/L $HgCl_2$.

### 2.2   Culture Medium

1. For in vitro culture initiation: Murashige and Skoog (MS) [9] tissue culture medium supplemented with 200 mg/L glutamine, 30 g/L sucrose, and 0.1 mg/L 2,4-D (Table 1).

2. Establishment of embryogenic cell suspension culture: Half-strength MS medium containing 0.1 mg/L 2,4-D and 300 mg/L activated charcoal.

3. Germination of somatic embryos and plantlet hardening: MS medium without plant growth regulators (PGRs).

### 2.3   Equipment

1. Horizontal laminar flow hood.

2. Orbital shaker.

## 3     Methods

### 3.1   Explant Disinfection

1. Excise offshoot leaves one by one to reach leaf primordia surrounding the apical dome in the shoot tip. Immerse excised juvenile leaves (less than 0.5 cm in length) in 0.1 g/L $HgCl_2$ solution for 1 h.

2. Wash the leaves three times in sterile distilled water.

**Table 1**
**MS medium composition [9]**

| Medium composition | Concentration (mg/L) |
| --- | --- |
| *Major inorganic nutrients* | |
| $Mg\ SO_4.7H_2O$ | 370 |
| $KH_2PO_4$ | 170 |
| $KNO_3$ | 1900 |
| $CaCl_2,2H_2O$ | 440 |
| $NH_4NO_3$ | 1650 |
| *Minor inorganic nutrients* | |
| $MnSO_4.4H_2O$ | 22.300 |
| $ZnSO_4.7H_2O$ | 8.600 |
| $H_3\ BO_3$ | 6.200 |
| $KI$ | 0.830 |
| $Na_2\ MoO_4.2H_2O$ | 0.350 |
| $CuSO_4.5H_2O$ | 0.025 |
| $CoCl_2.6H_2O$ | 0.025 |
| *Iron source* | |
| $Fe\ SO_4.7H_2O$ | 27.84 |
| $Na_2\ EDTA$ | 37.24 |
| *Vitamins and amino acids* | |
| Nicotinic acid | 0.5 |
| Pyridoxine HCl | 0.5 |
| Thiamine HCl | 0.1 |
| Myo-inositol | 100 |
| Glycine | 2 |
| *Plant growth regulators* | |
| 2,4-D | 0.1 |
| *Other compounds* | |
| Sucrose | 30,000 |
| Activated charcoal | 300 |
| Agar | 8000 |

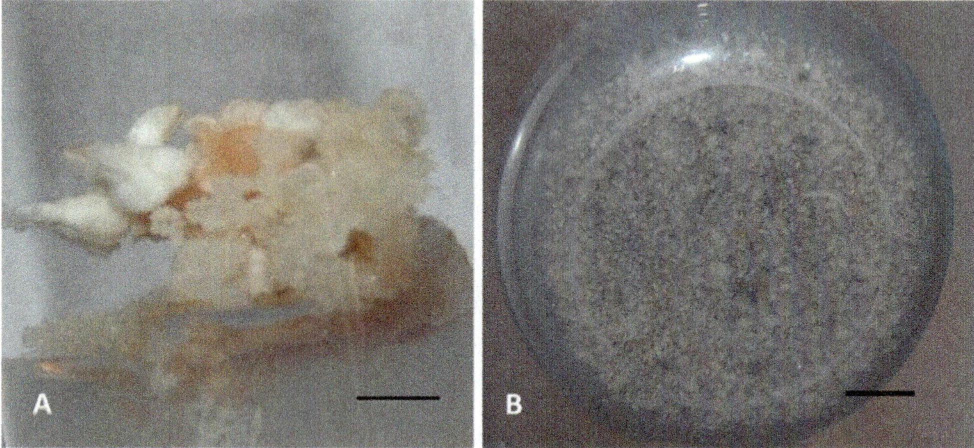

**Fig. 1** Date palm somatic embryogenesis: (**a**) embryogenic callus from a juvenile leaf (scale bar: 0.5 cm), (**b**) date palm proembryos in their exponential proliferation phase (scale bar: 1 cm)

***3.2 Callogenesis***

1. Culture leaf explants on MS medium supplemented with 0.1 mg/L 2,4-D.

2. Maintain in vitro cultures in the dark at 28 ± 2 °C for callus initiation.

3. After 5 months, transfer the explants onto the fresh culture medium and subculture them for 5 to10 months at 60-day intervals under the same culture conditions (*see* **Note 2**).

4. Isolate embryogenic callus after 2 years, and multiply them by successive subcultures at a 60-day interval using the same medium composition (Fig. 1a).

***3.3 Production of Mature Somatic Embryos***

1. Chop with a scalpel 0.5 g friable callus, 10 days after subculture, into small pieces (approx. 100–500 μm), and transfer them aseptically into 250 mL Erlenmeyer flasks containing 50 mL half-strength MS liquid medium containing 0.1 mg/L 2,4-D and 300 mg/L activated charcoal.

2. Sieve immediately after the operation the suspension cultures through a 500 μm mesh filter.

3. Maintain the cultures for 18 months on a rotary shaker at 120 rpm at 28 ± 2 °C under a 16-h photoperiod of 28 μmol/m$^2$/s (Fig. 1b; *see* **Note 3**). Subcultures are realized by renewing medium after cell decantation.

***3.4 Germination of Somatic Embryos***

1. Collect mature somatic embryos from the suspension cultures (Fig. 2a; *see* **Note 4**).

2. Put somatic embryos that are generally hyperhydrated in 250 mL Erlenmeyer flask on three layers of filter paper, and cap and maintain in the growth chamber for 4–6 days to ensure

**Fig. 2** Mature date palm somatic embryos: (**a**) before germination (scale bar: 2 cm), (**b** and **c**) after desiccation and germination (scale bar: 1 cm)

their partial desiccation. Desiccation should be ended when hyperhydrated embryos turn white in color.

3. Culture partially desiccated somatic embryos (water content is decreased from 90 before desiccation down to 75% after desiccation) on MS medium without plant growth regulators (PGRs).

4. Water content is determined by weighting plant material (P1), placing the tissue at 60 °C for 48 h, and then weighting the tissue for the second time (P2), and the water content is calculated as follows: water content (%) = [(P1 − P2)/P1] × 100.

5. Observe the germination of somatic embryos which will occur after a few weeks (Fig. 2b, c).

*3.5 In Vitro Hardening and Acclimation*

1. Discard the roots of the non-vigorous somatic embryo-derived plants, and induce new root on MS medium supplemented with 2 mg/L indolebutyric acid (IBA).

2. Transfer somatic embryo-derived plants to glass tubes containing MS liquid medium, and maintain them in the growth chamber for 6–12 months at 16-h photoperiods of 28 μmol/m$^2$/s photon flux (*see* **Note 5**).

3. Transfer four plantlets into a plastic vessel (500 mL volume) containing growth substrate consisting of 2 peat/1 sand (*see* **Note 6**).

4. Cover the plastic vessels with a plastic jar.

5. Acclimatize in vitro plants in the greenhouse under a moderate photon flux (30 μmol/m$^2$/s) at 25–30 °C.

6. After 3 months, remove the plastic jar and irrigate plants at 10-day intervals.

**Fig. 3** Date palm somatic embryo-derived plants: (**a**) acclimatized soma-plants (scale bar: 10 cm), (**b**) somatic embryo-derived plant growing in the field (scale bar: 40 cm)

7. After 4 months transfer plants to a 2 L plastic bag containing another substrate (peat/sand/compost: 1:1:1) (Fig. 3a).

8. Maintain the plants in the greenhouse for 10–15 months. Irrigation should be at 7-day intervals and not excessive.

*3.6  Field Transplantation*

1. Transplant acclimatized in vitro plants (30 cm in height) to the field in February.

2. Surround the plants with palm leaves to protect them from weather fluctuation.

3. Irrigate weekly during 12 months (Fig. 3b; *see* **Notes** **7** and **8**).

## 4  Notes

1. Juvenile leaves, surrounding the apical meristem, are excised from date palm offshoots growing at the base of adult date palms.

2. Embryogenic callus starts to appear after 12–36 months. The period depends on the physiological status of the explants and their position on the culture media. The basal part of the leaves is more responsive.

3. Proembryonic masses (PEMs) are observed after 20–30 days of culture without any transfer (Fig. 1b). Proliferation and maturation of somatic embryos occur simultaneously in the same culture vessel. PEMs differentiate into somatic embryos after weekly transfer of suspension cultures on the fresh culture medium (Fig. 2a).

4. Embryogenic suspensions which produce immature and consequently non-vigorous plants should be treated with 4 mg/L ABA for 2 weeks. This is when most of embryos are at the post-globular stage. The overall production of somatic embryos generally exceeds 10,000 units per liter per month.

5. Acclimatization is the final stage of micropropagation; when not properly controlled, it can cause high economic losses [10]. In vitro hardening of 12 months is necessary for an 83.3% survival rate during acclimatization. Plants transplanted into the greenhouse should progressively adapt to (1) higher photon flux, (2) lower relative humidity, (3) fluctuation of temperature, and (4) biotic stresses.

6. Scanning electron microscopy examination shows that stomatal opening is almost totally regulated in vitro plants after hardening for 12 months. Zaid and Hughes [10] reported that a polyethylene glycol treatment of vitro plants increased the amount of wax deposition on leaf epidermis, and as a consequence this wax layer is able to limit water loss during acclimatization. Several reports indicated that non-acclimatized vitro plants showed permanent stomatal opening or poor control of water loss [11, 12].

7. Three major constraints in date palm somatic embryogenesis are endophytic bacterial contamination [13], abnormal embryo differentiation (malformed or immature), and somaclonal variation [10]. Concerning endophytic bacterial contamination, only juvenile explants can be used to establish clean in vitro tissue culture, since antibiotics such as cefotaxime have only a bacteriostatic effect [14]. Immaturity of vascular tissue in these explants may explain the absence of endophytic contaminants.

8. Both abnormal somatic embryo differentiation and somaclonal variation are mainly associated with the utilization of high concentrations of 2,4-D. Reducing its concentration significantly minimizes the number of abnormal somatic embryos and somaclonal variants [8, 15].

## Acknowledgments

This work was supported financially by the Ministry of Higher Education and Scientific Research in Tunisia; the International Atomic Energy Agency (IAEA); the Arab League Educational, Cultural and Scientific Organization (ALECSO); The European Cooperation in Science and Technology (COST); the Swiss National Science Foundation (SNSF); and the Technical Centre of Dates in Tunisia (TCDT).

## References


1. Fki L, Masmoudi R, Kriaa W, Mahjoub A, Sghaier B, Mzid R et al (2011) Date palm micropropagation via somatic embryogenesis. In: Jain SM, Al-Khayri JM, Johnson DV (eds) Date palm biotechnology. Springer, The Netherlands, pp 47–68

2. Fki L, Bouaziz N, Kriaa W, Benjemaa-Masmoudi R, Gargouri-Bouzid R, Rival A, Drira N (2011) Multiple bud cultures of 'Barhee' date palm (*Phoenix dactylifera*) and physiological status of regenerated plants. J Plant Physiol 168:1694–1700

3. Al-Khayri JM (2013) Factors affecting somatic embryogenesis in date palm (*Phoenix dactylifera* L.) In: Aslam J, Srivastava PS, Sharma MP (eds) Somatic embryogenesis and genetic transformation in plants. Narosa Publishing House, New Delhi, pp 15–38

4. Fki L, Masmoudi R, Drira N, Rival A (2003) An optimised protocol for plant regeneration from embryogenic suspension cultures of date palm (*Phoenix dactylifera* L.) cv. Deglet nour. Plant Cell Rep 21:517–524

5. Drira N, Benbadis A (1985) Multiplication végétative du palmier dattier (*Phoenix dactylifera* L.) par réversion, en culture in vitro, d'ébauches florales de pieds femelles adultes. J Plant Physiol 119:227–235

6. Sharma DR, Deepak S, Chowdhury JB (1986) Regeneration of plantlets from somatic tissues of date palm (*Phoenix dactylifera* L). Indian J Exp Biol 24:763–766

7. Naik PM, Al-Khayri JM (2016) Somatic embryogenesis of date palm (*Phoenix dactylifera* L.) through cell suspension culture. In: Jain SM (ed) Protocols for in vitro cultures and secondary metabolite analysis of aromatic and medicinal plants, Methods in molecular biology, 2nd edn. Springer, New York, pp 357–366

8. Boufis N, Khelifi-Slaoui M, Djillali Z, Zaoui D, Morsli A, Bernards MA et al (2014) Effects of growth regulators and types of culture media on somatic embryogenesis in date palm (*Phoenix dactylifera* L. cv. Degla Beida). Sci Hort 172:135–142

9. Murashige T, Skoog F (1962) A revised medium for rapid growth and bioassays with tobacco tissue culture. Physiol Plant 15:473–479

10. Zaid A, Hughes H (1995) Water loss and polyethylene glycol-mediated acclimatization of in vitro grown Seedlings of 5 cultivars of date palm (*Phoenix dactylifera* L.) plantlets. Plant Cell Rep 14:385–388

11. Brainerd KE, Fuchigami LH (1982) Stomatal functioning of in vitro and greenhouse apple leaves in darkness, manitol, ABA and CO2. J Exp Bot 33:388–392

12. Zaid A, Hughes H (1995) In vitro acclimatization of date palm (*Phoenix dactylifera* L.) plantlets: a quantitative comparison of epicuticular leaf wax as a function of polyethylene glycol treatment. Plant Cell Rep 15:111–114

13. Fki L, Bouaziz N, Sahnoun N, Swennen R, Drira N, Panis B (2011) Palm cryobanking. CryoLetters 32(6):451–462

14. Reed BM, Tanprasert P (1995) Detection and control of bacterial contaminants of plant tissus culture. A review of recent literature. Plant Tiss Cult Biotechnol 1(3):137–142

15. Salma M, Fki L, Engelmann-Sylvestre I, Niino T, Engelmann F (2014) Comparison of droplet-vitrification and D-cryoplate for cryopreservation of date palm (*Phoenix dactylifera* L.) polyembryonic masses. Sci Hort 179:91–97

# Chapter 10

# Desiccation-Enhanced Maturation and Germination of Date Palm Somatic Embryos Derived from Cell Suspension Culture

Nazim Boufis, Khayreddine Titouh, and Lakhdar Khelifi

## Abstract

In vitro plant regeneration via somatic embryogenesis is a powerful tool for rapid, large-scale production of healthy true-to-type plants. This approach is suitable to preserve existing natural genetic variability and propagation of variability generated from genetic improvement programs, including crossing, somaclonal variation, mutagenesis, and somatic hybridization. This chapter outlines a simplified protocol for date palm regeneration via somatic embryogenesis induced in cell suspension cultures. In this protocol, culture medium composition is manipulated, including plant growth regulators and solid (addition of agar) and liquid media to achieve reduction of production cycle of somatic embryogenesis, which increases the multiplication rate of embryogenic callus and improves the quantity and quality of somatic embryos.

**Key words** Cell suspension culture, Embryogenic callogenesis, In vitro regeneration, *Phoenix dactylifera*, Somatic embryogenesis

## 1 Introduction

The date palm (*Phoenix dactylifera* L.) is among the most important economic species in the Palm family (Arecaceae). It is a long-lived dioecious monocotyledon adapted to local conditions of arid and semiarid regions and represents a cornerstone of the economy in North Africa and the Middle East [1]. The world's date palm heritage is characterized by a wide diversity of cultivars, over 5000 cultivars more or less described [2], and the palm is subject to many biotic and abiotic threats. This situation is compounded by genetic erosion and likely exacerbated by global climate change [1, 3–5]. Therefore, in order to preserve and/or improve this varietal diversity, control of the date palm propagation is preeminent. This objective can be achieved using plant tissue culture techniques to overcome the deficiencies in conventional propagation methods using either seeds or offshoots [4, 6].

Jameel M. Al-Khayri et al. (eds.), *Date Palm Biotechnology Protocols Volume 1: Tissue Culture Applications*,
Methods in Molecular Biology, vol. 1637, DOI 10.1007/978-1-4939-7156-5_10, © Springer Science+Business Media LLC 2017

Plant tissue culture techniques allow a rapid large-scale production of healthy true-to-type plants, regardless of the season of the year [7]. Two techniques of in vitro propagation are used for date palm regeneration. The direct-organogenesis based on shoot formation ability of different types of explants and somatic embryos formation from somatic cells [8]. Between these techniques, somatic embryogenesis is the most advantageous because it is time-saving and low in cost and gives high rates of regeneration despite a risk of somaclonal variation [9, 10]. However, date palm remains a recalcitrant species to in vitro regeneration because of genotypic factors that affect the explant response in vitro culture and the poor germination rate of somatic embryos [4, 6]. However, significant improvement was made in date palm somatic embryogenesis by using embryogenic cell suspension cultures [11–15], which has an advantage to promote the uniform absorption of nutrients and reduce the inhibitory effect of polyphenols on cell growth. It improves considerably the yield of viable somatic embryos [16] by maintaining genetic fidelity of regenerated plants and is useful in breeding programs [17–19].

This chapter describes a simplified protocol for date palm plant regeneration through somatic embryogenesis by using embryogenic cell suspension culture of Algerian date palm cultivars. The protocol consists of five main steps: induction and maintenance of embryogenic callus, multiplication of embryogenic cultures in agitated liquid medium, initiation of somatic embryogenesis in liquid medium, maturation, and germination of proembryos on solid media.

## 2  Materials

***2.1  Plant Material and Surface Sterilization***

1. Offshoots of date palm having an average weight 4–6 kg (Fig. 1a).

2. Disinfection solution: Mercuric chloride (HgCl$_2$), 150 mg/L.

***2.2  Equipment***

1. Explant isolation tools: Tree saw and hand pruner.

2. Surgical tools: Scalpels, forceps, and spatulas.

3. Glassware and culture vessels: Beakers, Erlenmeyer flasks, boiling flasks, graduated cylinders, glass laboratory bottles, glass funnels, graduated pipettes, test tubes (160 × 24 × 1.2 mm), sterile rectangular flasks (650 mL), glass, and sterile plastic Petri dishes (∅ 9 cm).

4. Instruments: Water distillation unit, autoclave, dry-heat sterilizer, laboratory ice machine, refrigerator, analytical balance, magnetic stirrer with heating, pH meter, rotary shaker, laminar flow bench, gas burners, and strainers (with sieves of 500 μm mesh size).

**Fig. 1** Illustration of the different date palm regeneration stages via somatic embryogenesis induced in cell suspension culture: (**a**) offshoot of date palm, (**b**) offshoot core, (**c**) sterilization of offshoot core with HgCl₂, (**d**) the sterilized apical part of offshoot, (**e**) explant in test tube cultures for callogenesis initiation, (**f**) initiation of embryogenic callogenesis, (**g**) embryogenic calli maintained on solid medium, (**h**) embryogenic calli after 8 weeks of multiplication in liquid medium, (**i**) general appearance of suspensions in liquid medium, (**j**) clusters of proembryos obtained in liquid medium, (**k**) development of somatic embryos on solid medium, (**l**) structured somatic embryos after maturation on solid media, (**m**) development of somatic embryos into in vitro plants. Source: Photos **f, g, i, k, l,** and **m** are taken from Boufis et al. [15]

5. Supplies: Absorbent paper, stretch cling film or parafilm, cotton, and aluminum foil.

6. Plant tissue culture room set at 25 °C ± 2 and 16 h photoperiod provided by fluorescent lamps (28 μmol/m²/s).

**2.3 Culture Medium**

1. Basal culture medium: Stock solutions of inorganic salts and organic elements of modified MS medium (Murashige and Skoog [20]) (*see* Table 1).

2. Hormonal stock solutions: 2,4-dichlorophenoxyacetic acid (2,4-D) and 6-benzylaminopurine (BAP), each at 1 mg/mL.

3. Culture medium additives: Myoinositol, L-glutamine, adenine, sucrose, activated charcoal, polyphenols, and agar (*see* Table 2).

4. Culture media additives for each culture stage: Initiation and maintenance of embryogenic callus (CM), multiplication of embryogenic callus (AM), induction of embryogenesis (EM), maturation of proembryos (MM), and germination of somatic embryos (GM) (*see* Table 2).

5. pH adjustment solutions: 1 M NaOH and 1 M HCl.

**Table 1**
**Composition of stock solutions based on modified Murashige and Skoog (MS) medium [20]**

| Components | Final concentration of culture medium (mg/L) | Stock solution strength[a] | Concentration of the stock solutions (mg/L) |
|---|---|---|---|
| *Macroelements* | | ×20 | |
| $KNO_3$ | 1900 | | 38,000 |
| $NH_4NO_3$ | 2000 | | 40,000 |
| $CaCl_2, 2H_2O$ | 440 | | 8800 |
| $MgSO_4, 7H_2O$ | 370 | | 7400 |
| $KH_2PO_4$ | 270 | | 5400 |
| $Na\ H_2PO_4$ | 170 | | 3400 |
| *Microelements* | | ×100 | |
| $MnSO_4, 4H_2O$ | 22.3 | | 2230 |
| $ZnSO_4, 4H_2O$ | 8.6 | | 860 |
| $H_3BO_3$ | 6.2 | | 620 |
| KI | 0.83 | | 83 |
| $Na_2MoO_4, 2H_2O$ | 0.25 | | 25 |
| $CuSO_4, 5H_2O$ | 0.025 | | 2.5 |
| $CoCl_2, 6H_2O$ | 0.025 | | 2.5 |
| *Fe-EDTA* | | ×100 | |
| $Na_2EDTA$ | 37.25 | | 3725 |
| $FeSO_4, 7H_2O$ | 27.85 | | 2785 |
| *Vitamins* | | ×100 | |
| Glycine | 2 | | 200 |
| Pyridoxine | 0.5 | | 50 |
| Nicotinic acid | 0.5 | | 50 |
| Thiamine HCl | 0.1 | | 10 |

[a]All stock solutions are prepared with distilled water and are stored at 4 °C in total darkness

**Table 2**
**Culture media and additives used for each stage of date palm tissue culture (amounts are per final volume to 1 L medium)**

| Components | Initiation and maintenance of embryogenic callus (CM) | Multiplication of embryogenic callus (AM) | Induction of somatic embryogenesis (EM) | Maturation of proembryos (MM) | Germination of somatic embryos (GM) |
|---|---|---|---|---|---|
| Macroelements (mL/L) | 50 | 25 | 25 | 25 | 25 |
| Microelements (mL/L) | 10 | 5 | 5 | 5 | 5 |
| Fe-EDTA (mL/L) | 10 | 5 | 5 | 5 | 5 |
| Vitamins (mL/L) | 10 | 10 | 10 | 10 | 10 |
| Myoinositol (mg/L) | 100 | 100 | 100 | 100 | 100 |
| L-Glutamine (mg/L) | 100 | 100 | 100 | 100 | 100 |
| Adenine (mg/L) | 40 | 40 | 40 | 40 | 40 |
| Sucrose (g/L) | 45 | 30 | 30 | 30 | 30 |
| Activated charcoal (mg/L) | 300 | 300 | 300 | 300 | 300 |
| 2,4-D (mg/L) | 10 | 1 | 0 | 0 | 0 |
| BAP (mg/L) | 1 | 0 | 0 | 0 | 0 |
| Agar (g/L) | 7 | 0 | 0 | 7 | 7 |

## 3  Methods

All plant tissue culture steps are carried out under aseptic conditions by sterilization of the culture media, containers and instruments, and surface disinfection under a laminar flow bench with 70% ethanol before use.

### 3.1  Preparation of Culture Medium

1. Prepare the culture media in distilled water as described in Table 2. The composition of the culture media used depends on culture stages: initiation and maintenance of embryogenic callus (CM), multiplication of embryogenic callus (AM), induction of embryogenesis (EM), maturation of proembryos (MM), and germination of somatic embryos (GM) (Table 2).

2. Adjust the pH of the media to 5.6–5.8 with either 1 M NaOH or 1 M HCl.

3. Distribute the media into suitable containers for each culture stage of the described protocol (Table 3) (*see* **Note 1**).

**Table 3**
**Volumes of culture media and containers used for each stage of described protocol**

| Culture stage | Initiation and maintenance of embryogenic callus (CM) | Multiplication of embryogenic callus (AM) | Induction of somatic embryogenesis (EM) | Maturation of Proembryos (MM) | Germination of somatic embryos (GM) |
|---|---|---|---|---|---|
| Type of culture medium | Solid medium | Liquid medium | Liquid medium | Solid medium | Solid medium |
| Volume of culture medium | 20 mL | 150 mL | 150 mL | 150 mL | 20 mL |
| Tissue culture containers | Test tubes (160 × 24 × 1.2 mm) | Sterile rectangular flasks (650 mL) | Erlenmeyer flasks (500 mL) | Erlenmeyer flasks (500 mL) | Test tubes (160 × 24 × 1.2 mm) |

4. Sterilize the media by autoclaving at 121 °C for 20 min (*see* **Note 2**).

5. Store the culture medium at 25 °C in the dark for a few days or at 4 °C for a prolonged storage time (*see* **Note 3**).

**3.2 Dissection and Disinfection of Plant Material**

1. Use a tree saw and a hand pruner to remove the root parts, the spiny palm leaves, and their fibrous sheaths until the last two layers that protect the fleshy parts of the offshoot core (the meristematic area) (*see* **Note 4**).

2. Rinse extracted offshoot core with tap water (Fig. 1b).

3. Prepare the disinfection solution containing 150 mg/L $HgCl_2$ in sterile distilled water under a laminar flow bench (*see* **Note 5**).

4. Immerse the offshoot core in the sterilizing solution for 1 h on a laminar flow bench (Fig. 1c).

5. Clean and remove the last two apparent meristem layers with sterile forceps and scalpel.

6. Immerse the offshoot core in the sterilizing solution a second time for 1 h.

7. Rinse three consecutive times with sterile distilled water.

8. Remove necrotic tissue with forceps and scalpel, which is caused by $HgCl_2$ (*see* **Note 6**).

**3.3 Initiation and Maintenance of Embryogenic Callus**

1. Divide the sterilized apical part of offshoot core in small explants, 0.5 $cm^3$ size, with forceps and scalpel (Fig. 1d).

2. Place the explants 0.5 $cm^3$ with forceps in test tubes containing 20 mL solid medium (CM) containing 10 mg/L 2,4-D and 1 mg/L BAP (*see* **Note 7**). The explants are cultured separately (1 explant per test tube) (Fig. 1e; *see* **Note 8**).

3. Incubate cultures in total darkness for the entire period of the initiation and maintenance of embryogenic callus at 25 °C ± 2.

4. Subculture explants on a fresh medium at 8-week intervals.

5. Eliminate brown or necrotic parts of explants during the subculture (*see* **Note 9**).

6. Excise explants showing prolific embryogenic callus growth during the renewal of solid medium (CM) (Fig. 1f, g; *see* **Note 10**).

**3.4 Multiplication of Embryogenic Callus**

1. Take 0.5 g explants embryogenic callus, and cut it into small pieces on a sterile Petri dish with a scalpel.

2. Transfer fragmented callus into sterile rectangular flasks (650 mL) containing 150 mL liquid medium (AM).

3. Incubate the suspension cultures on a rotary shaker, 100 rpm, at 25 ± 2 °C and 16 h photoperiod (28 $\mu mol/m^2/s$).

4. Renew the liquid medium (AM) every 4 weeks of culture, after cell decanting.

5. After 8 weeks of culture, re-transfer embryogenic calli on the solid medium (CM) (Fig. 1h), and incubate at 25 °C in total darkness, after partial drying for 2 h under a laminar flow bench.

**3.5  Induction of Somatic Embryogenesis**

1. Cut 0.4 g embryogenic calli (*see* **Note 11**) into small pieces in a sterile Petri dish using a sterile scalpel.

2. Place a funnel over a sterile Erlenmeyer 500 mL flask.

3. Place over the funnel a sieve, 500 μm mesh size.

4. Transfer and crush the chopped embryogenic callus on the sieve with a spatula.

5. Add gradually 150 mL liquid medium (EM) in order to retrieve all cell aggregates.

6. Incubate the suspensions of cell aggregates in a rotary shaker set at a speed of 100 rpm, at 25 °C ± 2 and 16 h photoperiod (28 μmol/m$^2$/s).

7. Renew the liquid medium (EM) at each 4-week interval after decanting the cell suspension cultures for 2 h.

8. The achievement of this stage requires 12 weeks of culture (Fig. 1i).

**3.6  Maturation of Proembryos**

1. Collect the proembryos after decanting (*see* **Note 12**).

2. Eliminate the excessive liquid medium (EM) using a sterile absorbent paper, and transfer proembryos onto a sterile Petri dish containing sterile absorbing paper (Fig. 1j).

3. Incubate the closed Petri dish for 24 h for desiccation in total darkness at 25 °C ± 2.

4. Collect and plate with a spatula the desiccated proembryos into 500 mL Erlenmeyer flasks containing 150 mL maturation solid medium (MM).

5. Incubate the plated proembryos at 25 °C ± 2, 16-h provided by fluorescent lamps (28 μmol/m$^2$/s) for 4 weeks (Fig. 1k; *see* **Note 13**).

**3.7  Germination of Somatic Embryos**

1. Desiccate well-developed somatic embryos (*see* **Note 14**) with sterile absorbing paper for 2 h under a laminar flow bench (Fig. 1l).

2. Put the somatic embryos with forceps into test tubes containing 20 mL solid medium (GM).

3. Incubate the somatic embryos at 25 °C ± 2 and a 16 h photoperiod (28 μmol/m$^2$/s) (*see* **Note 15**).

4. Transfer the cultured somatic embryos after 4 weeks to a fresh medium (Fig. 1m; *see* **Note 16**).

## 4  Notes

1. The test tubes and Erlenmeyer flasks are capped with cotton plugs and aluminum foil.

2. Autoclaved solid culture media contained in test tubes or Erlenmeyer flasks are stirred then cooled in ice to maintain the uniformity of activated charcoal in the solid culture media.

3. The storage of culture media in the dark at low temperature prevents the modification of their composition before use.

4. Only the apical part of the offshoot core is used. This part includes the apical meristem, leaf primordia, axillary buds, and base of young leaves.

5. As $HgCl_2$ is highly toxic to human health, while handling, take all appropriate precautions: glasses, mask, and gloves.

6. A good physiological and phytosanitary status of date palm offshoots is necessary for the success of callus initiation since the surface or contact sterilization is achieved gradually with apical part isolation.

7. For cultivars containing high levels of polyphenols, it is necessary to increase the quantity of activated charcoal in the culture medium. Thus, a culture medium containing 1.5 g/L activated charcoal combined with 50 mg/L 2,4-D and 1 mg/L BAP is most appropriate.

8. The explants derived from the apical part of the offshoot core are grown separately in test tubes to reduce the risk of contamination.

9. The secretion of polyphenols induces browning of explants and culture media. This phenomenon occurs within the first week of culture. The unfavorable impact of polyphenols in the first step can be reduced by removing brown or necrotic portions of explants.

10. The initial response of explants is either complete or partial swelling followed by the appearance of the primary callus, which is hyperhydric, compact, or friable. The nodular or embryogenic callus is the result of the development of compact primary callus. The required time for obtaining embryogenic callus is around 16 weeks but may take even longer than 48 weeks for recalcitrant cultivars.

11. Nodular callus is a potentially embryogenic callus. Its texture is more or less friable, and its color is white without browning.

12. After 12 weeks in the liquid culture medium (EM), a number of structured somatic embryos are produced together with a large number of proembryos: somatic embryos at the globular and post-globular stage of somatic embryogenesis.

13. The plated proembryos are preliminarily incubated in total darkness during the first 48 h to promote adaptation to new growing conditions.

14. Well-developed somatic embryos have elongated and bipolar forms, approximately 6–7 mm long and a slightly curved shape on one side and pointed on the other.

15. Well-developed somatic embryo germination is manifested by the emergence of the aerial part (cotyledonary leaf) followed by the appearance of a radicle that will develop later into the taproot system. The germination somatic embryos can be observed from the first week of culture.

16. Transfer germinated somatic embryos to the same composition solid medium (GM) with reduced gelling agent concentration (3.5 g/L agar) to promote rooted in vitro plant growth.

## References


1. El Hadrami A, Al-Khayri JM (2012) Socioeconomic and traditional importance of date palm. Emir J Food Agric 24:371–385

2. El Hadrami I, El Hadrami A (2009) Breeding date palm. In: Jain SM, Priyadarshan PM (eds) Breeding plantation tree crops. Springer, New York, pp 191–216

3. Chao CT, Krueger RR (2007) The date palm (*Phoenix dactylifera* L.): overview of biology, uses, and cultivation. HortScience 42 (5):1077–1082

4. Jain SM (2012) In vitro mutagenesis for improving date palm (*Phoenix dactylifera* L.) Emir J Food Agric 24(5):400–407

5. Shabani F, Kumar L, Taylor S (2012) Climate change impacts on the future distribution of date palms: a modeling exercise using CLIMEX. PLoS One 7(10):e48021

6. Zaid A, de Wet PF (2002) Date palm propagation. In: Zaid A (ed) Date palm cultivation. Food and Agriculture Organization Plant Production and Protection Paper No, vol 156. FAO, Rome, Italy, pp 73–105

7. Bhojwani SS, Dantu PK (2013) Plant tissue culture: an introductory text. Springer, India

8. Loyola-Vargas VM, De-la-Peña C, Galaz-Ávalos RM, Quiroz-Figueroa FR (2008) Plant tissue culture. In: Walker JM, Rapley R (eds) Molecular biomethods handbook, 2nd edn. Humana Press, Totowa, NJ, pp 875–904

9. Ree JF, Guerra MP (2015) Palm (Arecaceae) somatic embryogenesis. In Vitro Cell Dev Plant 51(6):589–602

10. Al-Khalifah NS, Askari E (2011) Growth abnormalities associated with micropropagation of date palm. In: Jain SM, Al-Khayri JM, Johnson DV (eds) Date palm biotechnology. Springer, The Netherlands, pp 205–219

11. Fki L, Masmoudi R, Drira N, Rival A (2003) An optimised protocol for plant regeneration from embryogenic suspension cultures of date palm, *Phoenix dactylifera* L. cv. Deglet Nour. Plant Cell Rep 21(6):517–524

12. Sané D, Aberlenc-Bertossi F, Gassama-Dia YK, Sagna M, Trouslot MF, Duval Y, Borgel A (2006) Histocytological analysis of callogenesis and somatic embryogenesis from cell suspensions of date palm (*Phoenix dactylifera* L.) Ann Bot 98(2):301–308

13. Othmani A, Bayoudh C, Drira N, Marrakchi M, Trifi M (2009) Somatic embryogenesis and plant regeneration in date palm *Phoenix dactylifera* L. cv. Boufeggous is significantly improved by fine chopping and partial desiccation of embryogenic callus. Plant Cell Tissue Organ Cult 97(1):71–79

14. Al-Khayri JM (2012) Determination of the date palm cell suspension growth curve, optimum plating efficiency, and influence of liquid medium on somatic embryogenesis. Emir J Food Agric 24(5):444–455

15. Boufis N, Khelifi-Slaoui M, Djillali Z, Zaoui D, Morsli A, Bernards MA, Khelifi L (2014) Effects of growth regulators and types of culture media on somatic embryogenesis in date palm (*Phoenix dactylifera* L. cv. Degla Beida). Sci Hort 172:135–142

16. Preil W (2005) General introduction: a personal reflection on the use of liquid media for in vitro culture. In: Hvoslef-Eide AK, Preil W (eds) Liquid culture systems for in vitro plant propagation. Springer, The Netherlands, pp 1–18

17. Aslam J, Khan SA, Azad MAK (2015) Agrobacterium-mediated genetic transformation of date palm (*Phoenix dactylifera* L.) cultivar "Khalasah" via somatic embryogenesis. Plant Sci Today 2(3):93–101

18. Titouh K, Khelifi L, Slaoui M, Boufis N, Morsli A, HadjMoussa K, Makhzoum A (2015) A simplified protocol to induce callogenesis in protoplasts of date palm (*Phoenix dactylifera* L.) cultivars. Iran J Biotech 13(1):26–35

19. El Hadrami A, Daayf F, Elshibli S, Jain SM, El Hadrami I (2011) Somaclonal variation in date palm. In: Jain SM, Al-Khayri JM, Johnson DV (eds) Date palm biotechnology. Springer, The Netherlands, pp 183–203

20. Murashige T, Skoog F (1962) A revised medium for rapid growth and bioassays with tobacco tissue cultures. Physiol Plant 15:473–497

# Desiccation and Cold Hardening of Date Palm Somatic Embryos Improve Germination

## Hussein J. Shareef

## Abstract

Embryogenic suspension cultures of date palm are ideal for mass propagation of somatic embryos; however, the low percentage of germination of somatic embryos (SE) remains an impediment. This chapter focuses on two important physical factors to improve germination of date palm somatic embryos: the use of partial desiccation (3 h) of somatic embryos and the exposure to low temperature (4 °C for 24 h). High germination percentage (41%) is achieved by desiccation for 3 h. Moreover, adding 0.3 g/L activated charcoal (AC) to the liquid medium further improves somatic embryo number and weight as well as the percentage of germination. Moreover, partial desiccation and low temperature exposure tend to increase proline content. This improved protocol for somatic embryo germination is potentially applicable for commercial micropropagation of date palm.

**Key words** Cell suspension, Germination, Low temperature, Desiccation, Somatic embryos

## 1 Introduction

Micropropagation of date palm has been achieved from several genotypes through organogenesis and somatic embryogenesis using various meristematic explants including zygotic embryos, shoot tips, and lateral buds [1]. However, the date palm remains a recalcitrant species to in vitro techniques because of the influence of genotypic factors that affect the explant response and the frequency of maturation and germination of embryos, thus hindering the establishment of simple, reliable, and reproducible protocols [2, 3]. In fact, date palm tissue cultures grow very slowly; thus, the initiation phase may require more than 24 months, especially when low concentrations of plant growth regulators are added to the culture medium to prevent somaclonal variation [4].

Somatic embryogenesis is the most efficient regeneration process for date palm micropropagation [5]. It is considered a rapid, efficient method for large-scale micropropagation of date palm and highly useful for breeding programs [6]. Several researchers

Jameel M. Al-Khayri et al. (eds.), *Date Palm Biotechnology Protocols Volume 1: Tissue Culture Applications*, Methods in Molecular Biology, vol. 1637, DOI 10.1007/978-1-4939-7156-5_11, © Springer Science+Business Media LLC 2017

successfully obtained in vitro plant regeneration using suspension cultures established from date palm embryogenic callus [7, 8]. Moreover, adding low concentrations of auxin and activated charcoal (AC) enhanced the number of date palm somatic embryos induced in the liquid cultures [9]. Meanwhile, Othmani et al. [8] enhanced plant regeneration of date palm in vitro cultures through partial desiccation of somatic embryos. In addition, cold hardening increased the somatic embryo germination rate [10], and a significant increase in somatic embryo germination was achieved using desiccation [11].

This chapter describes two effective approaches to enhance the germination of date palm somatic embryos by applying partial desiccation and low temperature treatments along with the role of activated charcoal (AC).

## 2   Materials

### 2.1   Plant Material and Explant Sterilization

1. Explant source: Shoot tips of date palm Barhee cv., separated from healthy offshoots (3–4 years old), weighing 5–7 kg.

2. Chilled antioxidant solution: 150 mg/L ascorbic acid and 150 mg/L citric acid.

3. Disinfectant solution: 0.3% $HgCl_2$ containing 3 drops of Tween 20 per 100 mL solution.

### 2.2   Culture Medium

1. Basal culture medium: Murashige and Skoog (MS) [12] medium stock solutions (Table 1).

2. Hormone stock solutions: 2,4-dichlorophenoxyacetic acid (2,4-D, 5 mg/mL), 2-sopentenyladenine (2iP, 1 mg/mL), and naphthalene acetic acid (NAA, 1 mg/mL) (*see* **Note 1**).

3. pH adjustment solutions: NaOH and HCl at 0.1 and 1 N each.

4. Medium additives for different culture phases: The basal culture medium (Table 1) containing hormones, agar, and AC additives according to the culture phase as specified in (Table 2). The media needed for various culture phases are culture initiation (CI), callus proliferation (CP), callus maintenance (CM), cell suspension (CS), somatic embryo maturation (EM), somatic embryo germination (EG), and rooting (RT) media (*see* Table 2).

### 2.3   Proline Analysis

1. Ninhydrin acid solution: 1.25 g ninhydrin, 30 mL glacial acetic acid, and 20 mL 6M phosphoric acid.

2. Proline standard: 0.1 g proline, 1 mL sulfosalicylic acid, 2 mL glacial acetic acid, and 2 mL acid ninhydrin in 100 mL distilled water.

3. Standards blank: Distilled water and toluene.

**Table 1**
**Chemical composition of the MS medium used for date palm tissue culture [12]**

| Chemical combination | Concentration in stock solution (mg/L) | Final concentration in culture medium (mg/L) |
| --- | --- | --- |
| *Stock I: major inorganic nutrients (20×) use 50 mL for preparing 1 L of medium* | | |
| NH4NO3 | 33,000 | 1650 |
| KNO3 | 38,000 | 1900 |
| CaCl2·2H2O | 8800 | 440 |
| MgSO4·2H2O | 7400 | 370 |
| KH2PO4 | 3400 | 170 |
| NaH2PO4.H2O | 3400 | 170 |
| *Stock II: minor inorganic nutrients (200×) use 5 mL for preparing 1 L of medium* | | |
| KI | 166 | 0.83 |
| H3BO3 | 1240 | 6.2 |
| MnSO4·2H2O | 4460 | 22.3 |
| ZnSO4·7H2O | 1720 | 8.6 |
| Na2.MoO4·2H2O | 50 | 0.25 |
| CuSO4·5H2O | 5 | 0.025 |
| CoCl2·6H2O | 5 | 0.025 |
| *Stock III: iron source(200×) use 5 mL for preparing 1 L of medium* | | |
| FeSO4·7H2O | 5560 | 27.8 |
| Na2EDTA·2H2O | 7460 | 37.3 |
| *Stock IV: organic supplements (200×) use 5 mL for preparing 1 L of medium* | | |
| Myo-Inositol | 25,000 | 125 |
| Pyridoxine·HCl | 200 | 1 |
| Thiamine·HCl | 500 | 2.5 |
| Glutamine | 200 | 2 |
| Biotin | 200 | 1 |
| *Carbon source* | | |
| Sucrose | | 30,000 |
| pH | | 5.7 |

**2.4 Acclimatization Stage**

1. Potting mixture: peat moss and perlite (2:1 v/v) in 10 cm polyethylene nursery pots (Fig. 2f).

2. Fungicide solution: 0.5 g Benlate in 1 L distilled water.

**Table 2**
**Media used for various culture stages and their corresponding hormonal and activated charcoal additives**

| Culture stage | Medium | Hormones, agar, and activated charcoal additives | | | | |
| --- | --- | --- | --- | --- | --- | --- |
| | | 2,4-D (mg/L) | 2iP (mg/L) | NAA (mg/L) | Agar (g/L) | Activated charcoal (g/L) |
| Culture initiation (CI) | MS | 10 | 3 | – | 7 | 1.5 |
| Callus proliferation (CP) | MS | – | 3 | 10 | 7 | 1.5 |
| Callus maintenance (CM) | MS | – | 1.5 | 10 | 7 | – |
| Cell suspension (CS) | ½ MS | 1 | 1.5 | – | – | 0.3 |
| Somatic embryo maturation (EM) | MS | – | 1 | 0.1 | 7 | 0.1 |
| Somatic embryo germination (EG) | MS | – | – | 0.1 | 7 | 0.1 |
| Rooting (RT) | MS | – | – | 0.2 | 7 | 0.1 |

### 2.5 Equipment

1. Rotary shaker.
2. Refrigerator.
3. Flasks (150 mL).
4. Centrifuge.
5. Water bath.
6. Mortar and pestle.
7. Spectrophotometer.
8. Ice bucket.
9. pH meter.
10. Petri dishes.
11. Magenta vessels (100 mL) with Magenta B-cap.
12. Autoclave.

## 3  Methods

### 3.1 Explant Preparation

1. Remove outer leaves, exposing the heart of the offshoot, 15–20 cm long, 6–8 cm wide.
2. Remove outer leaves of the offshoot heart under aseptic conditions exposing the shoot tip region, 3–4 cm long, 1–1.5 cm wide, with 3–4 primordial leaves.
3. Disinfect the shoot tips by immersion in the disinfection solution for 5 min under agitation, and then wash three times in sterile distilled water.
4. Carefully remove the tissue surrounding the shoot tip until you reach the shoot. Use whole shoot.

***3.2 Induction and Maintenance of Embryogenic Callus***

1. Culture the shoot tip on CI medium (Table 2) for 2–3 months [9] (Fig. 1a), and incubate in the dark at 24 ± 2 °C.

2. Transfer the resultant callus to CM medium (Table 2) (Fig. 1b), and incubate in the dark at 24 ± 2 °C for 12 weeks and subculture at a 4-week interval.

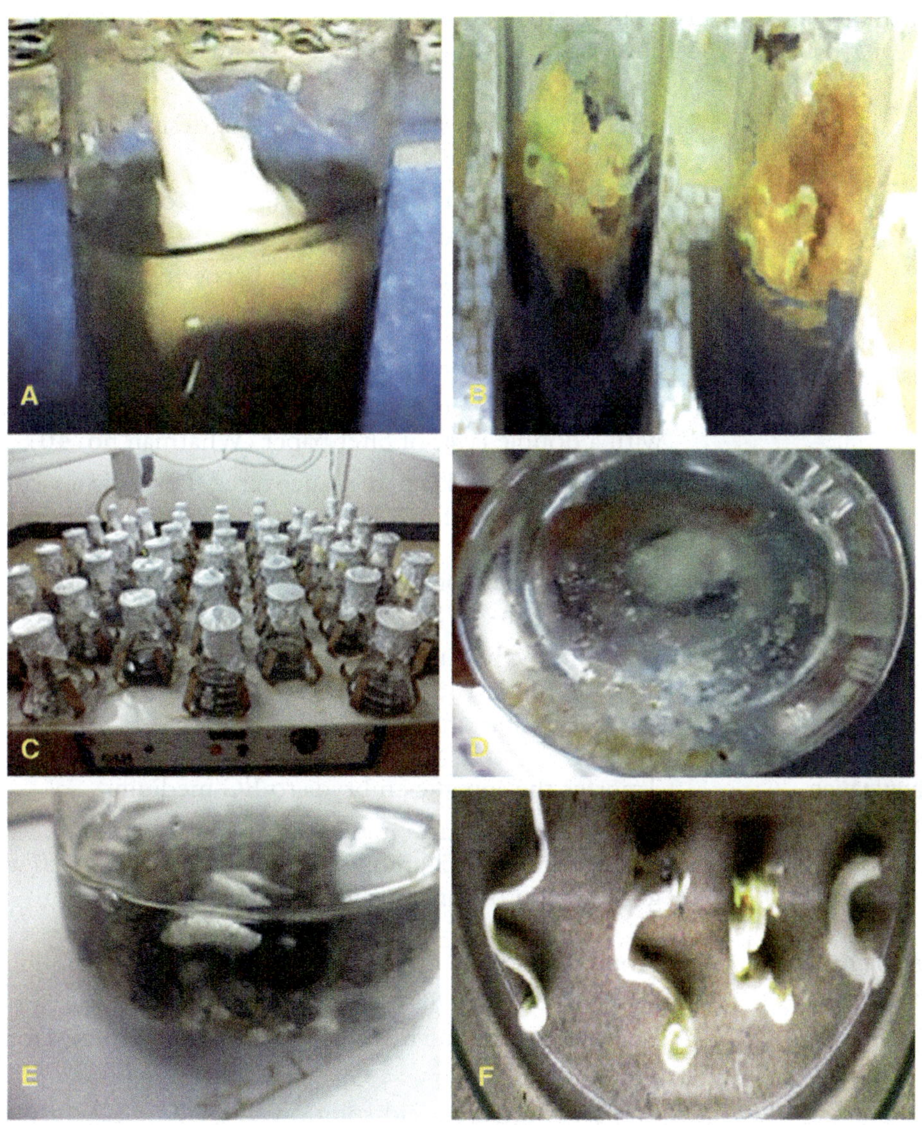

**Fig. 1** The different somatic embryogenesis phases in Barhee cultivar from shoot tip culture to somatic embryos: (**a**) shoot tip on culture initiation medium, (**b**) embryogenic friable callus, (**c**) suspensions system (a rotary shaker), (**d**) somatic embryos derived from embryogenic friable callus, (**e**) somatic embryos in liquid medium with AC, (**f**) somatic embryos for desiccation

| | |
|---|---|
| ***3.3 Cell Suspension Establishment*** | 1. Separate the portions of embryogenic friable callus (200 mg) and culture in CS medium dispensed in 150 mL flasks (*see* **Note 2**). |
| | 2. Cap the flasks with cotton and aluminum foil (*see* **Note 3**). |
| | 3. Incubate the suspension cultures on a rotary shaker set at 100 rpm (Fig.1c) under 16-h photoperiods of cool-white florescent light (35 $\mu$mol/m$^2$/s) at $24 \pm 2$ °C for 4 weeks, and subculture at a 2-week interval (Fig. 1d, e). |
| ***3.4 Maturation of Somatic Embryos*** | 1. Transfer the resultant clumps of globular and mature somatic embryos to EM medium (Table 2) for 6 weeks, and subculture at a 4-week interval (Fig. 2a, b). |
| | 2. Select the mature somatic embryos to desiccation or cold hardening process (*see* **Note 4**). |
| | 3. Transfer the mature somatic embryos onto Petri dishes for desiccation process. |
| | 4. Culture the mature somatic embryos in the tubes for cold hardening process. |
| ***3.5 Desiccation of Embryo Cultures*** | 1. Transfer the resultant clumps of somatic embryos, each clump with 3–5 non-germinated embryos (Fig. 1f) to Petri dishes with a double layer of filter paper (Whatman no. 40). |
| | 2. For desiccation, keep the uncovered Petri dishes containing somatic embryos for 3 h inside the laminar air flow cabinet. |
| ***3.6 Cold Hardening Procedure*** | 1. Keep the tubes containing somatic embryos in EM medium at 4 °C for 24 h (*see* Subheading 3.4, **step 4**). |
| | 2. Incubate the cultures under a 16-h photoperiod of cool-white florescent light (35 $\mu$mol/m$^2$/s) at $24 \pm 2$ °C for 6 weeks for the maturation of somatic embryos. |
| ***3.7 Germination of Somatic Embryos*** | 1. Transfer the somatic embryos to EG medium. |
| | 2. Incubate the cultures under a 16-h photoperiod of cool-white florescent light (35 $\mu$mol/m$^2$/s) at $24 \pm 2$ °C for 10 weeks for germination of somatic embryos. |
| ***3.8 Determination of Proline Content*** | 1. Homogenize 100 mg plant material in 1.5 mL 3% sulfosalicylic acid (*see* **Notes 5** and **6**). |
| | 2. Remove the residue by centrifugation at 2,415 $\times$ *g* for 10 min at 4 °C. |
| | 3. Prepare ninhydrin acid solution by mixing 1.25 g ninhydrin, 30 mL glacial acetic acid, and 20 mL 6M phosphoric acid, and heat in water bath at 100 °C until dissolved. |
| | 4. Mix 100 mL extract with 2 mL glacial acetic acid and 2 mL ninhydrin acid solution, and incubate in a water bath at 100 °C for 1 h. |

**Fig. 2** The different somatic embryogenesis phases in Barhee cultivar from germination of somatic embryos to in vitro plantlets: (**a**) development of embryos on solid medium (maturation medium), (**b**) structured somatic embryos after maturation on solid media, (**c**) germination of somatic embryos, (**d**) development of somatic embryos into plantlets, (**e**) plantlets, (**f**) plantlets in polyethylene nursery pots with the potting mixture for acclimatization stage

5. Terminate the extract reaction by placing the mixture in an ice bucket for 15 min.

6. Add 1 mL toluene to the extract reaction and mix.

7. Separate the toluene layer.

8. Warm the chromophore containing toluene at room temperature.

9. Measure the optical density at 520 nm by spectrophotometer.

10. Determine the amount of proline from a standard curve (*see* **Note 7**).

11. Express the proline content on fresh-weight basis (*see* **Note 8**).

### 3.9 Rooting and Acclimatization

1. After the appearance of mature shoots from EG medium, transfer them to RT medium for 9 weeks to stimulate root induction and shoot elongation for the complete plant formation (Fig. 2c, d; *see* **Note 9**).

2. Collect the plantlets, 10–12 cm long, from RT medium, and gently rinse under a slow stream of tap water to remove residual medium sticking to the rooting region.

3. Dip the roots into the fungicide solution, and transfer to 10 cm polyethylene nursery pots with the potting mixture (Fig. 2f). Water the plantlets with 1/2 MS and subsequently as needed.

4. Cover the plantlets with a glass bottle to maintain humidity preventing the plants from dehydration. Maintain the plants in a greenhouse under natural sunlight at 27 ± 2 °C and 70% relative humidity.

5. Transfer the plantlets to a larger polyethylene pots. After 2–4 months of culture, transfer the plants to a shade house and maintain for 12–24 months, and finally transfer them to the field for further growth.

## 4  Notes

1. Prepare the stock solutions of plant growth regulators by dissolving the 2, 4-D and NAA in 95% ethanol or 1 N NaOH and 2iP using 1 N HCl, and make up the required volume by adding double-distilled water. Store in the refrigerator at 4 °C for up to 1 month.

2. Liquid culture system is a promising technique for rapid mass propagation of date palm. Somatic embryo production in liquid medium is about ten times greater than that on the solid medium. Furthermore, the suspension cultures are technically easier and more economical than the bioreactors [9].

3. It is necessary to link cap of flasks with duct tape to prevent contamination during the rocking movement.

4. Somatic embryos should be washed well several times with sterile distilled water before the desiccation or cold hardening process to remove the traces of plant growth hormones.

5. Recent proteomic, genomic, and metabolic studies have revealed that the function of proline is not as straightforward as initially believed. Research studies on plants, especially those on proline synthesis and catabolic genes, have demonstrated that the proline produced under stressful conditions can act as a compatible solute in osmotic adjustment; a free radical scavenger; a metal chelator; an activator of detoxification pathways; a cell redox balancer; a cytosolic pH buffer; a source of energy, nitrogen, and carbon; a stabilizer for subcellular structures and membranes including photosystem II (PS II); or a signaling molecule [13].

6. Proline analysis is based on procedures described by Bates et al. [14].

7. Proline standard: Dissolve 0.1 g proline in 100 mL distilled water, make 10 concentrations by taking 1–10 mL from proline solution and making volume up to 100 mL by adding distilled water, and take 1 mL from each above concentrations. Add 1 mL sulfosalicylic acid to this. Add 2 mL glacial acetic acid and 2 mL acid ninhydrin, and heat in the boiling water bath for 1 h. A sample of only distilled water is also run doing the same process as for proline samples. For standards blank is distilled water; for samples, blank is toluene.

8. Express the proline content on fresh-weight basis as follows (115.5 is the molecular weight of proline):

$$\mu\text{moles per g tissue} = \frac{\text{g proline/mL} \times \text{mL toluene}}{115.5} \times \frac{5}{\text{g sample}}$$

9. Somatic embryos are considered germinated as soon as radical and roots emerge, and the full plantlet develops as shoots become greener and elongate, and roots are well established [8].

## References


1. Al-Khayri JM (2010) Somatic embryogenesis of date palm (*Phoenix dactylifera* L.) improved by coconut water. Biotechnol 9:477–484

2. Zaid A, Arias-Jimenez EJ (2002) Date palm cultivation. FAO plant production and protection paper. FAO, Rome

3. Jain SM, Al-Khayri JM, Johnson DV (eds) (2011) Date palm biotechnology. Springer, The Netherlands, pp 47–68

4. Cohen Y (2011) Molecular detection of somaclonal variation in date palm. In: Jain SM, Al-Khayri JM, Johnson DV (eds) Date palm biotechnology. Springer, The Netherlands, pp 221–235

5. Fki L, Masmoudi R, Drira N, Rival A (2003) An optimised protocol for plant regeneration from embryogenic suspension cultures of date palm, *Phoenix dactylifera* L., cv. Deglet Nour. Plant Cell Rep 21:517–524

6. El-Hadrami I, El-Bellaj M, El-Idrissi A, J'Aiti F, El-Jaafari S, Daaf F (1998) Biotechnologie végétalers et amelioration du palmier dattier (*Phoenix dactylifera* L.) Pivot de l'agriculture oasienne Marocaine. Cah Agric 7:463–468

7. Sharma DR, Kumari R, Chowdhury JB (1980) *In vitro* culture of female date palm (*Phoenix dactylifera* L.) tissues. Euphytica 29:169–174

8. Othmani A, Bayoudh C, Drira N, Trifi M (2009) *In vitro* cloning of date palm *Phoenix dactylifera* L. cv. Deglet Bey by using embryogenic suspension and temporary immersion bioreactor (TIB). Biotechnol Biotechnol Equip 23(2):1181–1188

9. Ibraheem Y, Pinker I, Böhme M (2013) A comparative study between solid and liquid cultures relative to callus growth and somatic embryo formation in date palm (*Phoenix dactylifera* L.) cv. Zaghlool. Emir J Food Agric 25 (11):883–898

10. Fki L, Bouazizi N, Chkiri O, Benjemaa-Masmoudi R, Rival A, Swennen R et al (2013) Cold hardening and sucrose treatment improve cryopreservation of date palm meristems. Biol Plant 57(2):375–379

11. Shareef HJ, Al-Mayahi AMW, Alhamd AD (2016) Effect of desiccation and cold hardening on germination of somatic embryos in date palm (*Phoenix dactylifera* L.) Barhee cultivar *in vitro*. Adv Appl Sci Res 7(3):58–64

12. Murashige T, Skoog F (1962) A revised medium for rapid growth and bioassays with tobacco tissue cultures. Physiol Plant 15:473–497

13. Hossain MA, Hoque MA, Burritt DJ, Fujita M (2014) Proline protects plants against abiotic oxidative stress: biochemical and molecular mechanisms. In: Ahmad P (ed) Oxidative damage to plants. Elsevier, Amsterdam, pp 477–522

14. Bates LE, Waldren RP, Teare ID (1973) Rapid determination of free proline for water stress studies. Plant and Soil 39:205–207

# Chapter 12

# Histological Evidence of Indirect Somatic Embryogenesis from Immature Female Date Palm Inflorescences

## Eman M.M. Zayed and Ola H. Abdelbar

## Abstract

Rapid production of somatic embryogenesis and date palm regeneration is achieved by culturing immature female inflorescence explants. Inflorescence explants are soft, creamy in color, average 6–7 cm in length, and cultured on Murashige and Skoog (MS) medium containing 1 mg/L thidiazuron (TDZ). Callus induction occurs after 4–5 weeks of culture on the callus induction medium. Subsequently, callus develops embryogenic calli on MS medium supplemented with 0.1 mg/L naphthalene acetic acid (NAA). Histological samples were collected successively at the culturing time and during morphogenetic changes throughout the developmental stages of somatic embryos. Initiation of callus and different successive developmental stages for somatic embryos including two-celled, four-celled, globular, bipolar, and fully developed cotyledonary somatic embryos were observed. Mature somatic embryos develop within 10–12 weeks after culture establishment.

**Key words** In vitro, Immature female inflorescence, Histology, Thidiazuron (TDZ), Somatic embryogenesis

## 1  Introduction

Indirect somatic embryogenesis is an important and appropriate method for the mass propagation of date palm [1]. This method has been successful with plant material containing meristematic cells that become crucial requirements for somatic embryogenesis. Hence, several meristematic sources have been used as explants in tissue culture of date palm, including axillary buds, shoot tips, immature inflorescences, and immature embryos [2, 3]. Young inflorescences are readily available as compared to an insufficient number of offshoots, the common source of explants in date palm [4].

Histological analysis revealed that immature female inflorescences appear as masses of meristematic cells and exhibit morphogenetic plasticity that stands in contrast with the morphogenetic rigidity characterizing vegetative tissue [5]. To avoid the risk of somaclonal variation, the use of minimal concentrations of plant

Jameel M. Al-Khayri et al. (eds.), *Date Palm Biotechnology Protocols Volume 1: Tissue Culture Applications*,
Methods in Molecular Biology, vol. 1637, DOI 10.1007/978-1-4939-7156-5_12, © Springer Science+Business Media LLC 2017

growth regulators in the culture medium is recommended [6]. These floral explants have a high level of endogenous gibberellins that promote inflorescence development. The explants need specific plant growth regulators to reverse flower meristematic cells to the vegetative growth. TDZ in the culture medium modulates endogenous levels of plant growth hormones, especially IAA/cytokinin ratio, which is responsible for callus growth, and subsequently the floral explant converts to a vegetative growth phase [4, 5, 7]. Gibberellins can be affected by TDZ; the latter could mediate endogenous GA and stimulate somatic embryogenesis in many species by GA-synthesis inhibitors [7, 8]. Exogenous TDZ is responsible for increasing zeatin, which indicates the active extent of cell division and metabolism of the plant [9, 10]. This greater effectiveness of TDZ may be due to its slow metabolism in tissue culture. Moreover, TDZ is a urea-based cytokinin and, therefore, is nondegradable by cytokinin oxidase enzymes in plant tissue. This characteristic causes TDZ to be persistent in tissue and to modify endogenous hormones either directly or indirectly and produce reactions in cells and tissue necessary for their division and regeneration [11].

This chapter presents a rapid and reliable protocol for somatic embryogenesis induction and plant regeneration from immature female inflorescences as meristematic explants, which requires only a short period of callus production and avoids the risk of somaclonal variation by exogenous application of cytokinin derivatives (TDZ). Furthermore, it describes the morphological and histological analysis of immature female inflorescences and the changes beginning from callus initiation and continuing until establishment of healthy bipolar somatic embryos.

## 2    Material

### 2.1    Plant Material, Sterilization, and Acclimatization Reagents

1. Immature female inflorescences (spathes) collected from selected adult female trees of date palm Siwy cv. (*see* **Note 1**).

2. Antioxidant solution: 100 mg/L ascorbic acid and 150 mg/L citric acid.

3. Disinfectant solution: 40% Clorox, commercial bleach 5.25% sodium hypochlorite solution containing three drops of Tween 20 per 100 mL solution.

4. Fungicide solution: 0.5% (w/v) Benlate.

### 2.2    Culture Medium

1. Basal culture medium: Murashige and Skoog (MS) salts and vitamins [12] (Table 1).

2. Plant growth regulators stock solutions: Thidiazuron (TDZ, 1 mg/mL), naphthaleneacetic acid (NAA, 1 mg/mL), and benzyladenine (BA, 1 mg/mL).

**Table 1**
**Chemical composition of modified Murashige and Skoog medium (MS) [12]**

| Components | Concentration (mg/L) |
| --- | --- |
| *Macronutrients* | |
| $KNO_3$ | 1900 |
| $NH_4NO_3$ | 1650 |
| $MgSO_4 \cdot 7H_2O$ | 370 |
| $KH_2PO_4$ | 170 |
| $NaH_2PO_4 \cdot H_2O$ | 170 |
| $CaCl_2 \cdot 2H_2O$ | 440 |
| *Micronutrients* | |
| $H_3BO_3$ | 6.2 |
| $MnSO_4 \cdot 2H_2O$ | 22.3 |
| $ZnSO_4 \cdot 7H_2O$ | 8.6 |
| $Na_2MoO_4 \cdot 2H_2O$ | 0.25 |
| $CuSO_4 \cdot 5H_2O$ | 0.025 |
| $CoCl_2 \cdot 6H_2O$ | 0.025 |
| KI | 0.83 |
| *Iron source* | |
| $FeSO_4 \cdot 7H_2O$ | 27.8 |
| $Na_2EDTA \cdot 2H_2O$ | 37.3 |
| *Vitamins and organic supplements* | |
| *myo*-Inositol | 100 |
| Glutamine | 200 |
| Nicotinic acid | 0.5 |
| Pyridoxine·HCl | 0.5 |
| Thiamine·HCl | 0.1 |
| Glycine | 2 |
| Adenine sulfate | 40 |
| Ascorbic acid | 100 |
| Citric acid | 150 |
| *Carbon source* | |
| Sucrose | 30,000 |

3. Induction medium: MS basal medium containing 1 mg/L TDZ (*see* **Note 2**).

4. Development medium: MS basal medium containing 0.1 mg/L NAA [13].

5. Maturation and germination medium: MS basal medium containing 0.1 mg/L NAA and 0.05 mg/L BA [14].

6. pH adjustment solutions: 0.1 and 1 N KOH and HCl, each.

### *2.3 Histological Examination*

1. Phosphate buffer (0.1 M): 19.5 mL of 0.2 M sodium phosphate monobasic (3.12 g $NaH_2PO_4$ in 100 mL) and 30.5 mL of 0.2 M sodium monohydrogen phosphate heptahydrate (3.56 g $Na_2HPO_4.7H_2O$ in 100 mL, pH 7). Store at 4 °C.

2. Fixative fluid: Glutaraldehyde 3% in 0.1 M phosphate buffer (25 mL 0.1 M phosphate buffer pH 7, 6 mL 25% glutaraldehyde, and 19 mL distilled water).

3. Dehydration: Gradient concentrations of ethanol series (5, 10, 15, 20, and 30% v/v).

4. Gradient concentration of tertiary butyl alcohol (TBA) series as shown in Table 2.

5. Embedding: Paraffin wax, melting point 54–56 °C.

6. Mounting gelatin adhesive: 2.5 g gelatin, 500 mL warm distilled water 35 °C, and 0.25 g chromium potassium sulfate.

7. Dewaxing reagent: Xylene.

8. Staining solutions: (a) Mordant reagent, ferric ammonium sulfate 4% w/v; (b) hematoxylin, 0.5% w/v.

### *2.4 Equipment*

1. Tissue culture glassware: Beakers (500–1000 mL), graduated cylinders (100, 500 and 1000 mL), culture jars (200 or 350 mL), and large test tubes (250 × 28 mm) (*see* **Note 3**).

2. Tissue culture tools and instruments: Forceps, scalpels, and blades. Precision balance, pH meter, refrigerator, autoclave,

**Table 2**
**Tertiary butyl alcohol (TBA) dehydration series**

| Mixture number | 95% Ethanol (%) | Absolute ethanol (%) | TBA (%) | Water (%) |
| --- | --- | --- | --- | --- |
| 1 | 50 | – | 10 | 40 |
| 2 | 50 | – | 20 | 30 |
| 3 | 50 | – | 35 | 15 |
| 4 | 50 | – | 50 | – |
| 5 | – | 25 | 75 | – |

laminar flow bench, growth chamber, pipettes, and magnetic stirrers.

3. Acclimatization of plantlets: Plastic pots 5 × 18 cm (torpedo), soil mixture (peat moss, vermiculite, and sand (1:1:1, v/v/v)), and greenhouse.

4. Microscopy preparation tools: Razor blades, filter paper, fine brush, glass vials (10–20 mL), pencil, origami dish or suitable mold, needle, wooden (metal or plastic) chucks, sharp blade or knife for microtome, clean black sheets, slides, long cover glass (24 × 45 mm), staining jars, forceps, and slide box.

5. Microscopy preparation instruments: Vacuum pump, desiccator, oven, differential heated and embedding hot plate, rotary microtome, and microscope fitted with a camera.

## 3  Methods

### 3.1  Medium Preparation

1. Prepare MS medium stock solutions (Table 1) using double-distilled water.

2. Prepare TDZ, NAA, and BA stock solutions (1 mg/mL each): Dissolve TDZ (0.1 g) in a few drops of absolute ethanol. To dissolve BA (0.1 g), use a few drops of 1 N HCl, and for NAA (0.1 g) use a few drops of 1 N KOH. Bring the final volume to 100 mL with distilled water and store the solutions at 4 °C.

3. Mix the components of MS culture medium (Table 1), and add 30/L sucrose and hormones according to the culture stage as specified in Subheading 2.2. Induction medium contains 1 mg/L TDZ; development medium contains 0.1 mg/L NAA; maturation and germination medium contains 0.1 mg/L NAA and 0.05 mg/L BA.

4. Adjust pH to 5.7 using KOH and HCl solutions, and then add 0.1 g/L activated charcoal and 6 g/L agar.

5. Heat the medium until agar is dissolved, and then dispense medium into 200 or 350 mL culture jars (40 mL per jar) and 250 × 28 mm large test tubes (30 mL per tube), cover with polypropylene caps, and autoclave for 20 min at 121 °C and 1.1 kg/cm$^2$.

### 3.2  Explant Preparation

1. Excise immature female inflorescences of date palm carefully from selected adult female trees in early spring (February). These spathes are creamy in color, soft, and 10–14 cm in length; store them in paper bags for 1–2 days in a refrigerator at 4 °C until use.

2. Wash the collected immature female inflorescences of date palm for 1 h under running tap water (Fig. 1a, b).

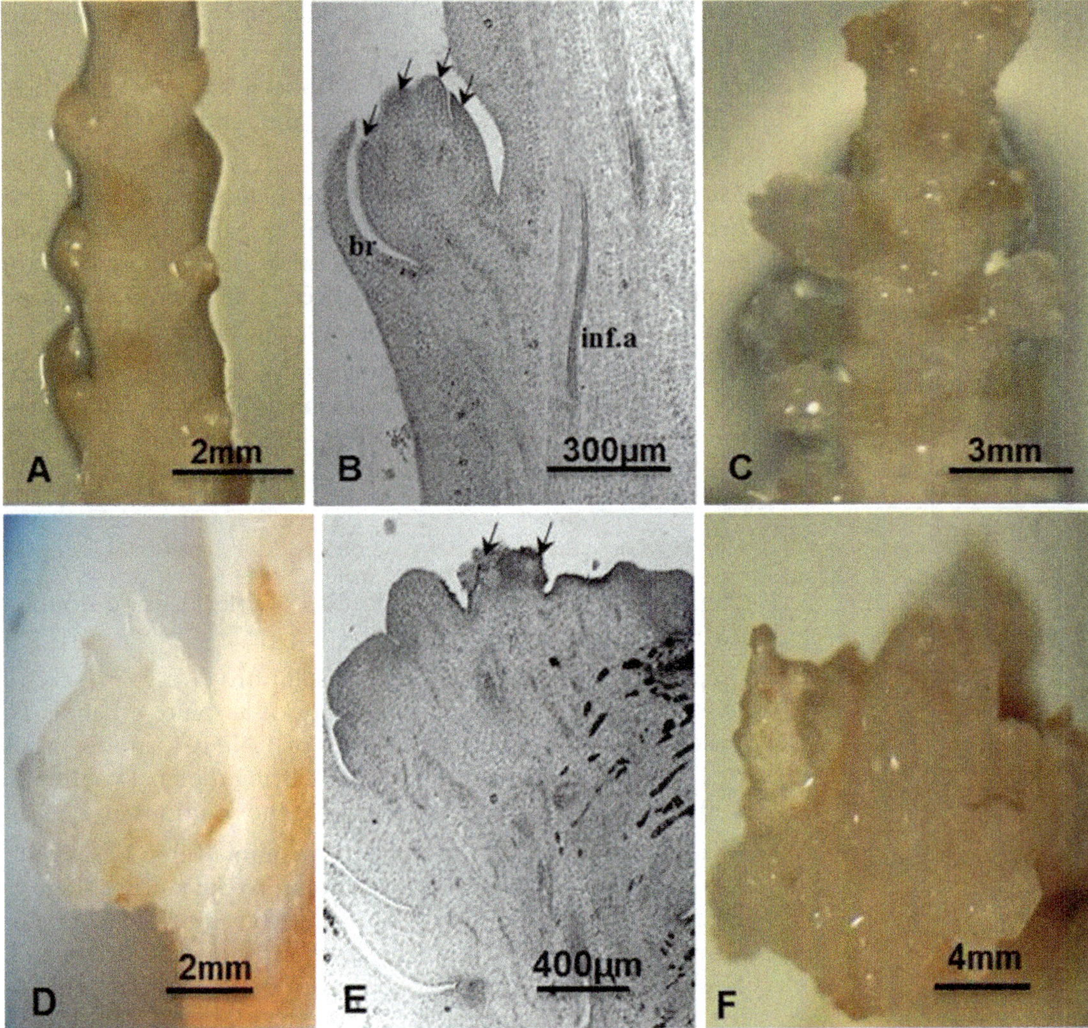



**Fig. 1** Morphology and anatomy of the immature female inflorescence explant of date palm. (**a**) Morphology of the immature floral buds on the soft inflorescence axis. (**b**) Longitudinal section of the same explant illustrates the pistillate floral bud consisting of meristematic tissue and subtending small bract (br), the *arrows* point to the primordial sepals and petals, and inf.a designates inflorescence axis. (**c**) At 2–3 weeks after culture on medium containing TDZ, the floral buds increased a little in size, and callus starts to develop all over the inflorescence axis and also the floral buds. (**d**) At 3–4 weeks, numerous bright callus spots arising all over the surface of the floral bud. (**e**) Longitudinal section of the same stage of callus formation showing the initiation of the callus from the epidermal and subepidermal layers of the floral bud (*arrows*); (**f**) 5–6 weeks later, development of callus into bulky translucent masses with irregular borders

3. Sterilize the laminar flow by UV radiation for 10 min and clean the surface with 70% ethanol before use.

4. Inside the laminar flow bench, disinfect the spathes by soaking in disinfectant solution while shaking for 20 min (*see* **Note 4**).

5. Rinse the spathes with sterilized distilled water three times for 2–3 min.

6. Soak the dissected inflorescence in the antioxidant solution to reduce browning.

7. Remove the protective sheath through a longitudinal incision from the bottom to the top of the protective sheath, and make another incision around the basal portion of the spathe with a scalpel and then peel it slowly.

8. Divide sterilized inflorescence longitudinally into 3–4 segments (spikes with part of base) completely intact for use as explants.

**3.3 Callus Development**

1. Culture one segment (2–3 cm) per jar horizontally, in close contact with the surface of the basal induction medium (Fig.1a, b).

2. Incubate all cultured explants at 25 ± 2 °C in the dark for 6 weeks observing the following culture development events:

   (a) Immature female flowers increase in size after 2–3 weeks of culture in the dark (Fig.1c) (*see* **Notes 5** and **6**).

   (b) Subsequently small masses of callus are formed on the surface of explants, which are translucent with irregular borders after 4–5 weeks (Fig. 1d, e) (*see* **Note 7**).

   (c) After 5–6 weeks of culture, most flowers have large masses of callus and easily separate from the inflorescence axis and fall onto the medium surface (Fig. 1f) (*see* **Note 8**).

**3.4 Somatic Embryogenesis Proliferation and Maturation**

1. Transfer the flowers with emerging callus into development media and incubate in the dark at 25 ± 2 °C.

2. Embryogenic callus starts to appear from callus after 6 weeks of culture. All the embryonic phases can be observed in the friable callus (Fig. 2a–f) (*see* **Notes 9** and **10**).

3. Transfer the embryogenic callus into maturation medium (Fig. 2c, g) (*see* **Note 11**).

4. Incubate cultures at 27 ± 2 °C and 16-h photoperiod (40 $\mu$mol/m$^2$/s).

5. Subculture at 6-week intervals and observe mature somatic embryo after 12–15 weeks of culture on the maturation medium (Fig. 2h, i) (*see* **Note 12**).

**3.5 Somatic Embryo Germination and Plant Formation**

1. Transfer well-matured somatic embryos to germination medium, and incubate cultures for 6–8 weeks at 27 ± 2 °C and 16-h photoperiod (40 $\mu$mol/m$^2$/s).

**3.6 Acclimatization**

1. Transfer plantlets to the greenhouse and immerse in 0.5% (w/v) Benlate fungicide solution for 1–2 min.

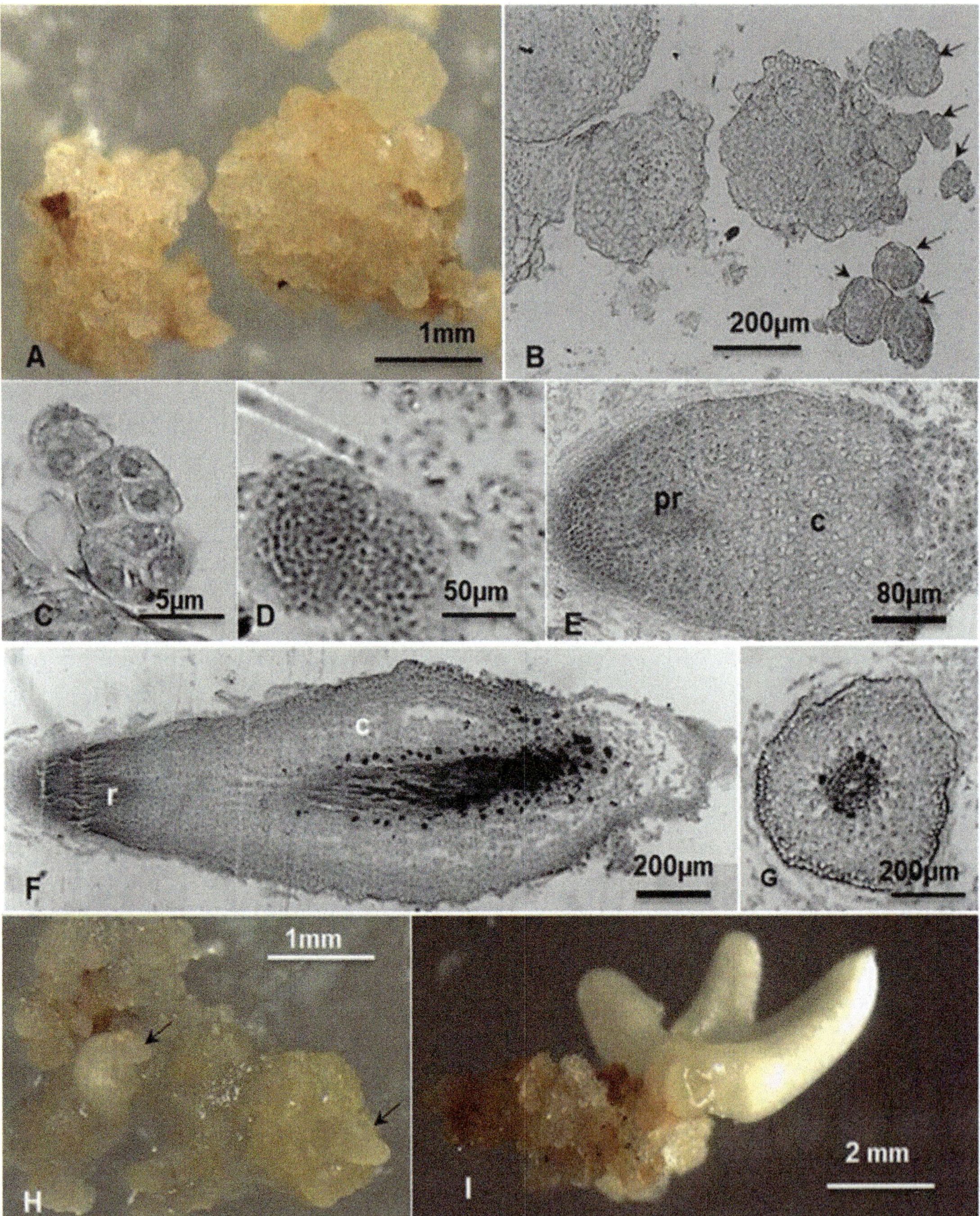

**Fig. 2** Initiation and developmental stages of the somatic embryos. (**a**) Formation of the embryogenic callus from the female floral buds; note the friable callus consists of granules with different sizes and *white* in color. (**b**) Transverse section through the embryogenic callus illustrates numerous confused callus aggregates which consist of two types. The first consists of thin, more vacuolated cells, while the second consists of compact meristematic aggregates (the *arrows* point to the meristematic aggregates of different sizes, and they are very rich in cytoplasm with small vacuoles and highly activated in cell divisions). (**c**) Two-celled and four-celled

2. Place plantlets in plastic pots 5 × 18 cm (torpedo) containing a mixture of peat moss, vermiculite, and sand (1:1:1, v/v/v).

**3.7  *Histological Examination***

This procedure is based on the schedule of the paraffin method described by Berlyn et al. [15].

*3.7.1  Fixation*

1. Cut the immature female inflorescence (at the culturing time) into small pieces 4–5 mm long using a sharp razor blade without pressing to minimize tissue damage.

2. Submerge samples into a vial containing a suitable volume of fixative fluid (glutaraldehyde 3% in 0.1 M phosphate buffer pH 7) instantly to avoid drying and to promote quick penetration of the fixative into the tissue.

3. Do the previous step with each ontogenesis stage occurring to the explant during the subculturing time (callus, embryogenic callus, and mature embryo).

4. Remove air from all samples with a vacuum pump.

5. After pumping, remove any floating pieces that do not immerse in glutaraldehyde solution with a fine brush because they will not be ready for the infiltration process.

6. Keep the pieces in glutaraldehyde for at least 24 h before continuing the process for embedding.

*3.7.2  Washing*

1. Rinse the fixative samples in phosphate buffer for 2–3 min to wash off the fixing fluid by decanting the fixative and adding a new amount of buffer immediately.

2. Remove the buffer solution and immerse samples in ascending concentrations of alcohol (5, 10, 20, and 30%) for 2 h at each concentration.

*3.7.3  Dehydration*

1. Dehydrate the samples gradually by immersing in ascending concentrations of TBA (as shown in Table 2) for 1 h at each concentration (*see* **Note 14**).

2. Make three changes of TBA and then continue with infiltration in paraffin (*see* **Note 15**).

**Fig. 2** (continued) embryonic stages. (**d**) The globular stage. (**e**) Bipolar embryo; note the polarity of this embryonic stage through a meristematic end, the root tip with a procambium strand (pr) and a more vacuolated tip cotyledon (c). (**f**) The fully developed embryo; note the root tip (r) with the cotyledon (c). (**g**) Transverse section at the same stage illustrates the procambium strand in the central position of the embryonic axis Note: the internal structure of this somatic embryo (illustrated in **f** and **g**) is similar to the zygotic embryo. (**h**) The *arrows* indicate two developed embryos arising from the embryogenic callus. (**i**) Cluster of somatic embryos

*3.7.4  Infiltration*

1. Prepare a mixture of pure TBA with paraffin oil (1:1 v/v).

2. Fill a new vial with a suitable amount of melting wax, let the wax solidify but not cool, put the samples on the wax surface, and then add a suitable amount of the previous mixture (TBA/paraffin oil) (*see* **Note 16**).

3. Place the open specimen vials into the wax oven at 35 °C.

4. Add more new melted wax when the solid layer dissolves. Continue adding melting wax until a thin layer of solidified wax remains on top of the solution. This means that the TBA is clearly saturated with paraffin at this temperature. This part of the process may be extended over 2–3 days (*see* **Note 17**).

5. Increase the oven temperature to 52–60 °C, so the solidified wax melts, and then allow the vial to stand for 4 h.

6. Pour off the homogenized solution of wax and TBA into a waste container and replace it with pure melted paraffin, and then quickly return the vial to the oven.

7. After 4 h, pour off all the paraffin wax and replace it with new pure paraffin. Repeat this step three times, 3–4 h for each change. Then the paraffin will be free of TBA.

*3.7.5  Embedding*

1. Make a suitable paper boat from slightly glossy smooth paper or use commercially available molds for the embedding with paraffin wax [15].

2. Place the embedding paper boat on the hot plate.

3. Take out the specimens vial from the oven; pour the vial contents (the specimens and the melted wax) into the boat. Add a suitable amount of melted wax to the top of the boat. Quickly arrange the specimens with a heated needle leaving enough space between each two pieces (*see* **Note 18**).

4. Move the boat toward the cold space. When the paraffin has hardened enough to keep the samples from moving, float the boat in a pan of cold water. Allow the surface of the wax to solidify, and slowly submerge the boat in the water, holding it under with a heavy object.

5. When the paraffin has cooled, discard the paper boat. Leave the wax blocks in the cold water for 30 min. Store the blocks away from dust and do not place one block upon another.

*3.7.6  Microtoming*

1. Using a sharp scalpel, cut the paraffin block into small pieces each containing an individual embryogenic stage.

2. Stick each paraffin piece on a wooden chuck or metal holder suitable for the microtome clamp by using a heated scalpel, passing between the paraffin block and chuck at the site of

contact to melt a surface layer of paraffin and press them together (*see* **Note 19**).

3. Trim the block to remove the excess of paraffin, leaving at least 2–3 mm of paraffin around the material (*see* **Note 20**).

4. Be sure that the clamping mechanism in the microtome is tightened securely before sectioning.

5. Fix the wooden chuck in the clamp of the microtome. Adjust the angle of the microtome knife so that the face of the paraffin block is parallel to the knife blade before sectioning (*see* **Note 21**).

6. Set the micrometer scale at 8 μm. This thickness is suitable for showing the organization of the meristematic tissue and their differentiations.

7. Make transverse sections for all the samples except for the fully developed embryos which need longitudinal sections.

8. Run the microtome and take sections to form a ribbon. Pick up the ribbon by a moisture brush; this is easier than using forceps and avoids blade damage.

9. Put the ribbon on a clean black paper, subdividing it into pieces with a razor blade to be placed on the glass slides (*see* **Note 22**).

*3.7.7 Mounting*

For best results clean the slides and cover slips before the mounting step even though they appear to be clean (*see* **Note 23**).

1. Coat the cleaned slide with few drops of gelatin adhesive. Allow it 1–2 min to dry.

2. Flood the slide with distilled water except for the labeled part for the slide handle.

3. Pick up the pieces of the ribbon by a moistened fine brush and float them on the water surface, and then arrange them uniformly to maintain a series of sections.

4. Place the slide on a warm plate to stretch and flatten the sections. The temperature of the warm plate should not be over 45 °C (*see* **Note 24**).

5. Transfer the slide from the hot plate; let it cool for 2–4 min.

6. Remove the excess water; let it dry and keep it away from dust until the staining process.

*3.7.8 Pre-Staining*

1. Immerse the slides in a staining jar containing enough xylene reagent to dissolve the paraffin; complete this step three times with 5 min intervals.

2. Transfer the slides in a series of staining jars containing descending concentrations of alcohol starting with anhydrous ethanol; then 95, 85, 70, 50, and 30%, for 5 min at each concentration; and then transfer slides to distilled water. The dewaxing procedure is illustrated in Fig. 4.

*3.7.9  Staining*

1. Prepare mordant reagent: Mix 500 mL ferric ammonium sulfate 4% (iron alum) with 5 mL acetic acid glacial and 6 mL sulfuric acid 10% (*see* **Note 25**).

2. Prepare hematoxylin stock solution: Add 0.01 g sodium bicarbonate to 1 L distilled water. Heat the water to the boiling point, remove the beaker from the heater, and add 5 g hematoxylin crystals. Immediately, cool the solution and store at 4 °C. Just before use, dilute the stock solution with twice its volume of distilled water.

3. Transfer the slide from the distilled water to a staining jar containing enough of the mordant reagent and keep for 4 h.

4. Rinse the slides in distilled water five changes at 1 min intervals.

5. Immerse the slides into 0.5% hematoxylin for 4 h.

6. Transfer the slides into a jar containing the de-staining reagent (mordant diluted with an equal volume of water) for 2–3 min until the dye is differentiated, i.e., until access staining is removed (*see* **Note 26**).

7. Transfer the slides in a series of jars containing ascending concentrations of ethanol start with 30, 50, 70, 85 and 95% to anhydrous ethanol, for 5–10 s each.

8. Immerse the slides into xylene jars, three changes at 2–3 min each.

9. Put a drop of mounting medium (Canada balsam or numerous synthetic resins) on the tissue, and cover the sections with cleaned cover glass without bubbles forming.

10. Place the preparations in a slide tray until dry. Keep it horizontal and away from dust. Now the slides are ready for microscopic observations. The hematoxylin staining procedures are summarized in Fig. 5.

---

# 4  Notes

1. Spray wounded plant parts with Thiram (fungicide) at 3 mL/L and Corazon (pesticide) at 3 mL/L to prevent diseases and pests, especially red palm weevil.

2. Prepare the induction media without activated charcoal due to low concentration of plant growth regulators.

3. Close all jars with polypropylene caps and all test tubes with aluminum foil.

4. During the soaking of explants in Clorox solution for 20 min, if the tissue starts to bleach and lacerate, immediately transfer them into sterilized distilled water.

5. Histological analysis of the immature female flower reveals small masses of meristematic cells.

6. Immature female inflorescence explants contain the highest level of endogenous gibberellic acid ($GA_3$) and indole acetic acid (IAA) and the lowest level of zeatin. Endogenous zeatin increases after culturing of floral explants on the induction medium containing 1 mg/L TDZ as it increased at the subsequent developmental stages. The highest concentration of endogenous zeatin is reached at the callus initiation and embryogenic callus formation and decreases in endogenous gibberellic acid ($GA_3$) [10].

7. Callus initiates from the epidermal and subepidermal layers of the immature female flowers.

8. Callus grows all over the tissue of immature female flowers and is characterized by translucent appearance. The inflorescent with the attached callus separates easily from the inflorescence axis and spontaneously drops onto the surface medium after 5–6 weeks of culture.

9. During this time, small parts of translucent calli differentiate into embryogenic calli. Morphological observation of this callus shows granules of different sizes, white in color of attached or separated cell clumps (Fig. 2a, b).

10. Histological observations indicate that the embryonic callus consists of two different types of cells, soft vacuolated cells and compact aggregate ones.

11. The friable callus is composed of disorganized masses of highly vacuolated cells 30–60 μm in diameter. This tissue does not originate from embryos; however, they are surrounded by the aggregate masses that serve as the source of embryoids. These masses are composed of meristematic and rich cytoplasmic cells and are 6–25 μm in diameter. All the embryonic phases can be observed in these aggregations from a single-cell, two-celled, four-celled, globular to bipolar structures (Fig. 2c–f).

12. The longitudinal sections of matured embryos show fully organized shoot and root apexes, in addition to well-differentiated procambial strands along the embryo axis and the cotyledon 800–1200 μm in length (Fig. 2e–g).

13. *See* Fig. 3 for a summary of in vitro propagation steps of date palm using immature female inflorescences to regenerate healthy plantlets.

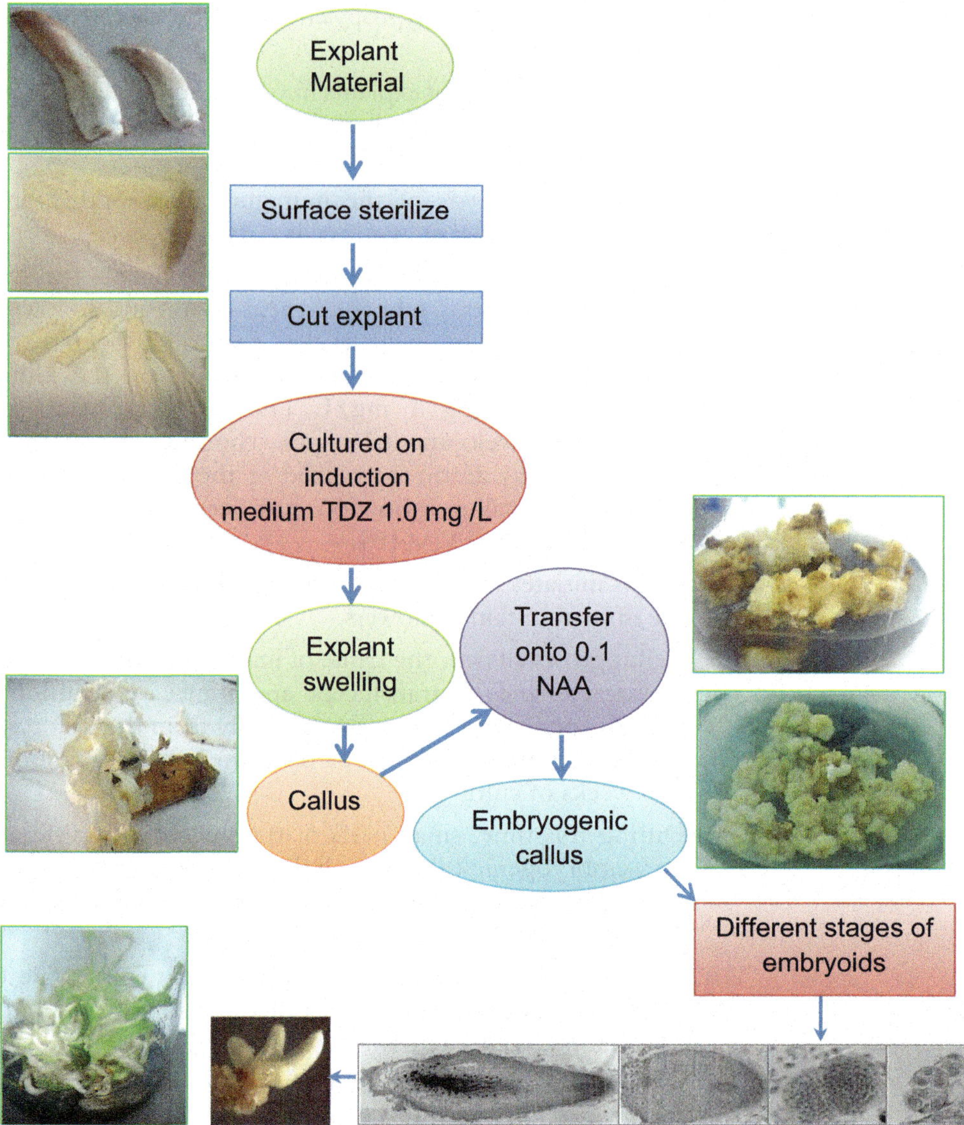


**Fig. 3** Summary of in vitro propagation steps of date palm using immature female inflorescences to regenerate healthy plantlets (*see* **Note 13**)

14. Samples must not be allowed to air dry during the dehydration steps.

15. Tertiary butyl alcohol is a dehydration reagent and also a paraffin solvent. It solidifies at 25 °C, a rather warm temperature for a laboratory or stock room. Alternatively, the container of TBA can be wrapped with a heat tape in conjunction with a regulatory rheostat set to deliver 26 °C; this has been successful to keep the reagent fluid.

16. Paraffin oil characterized with higher viscosity. For this, it will slowly be replaced with melted paraffin in the latter steps and reduce tissue shrinkage related to paraffin infiltration.

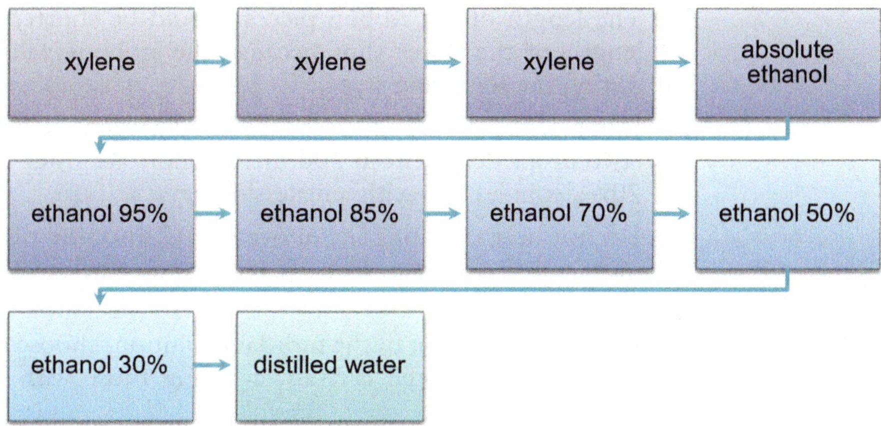


**Fig. 4** The pre-staining steps, each at 2–3 min interval

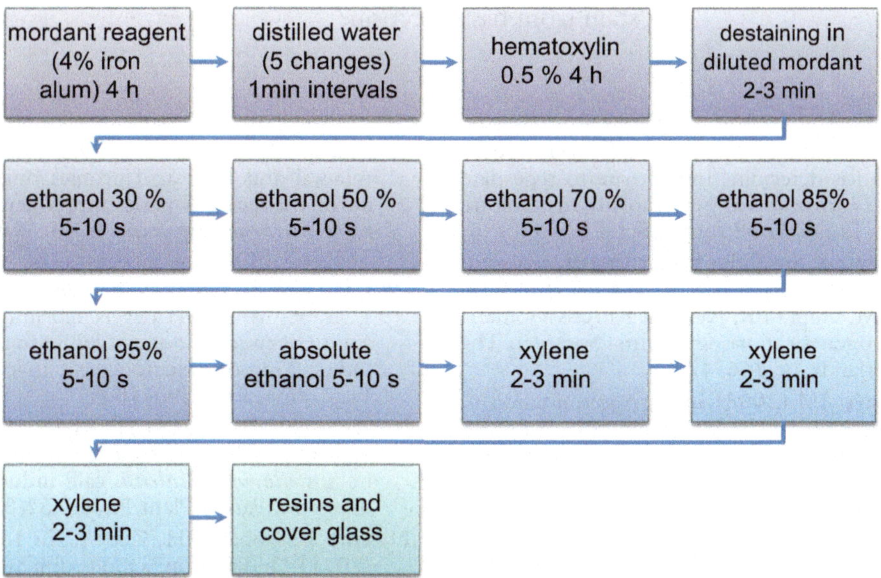


**Fig. 5** Summary of the hematoxylin staining procedures

17. If the vials become filled with solution, decant some of the solution into a waste container.

18. Never heat the bottom of the specimen vial because the tissue will become overheated and ruined.

19. Do not touch the sample with the hot scalpel when sticking the paraffin piece on the wooden chuck.

20. Make sure that the longest opposite edges of paraffin blocks are exactly parallel to each other for easy sectioning and to produce a straight ribbon.

21. Do not touch the cutting edge of the microtome knife/blades, as it is easily damaged.

22. The length of the ribbon pieces should be shorter than the length of the cover slips because the ribbons will elongate during the following step.

23. Immerse the slides and cover slips in a detergent in 70% alcohol (two drops detergent in 200 mL alcohol). Transfer them into 70% alcohol; clean with a furless tissue.

24. Do not heat the hot plate above 45 °C since the ribbons will melt and the samples lost. At temperature below 45 °C, the sections shrink.

25. During preparation of the mordant solution, choose the transparent violet-colored crystals, avoiding those with a yellow-rusted color.

26. Hematoxylin staining normally is followed by a dye differentiation step [16]. In this process, the stained sections are treated with iron alum (diluted mordant reagent) to remove excess stain from tissue section.

## References


1. Kunert KJ, Baaziz M, Cullis CA (2003) Techniques for determination of true-to-type date palm (*Phoenix dactylifera* L.) plants: a literature review. Emir J Food Agric 15:1–16

2. Bhaskaran S, Smith R (1995) Somatic embryogenesis in date palm (*Phoenix dactylifera* L.) In: Jain SM, Gupta PK, Newton RJ (eds) Somatic embryogenesis in woody plants. Springer, The Netherlands, pp 446–470

3. Al Khayri JM (2003) In vitro germination of somatic embryos in date palm: effect of auxin concentration and strength of MS salts. Curr Sci 84:680–683

4. Zayed EMM (2011) Propagation of *Phoenix dactylifera* L. *Chamaerops humilis* L. and *Hyophorbe verschaffeltii* L. Palms by using tissue culture technique. Ph.D. thesis. Fac Agric Cairo Univ, Cairo, Egypt

5. Zayed EMM, Abdelbar OH (2015) Morphogenesis of immature female inflorescences of date palm in vitro. Ann Agric Sci 60 (1):113–120

6. Cohen Y, Korchinsky R, Tripler E (2004) Flower abnormalities cause abnormal fruit setting in tissue culture-propagated date palm (*Phoenix dactylifera* L.) J Hort Sci Biotech 79 (6):1007–1013

7. Hutchinson MJ, Krishna RS, Saxena PK (1997) Inhibitory effect of $GA_3$ on the development of thidiazuron-induced somatic embryogenesis in geranium (*Pelargonium* x *hortorum* bailey) hypocotyl cultures. Plant Cell Rep 16:435–438

8. Murch SJ, Saxena PK (1997) Modulation of mineral and fatty acid profiles during thidiazuron mediated somatic embryogenesis in peanuts (*Arachis hypogaea* L.) J Plant Physiol 151:358–361

9. Casanova E, Valdés A, Fernández B, Moysset L, Isabel M (2004) Levels and immunolocalization of endogenous cytokinins in thidiazuron induced shoot organogenesis in carnation. J Plant Physiol 161:95–104

10. Zang CG, Li W, Mao YF, Zhao DL, Dong W, Guo GQ (2005) Endogenous hormonal levels in *Scutellaria baicalensis* calli induced by thidiazuron. Russ J Plant Physiol 52(3):345–351

11. Guo B, Abbasi BH, Zeb A, Xu LL, Wei YH (2011) Thidiazuron: a multi-dimensional plant growth regulator. Afr J Biotechnol 10 (45):8984–9000

12. Murashige T, Skoog F (1962) A revised medium for rapid growth and bioassays with tobacco tissue cultures. Physiol Plant 15:473–497

13. Mater AA (1986) In vitro propagation of *Phoenix dactylifera* L. Date Palm J 4(2):137–152

14. Omar MS (1988) In vitro response of various date palm explants. Date Palm J 6(2):371–388

15. Berlyn GP, Miksche JP, Sass JE (1976) Botanical microtechnique and cytochemistry. Iowa State University Press, Iowa, pp 24–100

16. Baker JR (1960) Experiments on the action of mordants, 1 'single-bath' mordant dyeing. Q J Microsc Sci 101:255–272

# Chapter 13

# Histological Analysis of the Developmental Stages of Direct Somatic Embryogenesis Induced from In Vitro Leaf Explants of Date Palm

Ola H. Abdelbar

## Abstract

Somatic embryogenesis is an ideal technique for the micropropagation of date palm using different explant tissue; however, histological studies describing the ontogenesis of plant regeneration are limited. This chapter provides a simple protocol for the histological analysis of the successive developmental stages of direct somatic embryogenesis induced from in vitro leaf explants. Direct somatic embryos are obtained from Murashige and Skoog (MS) medium containing 2 mg/L 6-benzylaminopurine. In order to observe the different developmental stages, histological analysis is carried out on samples at 15-day intervals for 60 days. Samples are fixed in formalin acetic alcohol and embedded in paraffin wax. Stain serial transverse and longitudinal sections, 8 μm thick, are stained with safranin-Fast Green. After 15 days on the induction medium, somatic embryos exhibit multicellular origin directly from the procambium cells, whereas the mesophyll and the epidermal cells are not involved in this process. After 2 months, several developmental stages (pre-globular, globular, early bipolar, bipolar, and cotyledonary-shaped) are observed. These embryos germinate after transferring to MS medium without plant growth regulators and rooting on 2 mg/L NAA-containing medium resulting in complete plantlets.

**Key words** Benzylaminopurine, Direct somatic embryogenesis, Histology, Leaf explant, Procambium cells

## 1 Introduction

Plant tissue culture techniques are used for large-scale propagation of date palm *Phoenix dactylifera* L. by either organogenesis [1, 2] or somatic embryogenesis [3, 4]. Different types of date palm explants such as shoot tips, leaves, immature inflorescences, and immature embryos have been used [4–7]. The induction of date palm somatic embryos was demonstrated by both indirect morphogenetic pathway mediated by a callus phase [8] and direct somatic embryogenesis without a callus phase [9]. The latter pathway, which is the focus of this chapter, is yet to be fully optimized for mass micropropagation of date palm.

Jameel M. Al-Khayri et al. (eds.), *Date Palm Biotechnology Protocols Volume 1: Tissue Culture Applications*, Methods in Molecular Biology, vol. 1637, DOI 10.1007/978-1-4939-7156-5_13, © Springer Science+Business Media LLC 2017

Recent discoveries of the cytomorphological features and biochemical and molecular markers controlling the process of somatic embryogenesis will enable improving in vitro propagation of date palm, which is a relatively slow process [10, 11]. Somatic embryogenesis, besides being one of the most widely used for mass propagation of date palm [12, 13], is considered an excellent morphogenetic system for studying in vitro differentiation processes. Several reports have provided accurate histological information on the origin and ontogeny of indirect somatic embryogenesis of date palm [7, 8, 14, 15]; however, only a few reports addressed the histological aspects of direct somatic embryogenesis [9, 11].

The gene *somatic embryogenesis receptor kinase 1 (SERK1)* is identified as a marker for embryonic competent cells, expressed in the procambium cells (immature vascular cells) in response to favorable plant growth regulators during the process of cell division and differentiation leading to somatic embryo formation [16, 17]. This prompted us to use young in vitro leaves as explants because of abundancy of procambium cells.

Histological analysis is a powerful tool to determine the tissues involved in various morphological processes, elucidate the various mechanisms of differentiation, and allow comparison of the internal structures between somatic embryos and zygotic embryos. This chapter describes a simple and rapid protocol for induction of direct somatic embryogenesis from young in vitro leaves, morphological and histological changes associated with the ontogeny of the direct somatic embryos from the initial cells until the establishment of the bipolar embryos and plantlet regeneration.

## 2  Materials

### 2.1  Plant Material

Leaf explants isolated from germinated mature somatic embryos of date palm Sakkoty cv. (Fig. 1a; *see* **Note 1**).

### 2.2  Culture Medium

1. Basal medium: Murashige and Skoog (MS) medium [18] (Table 1).

2. Induction medium: MS medium (Table 1) containing 2 mg/L 6-benzylaminopurine (BAP).

3. Germination medium: MS medium (Table 1) without plant growth regulators.

4. Rooting medium: ½ MS basal salts and vitamins containing 20 g/L sugar, 1 mg/L α-naphthaleneacetic acid (NAA), and 1 g/L activated charcoal.

5. Solutions to adjust pH: 0.1 and 1 N KOH and 0.1 and 1 N HCl.

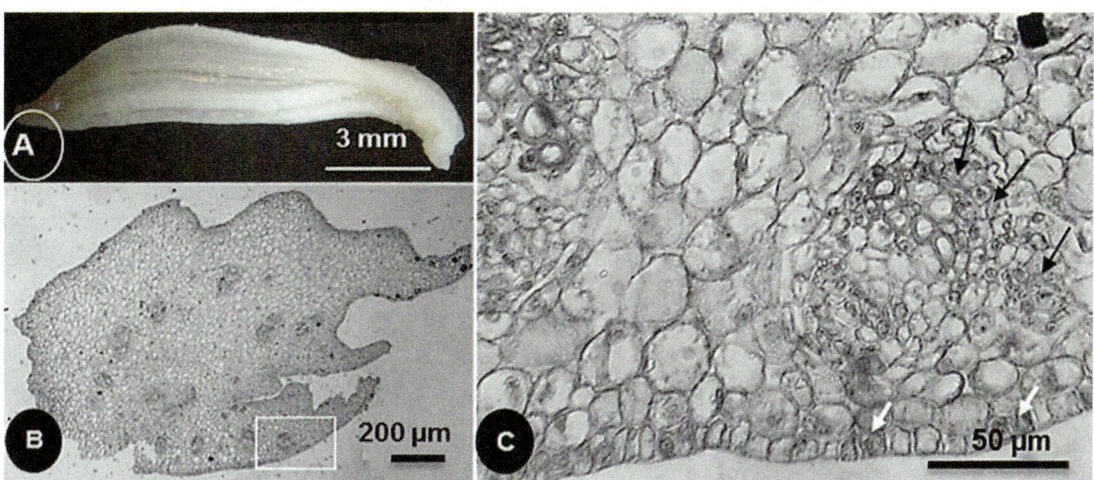

**Fig. 1** Morphology and histology of the in vitro small leaf of date palm cv. Sakoty before culture. (**A**) Morphology of the young leaf showing that the basal part is white and smooth; (**B**) cross section in the basal part of the young leaf showing uniseriate epidermis and parenchymatous ground tissue embedding the vascular bundles; (**C**) enlargement area of figure (**B**) showing anticlinal cell divisions in the epidermal layer (*white arrows*), the regular compact ground tissue, and procambial strands (*black arrows*). Source: Reproduced with permission [11]

***2.3 Microscopic Examination***

1. FAA solution: Formalin, glacial acetic acid, and 50% ethanol (5:5:90 v/v/v).

2. Series of ethanol solutions: 50, 70, 85, 95%, and absolute ethanol.

3. Xylene-absolute ethanol solutions in gradient concentrations: (1:3), (1:1), (3:1), and pure xylene.

4. Paraffin wax: Melting point 54–56 °C.

5. Haupt's adhesive: 1 g gelatin, 15 mL glycerin, 100 mL warm distilled water (35 °C), and 1 g phenol crystals (*see* **Note 2**).

6. Safranin-Fast Green stain: 1% safranin O and 0.1% Fast Green FCF.

***2.4 Equipment***

1. Glassware: Glass beakers (500–1000 mL), graduated cylinders (100, 500, and 1000 mL), and small jars (200 mL).

2. Surgical tools: Forceps and scalpels.

3. Tissue culture instruments: Autoclave, pH meter, laminar flow hood, growth chamber, and stereomicroscope.

4. Histological analysis instruments: Vacuum pump, desiccator, oven, differential heated hot plate, rotary microtome, and light compound microscope supplied with a digital camera.

**Table 1**
**Chemical composition of the modified MS medium [18]**

| Stock solutions and additives | Component | Concentration (mg/L) | Quantity in stock solution (g) | Volume of stock solution (mL) | Volume required for 1 L of medium (mL) |
|---|---|---|---|---|---|
| Macronutrients (20× stock) | $KH_2PO_4$ | 170 | 3.4 | 500 | 25 |
| | $NaH_2PO_4 \cdot H2O$ | 170 | 3.4 | | |
| | $KNO_3$ | 1900 | 38 | | |
| | $NH_4NO_3$ | 1650 | 33 | | |
| | $MgSO_4 \cdot 7H_2O$ | 370 | 7.4 | | |
| Calcium source (20× stock) | $CaCl_2 \cdot 2H_2O$ | 440 | 8.8 | 100 | 5 |
| Micronutrients (100× stock) | $MnSO_4 \cdot 2H_2O$ | 22.3 | 2.23 | 100 | 1 |
| | $H_3BO_3$ | 6.2 | 0.62 | | |
| | $ZnSO_4 \cdot 7H_2O$ | 8.6 | 0.86 | | |
| | $Na_2MoO_4 \cdot 2H_2O$ | 0.25 | 0.025 | | |
| | $CuSO_4 \cdot 5H_2O$ | 0.025 | 0.0025 | | |
| | $CoCl_2 \cdot 6H_2O$ | 0.025 | 0.0025 | | |
| | KI | 0.83 | 0.083 | | |
| Iron source (100× stock) | $FeSO_4 \cdot 7H_2O$ | 27.8 | 2.78 | 100 | 1 |
| | $Na_2EDTA \cdot 2H_2O$ | 37.3 | 3.73 | | |
| Vitamins (100× stock) | Nicotinic acid | 0.5 | 0.05 | 100 | 1 |
| | Pyridoxine·HCl | 0.5 | 0.05 | | |
| | Thiamine·HCl | 0.1 | 0.01 | | |
| | Glycine | 2.0 | 0.2 | | |
| | Myoinositol | 100 | 10 | | |
| | Adenine sulfate | 40 | 4 | | |
| Antioxidant, sugar, and agar | Glutamine | 200 | – | | |
| | Sucrose | 40 g/L | – | | |
| | Agar | 5.5 g/L | – | | |

5. Histological analysis supplies: Razor blades, filter paper, fine brush, glass vials (10 or 20 mL), pencil, origami dish or suitable mold, needle, wooden or (metal, plastic) chucks, sharp blade or knife for microtome, clean black sheets, slides, long cover glass (24 × 45 mm), Coplin jars, forceps, and slide box.

# 3    Methods

*3.1 Culture Media and Conditions*

1. Prepare MS stock solutions for macronutrients, micronutrients, iron, vitamins, and organic compounds (Table 1) by dissolving the components of each stock separately in distilled water using a magnetic stirrer. Transfer the stock solutions to reagent bottles, and store in a refrigerator at 4 °C until use.

2. Prepare BAP stock solution (1 mg/mL) by dissolving in few drops of 0.1 M KOH and NAA stock solution (1 mg/mL) by dissolving in a few drops of absolute ethanol and then make up the volume with distilled water.

3. Mix the stock solutions and other additives listed in Table 1 and then add the plant growth regulators as required for each medium according to the culture stages (*see* Subheading 2.2).

4. Adjust the pH of all media to 5.7 using KOH or HCl diluted solutions (0.1–1 N) and then add 5.5 g/L agar.

5. Dispense the culture medium in 200 mL jars (40 mL/jar) and autoclave for 20 min at 121 °C and 1.1 kg/cm$^2$.

*3.2 Induction of Direct Somatic Embryos*

1. In a laminar flow chamber, excise the young in vitro leaves, 1–1.5 cm in length, with sterile scalpel, white in color, and 1–2 weeks old (*see* **Note 3**).

2. Gently place the young in vitro leaves (5 leaves/jar) horizontally with the abaxial side in contact with the surface of MS induction medium.

3. Incubate the cultures in a growth chamber in the dark at 27 °C ± 2.

4. Again excise the in vitro leaves and prepare them for the microscopic examinations, to observe the internal structures of the explant (Fig. 1b, c) (*see* **Note 4**).

5. Screen the cultures for possible contamination at weekly intervals.

6. Use a stereomicroscope in a laminar flow chamber to observe any morphological changes occurring on the explants during the first 8 weeks.

*3.3 Ontogeny and Development of Direct Somatic Embryos*

1. Collect the histological samples at 15, 30, 45, and 60 days of culture to determine the tissue/cell(s) involved in the initiation of somatic embryos directly from the in vitro leaves and characterize their successive ontogeny stages (*see* **Note 5**).

2. After 15 days of culturing, a little swelling occurs at the basal part of the explant with few cell divisions in the undifferentiated vascular cells *procambial cells* as the first stage of embryo formation. Thus, these embryos have a multicellular origin based on histological evidence (Fig. 2a–d, *see* **Note 6**).

3. Subsequently, more divisions form a multicellular small ovate proembryos (Fig. 3a). These slight masses are fused to the vascular strands of *the parent tissue* (Fig. 3b).

4. After 30 days from culturing, intensive cell divisions form the globular embryos, ranging 200–250 μm in diameter, (*see* **Note 7**) embedded in loosened parenchyma inside the leaf tissues (Fig. 3b, c). The basal part of the young leaf is more swollen,

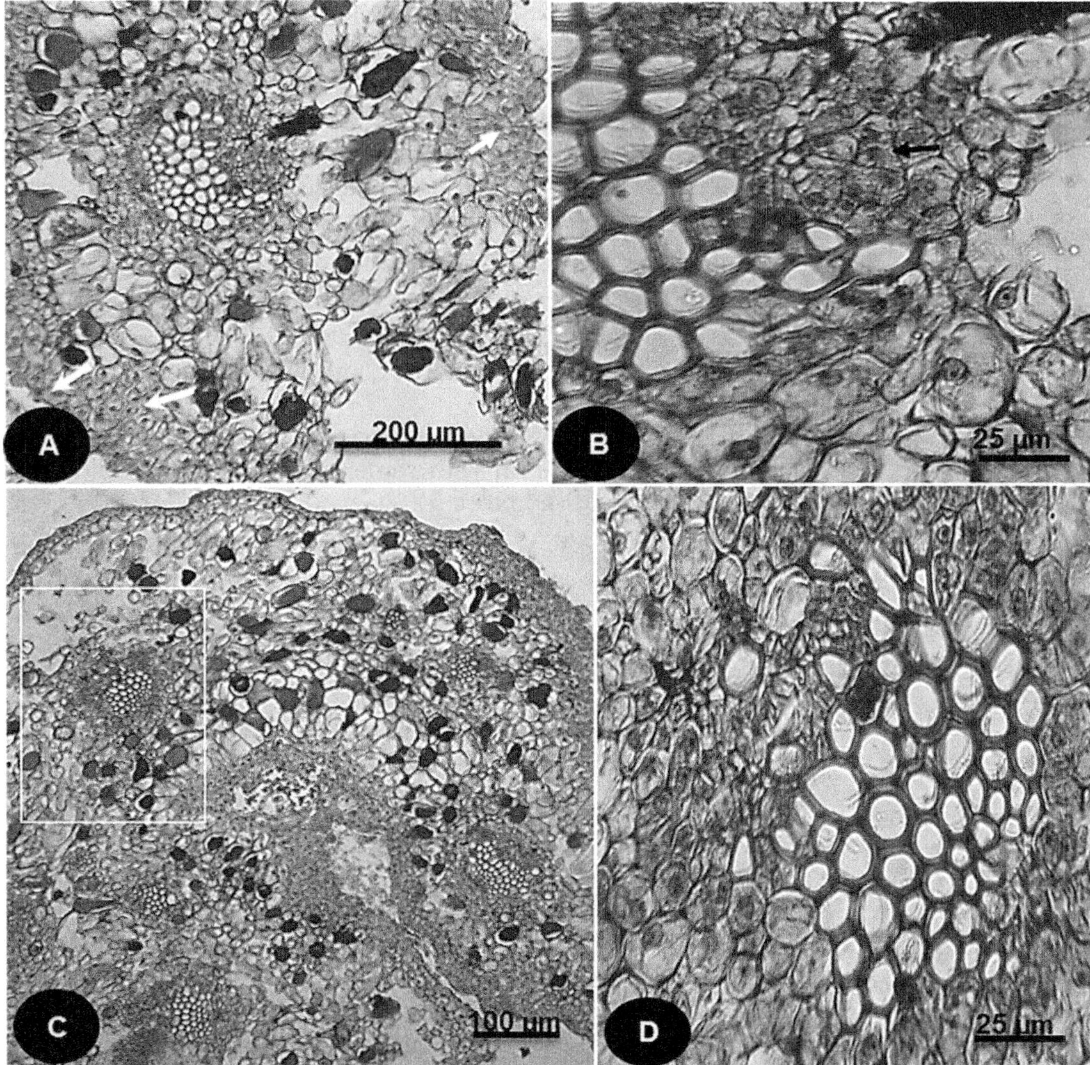

**Fig. 2** Ontogeny of direct somatic embryos. (**A**) After 15 days of culture showing a differentiated vascular bundle. Note the expanded ground tissue compared with those in the Fig. 1c, the arrows point to high mitotic activity occurred in the epidermal and subepidermal layers; (**B**) an enlarged view of the same bundle in the figure (**A**) showing cell divisions occurred on one side of this vascular bundle adjacent to the xylem elements (*arrow*); (**C**) transection in the culture young leaf; the rectangle reveals a vascular bundle, which was enlarged in the figure (**D**). Note the meristematic features in the cells surrounding the bundle. Source: Reproduced with permission [11]

and few small protuberances arise from the adaxial surface, i.e., upper side (Fig. 3d).

5. The globular embryo becomes somewhat prolonged to give an early bipolar shape, which is comprised of a meristematic pole and a more differentiated one. The cells of the later pole have large vacuoles (Fig. 4a, *see* **Note 8**).

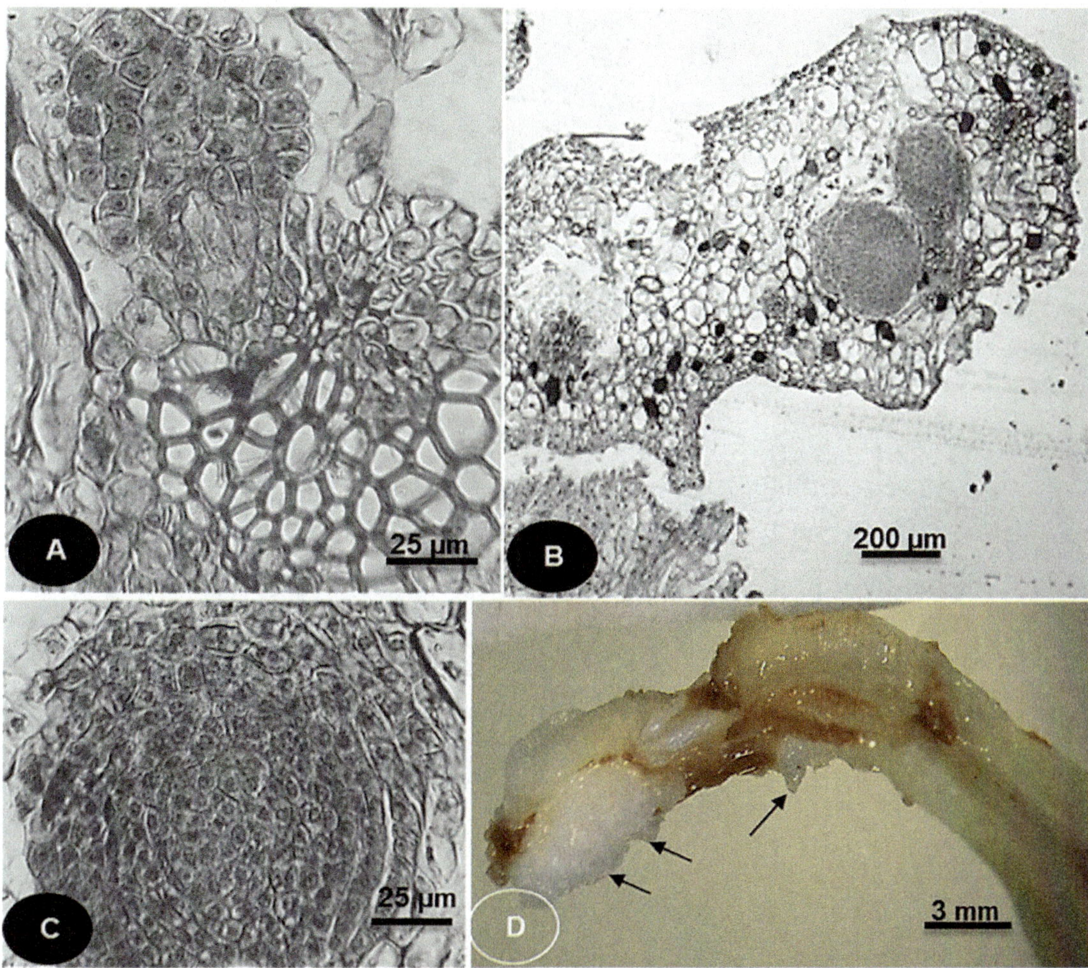

**Fig. 3** Development of direct somatic embryos. (**A**) The globular proembryo; (**B**) after 30 days from culturing, two globular proembryos. Note the degeneration of parenchymatous cells near the globular embryos; (**C**) an enlarged view of the globular embryo. Note the beginning of the differentiation of the meristematic tissues; (**D**) morphology of the in vitro young leaf 30 days from culturing showing the swollen leaf base, and few small protuberances arise from the adaxial surface (*arrows*). Source: Reproduced with permission [11]

6. After 45 days from culturing, a procambium strand is observed in the median plane of the bipolar-shaped embryo with formation of the cotyledon (Fig. 4b). With continued growth, the bipolar embryos become more visible outside the epidermis of the foliar explant. Thus, the epidermis turns brownish color and splits under the pressure of the growing inner embryos (Fig. 4a, d).

7. After 60 days from culturing, numerous direct embryos arise from the leaf surface with fully organized shoot and root apex, furthermore to well-differentiated procambial strands along the embryo axis and the cotyledon, similar to that of the zygotic embryo formation (Fig. 4c).

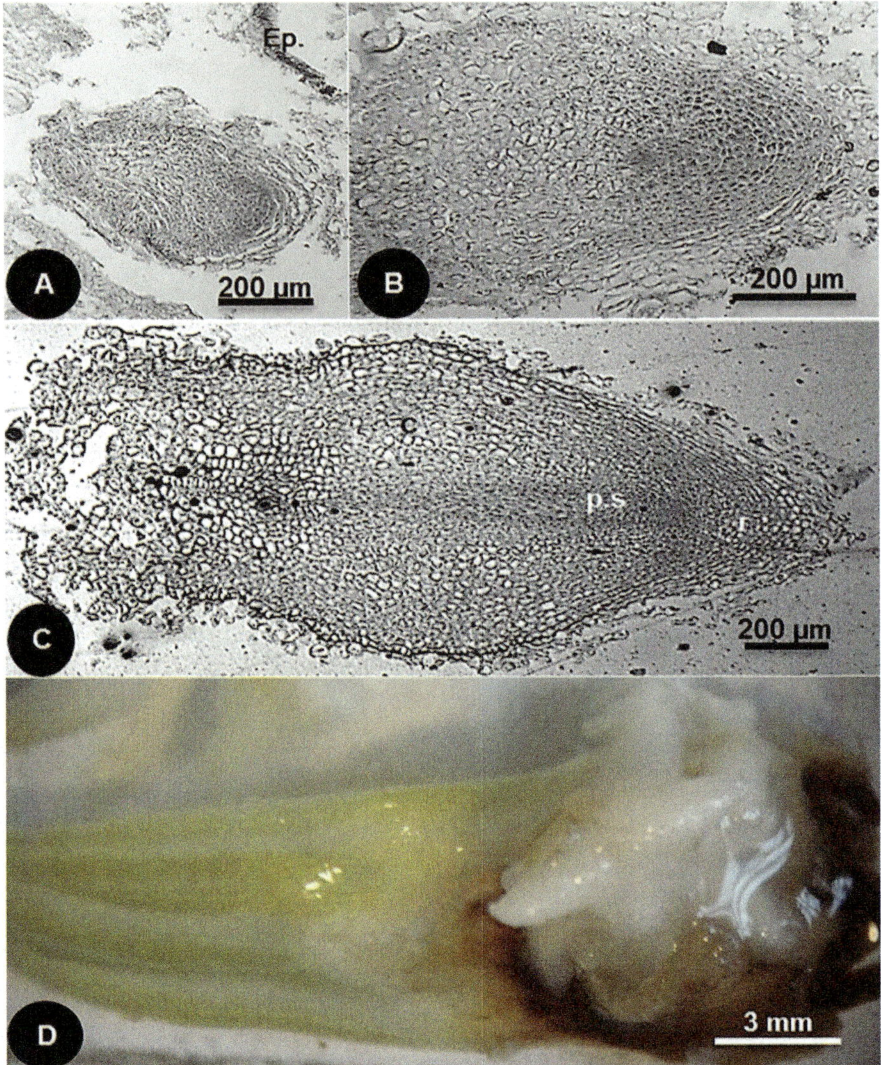

**Fig. 4** Development of direct somatic embryos. (**A**) After 45 days, the globular embryo expanded forming the early stage of the bipolar embryo which is characterized by a high dense meristematic region and a vacuolated one; note the degenerated parenchymatous ground tissue of the young leaf and the collapsed upper epidermis (ep); (**B**) the bipolar stage, note the cotyledon formation; (**C**) after 60 days, the complete bipolar embryo, the root meristem (r), procambial strands (p.s), and the cotyledon (**C**) were clearly visible; no vascular connection between somatic embryo and explant was observed; (**D**) morphology of the cultured small leaf after 60 days showing a cluster of direct somatic embryos arising from the basal part of the leaf surface Note that the color transformed into green compared to Fig. 1a, while the basal degenerative tissues became necrotic with a brownish color. Source: Reproduced with permission [11]

8. No vascular connection is seen between the vascular system of the somatic embryos and the parental tissue. Separate them from the foliar explant (Fig. 4d). The independent vascular systems in these somatic embryos indicate their ability to form normal somatic embryos and plantlets.

**Fig. 5** Different stages of plantlets. (**A**) Early stage of new shoots germinated from the direct somatic embryos; (**B**) fully developed date palm plantlets derived from direct somatic embryos. Source: Reproduced with permission [11]

9. Somatic embryo formation is an asynchronous process on the induction medium. Most of the early developmental stages are visible together with fully developed stage after 45 and 60 days of culturing.

### 3.4 Somatic Embryo Germination and Conversion

1. Transfer explants carefully with the emerging embryos from the induction medium onto the germination medium (Fig. 5a).

2. Incubate cultures at 27 °C $\pm$ 2 under light condition for 16 h (40 $\mu$mol/m$^2$/s) provided by cool white fluorescent lamps for 8 weeks.

3. Transfer the germinated embryos onto the rooting medium and incubate under the same conditions (Fig. 5b, *see* **Note 9**).

4. Subculture the materials at 8-week intervals three times.

### 3.5 Microscopic Examination [19]

#### 3.5.1 Fixation

1. Prepare FAA fixative solution: Mix together formalin, acetic acid, and 50% ethanol (5:5:90).

2. Remove the in vitro leaves from the germinated embryos at the culturing time by a sharp razor blade and also pick the explant samples after 15, 30, 45, and 60 days to follow development (*see* **Note 10**).

3. Put the leaves on a sheet of wet filter paper; subdivide into complete transverse pieces 3–4 mm thick using a razor blade. Keep the fully developed embryos intact without dividing.

4. Drop tissue pieces into a vial filled with suitable volume of the fixative fluid (FAA) by a fine brush to promote quick penetration of the fixative into the tissue.

5. Remove the air from samples using a vacuum pump and connect it with a glass desiccator. Decrease the pressure within the chamber gradually by giving the minimum pressure of the pump (approx. 5 m Bar). Turn off the vacuum and let the

samples equilibrate within the chamber. Replace a new amount of FAA fluid after pumping. Remove any floating pieces that do not immerse after pumping and submersion (*see* **Note 11**).

6. Keep tissue pieces in FAA for 24 h before continuing the processing for embedding.

*3.5.2 Dehydration*

1. Use 95% ethanol (ethyl alcohol) to prepare ascending concentration series of ethanol:

   (a)  For ethanol 50%, mix 50 mL ethanol and 45 mL distilled water.

   (b)  For ethanol 70%, mix 70 mL ethanol and 25 mL distilled water.

   (c)  For ethanol 85%, mix 85 mL ethanol and 15 mL distilled water.

   (d)  For ethanol 95%, use 95% ethanol directly without mixing with water.

   (e)  Finally, for absolute ethanol, use the anhydrous ethanol (*see* **Note 12**).

2. Dehydrate tissue in ethanol reagent by immersing in ascending concentration series of ethanol solutions (50, 70, 85, and 95%) and anhydrous ethanol. The interval of each of these solutions takes 30–60 min.

3. Make three changes of the anhydrous ethanol. Changes are done by decanting the ethanol from the tissues promptly flooding the leaf pieces with a generous volume of the next solution in the series, and this should be made quickly to avoid drying the tissues (*see* **Note 13**).

*3.5.3 Clearing*

1. Prepare the mixture of gradient concentrations of xylene-absolute ethanol:

   (a)  25% xylene mix, 25 mL xylene, and 75 mL absolute ethanol.

   (b)  50% xylene mix, 50 mL xylene, and 50 mL absolute ethanol.

   (c)  75% xylene mix, 75 mL xylene, and 25 mL absolute ethanol.

   (d)  100% pure xylene.

2. Transfer the samples to the clearing reagent xylene. Through this process the ethanol will be replaced by the paraffin solvent xylene, and this makes the leaf tissues in a translucent appearance. A series of ascending concentrations of xylene-ethanol mixtures is followed by 25, 50, 75%, and pure xylene. The interval in each mixture takes 90 min.

3. Make three changes of pure xylene to remove completely the remaining ethanol from the tissue samples before starting paraffin infiltration. The translucent appearance is an indicator that the tissues are free of moisture and ready for the next step (*see* **Note 14**).

*3.5.4 Infiltration with Paraffin wax*

1. Prepare a suitable container for supply of filter embedding paraffin wax with melting points between 54 and 56 °C.

2. Pour a teaspoonful of melted wax into the vial containing the specimens and the pure xylene along the side of the vial, where a solidified layer of wax will remain on the top of xylene. Add a further amount of wax equal to the xylene volume in the vial.

3. Place the specimen vial uncapped in the wax oven (35 °C). The paraffin layer will gradually dissolve and diffuse downward into the tissues.

4. Add more newly melted wax when the solid layer dissolves. When the vial is fully filled with solutions, decant some of the solutions into a waste container. Continue adding melted wax until a thin layer of solidified wax remains on top of the solution, which means that the solvent is obviously saturated with paraffin at this temperature. This part of the process may be extending over 2–3 days.

5. Transfer the specimen vials to a 55 °C oven, until the solidified paraffin melts. Continue adding newly melted wax and leave the vials in the oven for 4 h. Pour off the homogenized solution of wax and xylene into a waste container and replace it with pure melted paraffin, and then return the vials to the oven quickly.

6. After 4 h, pour off all the paraffin wax and replace with new pure paraffin. Repeat this step three times; 3 h is satisfactory for each exchange. Then the paraffin will be free of xylene.

*3.5.5 Embedding*

This step is achieved in suitable paper boats, which are made of slightly glossy smooth surface [19] (*see* **Note 15**)

1. Switch on a gradient heated hot plate (*see* **Note 16**). Place the paper boat on the hot side of the hot plate. Take out the specimen vials from the oven, shake well, and quickly pour the paraffin containing the tissue into the boat.

2. Add an adequate amount of paraffin to the boat to cover the materials. With a warmed needle, arrange the materials leaving enough space between each two pieces.

3. Move the boat toward the cold side of the plate. Sometimes by moving the boat from the melting zone toward the cooling edge, some pieces may move; in this case use a warmed needle to rearrange these pieces (*see* **Note 17**).

4. When the paraffin has cooled down and hardened enough to keep the pieces from moving, float the boat in a pan of cold water (*see* **Note 18**). Allow the surface of the wax to solidify and slowly submerge the boat in the water, holding it under with a heavy object.

5. When the paraffin has cooled, discard the paper boat. Leave the wax blocks in the cold water for 30 min. Store the blocks away from any dust and avoid stacking them on top of each other.

*3.5.6 Microtoming*

1. Remove a piece of material from the paraffin block using a sharp scalpel and stick it to a wooden chuck or metal holder suitable for the microtome clamp. Pass the hot scalpel between the paraffin block and chuck at the site of contact to melt the surface layer of paraffin and press them together (*see* **Note 19**).

2. Cut away the excess of paraffin leaving at least 2–3 mm of paraffin around the material. The longest opposite edges of paraffin should be exactly parallel to each other to produce a straight ribbon.

3. Insert the wooden block in the clamp of the microtome and make the face of the paraffin parallel with the knife blade. Be sure of the sample direction during the cutting process.

4. Make transverse sections for all the leaf samples except the fully developed embryos which need longitudinal sections. Set the micrometer scale at the desired thickness.

5. Sectioning of 8 μm may be a good compromise, showing the organization of the meristematic tissues and their differentiations.

6. Move the wheel of the rotary microtome after reassuring that the clamping mechanism is strongly secure and that the sections form a ribbon. Pick it up with a needle.

7. Put the ribbon on a clean black paper, subdividing it into pieces with a razor blade (*see* **Note 20**).

*3.5.7 Mounting*

Use Haupt's adhesive to glue sections on glass slides [19]. A magnetic heater mixer is used to dissolve the gelatin and glycerin in water. The phenol is included as a preservative (*see* **Note 21**).

1. Put a small drop of Haupt's adhesive on the slide, smear it evenly over the slide surface, and remove any excessive adhesive, leaving only a barely perceivable film.

2. Flood the slide with distilled water. With a moistened fine brush, pick up the portion of the ribbon on the flood and arrange it. Arrange ribbons uniformly to make a series of sections.

3. Put the slide on a warm plate to flatten out the wrinkles in the paraffin. The temperature of the warm plate should not be over 45 °C (*see* **Note 22**). After the sections have stretched, take the slide far away from the warm plate for a few minutes until the water cools, then drain off the excess water.

4. Set the mounted slides away in a dustproof place for a day or more, so the slides are ready for staining.

*3.5.8 Pre-Staining and Staining*

The pre-staining procedures are illustrated in Fig. 6. Follow the double staining system safranin-Fast Green combination. The staining procedures are illustrated in Fig. 7. In these procedures, use a Coplin jar, which is a vertical jar with grooves that hold the

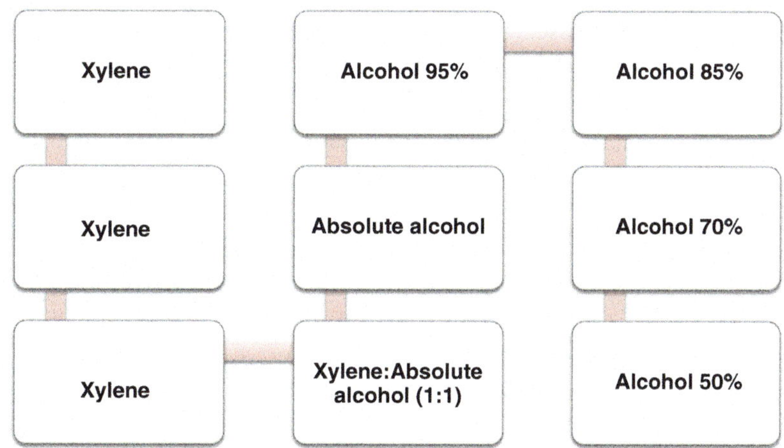



**Fig. 6** Pre-staining process steps at 3–5 min intervals

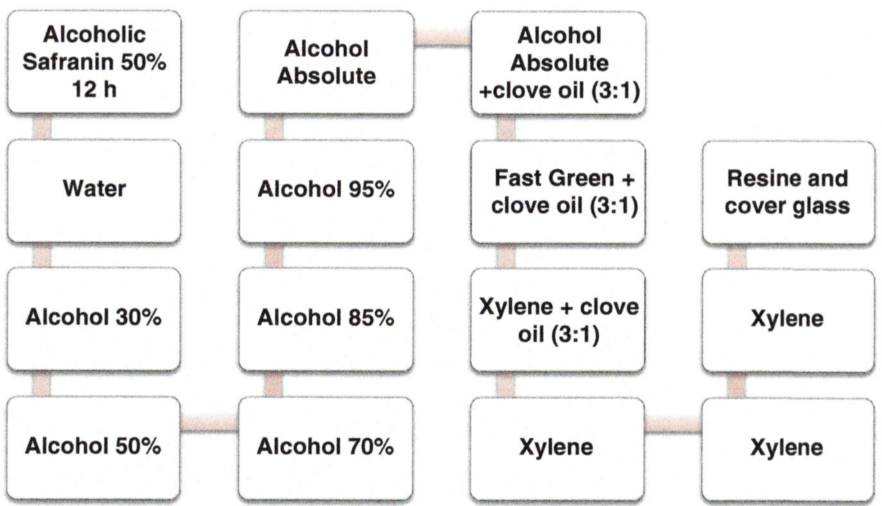



**Fig. 7** The staining process with safranin-Fast Green. Steps are at 5–10 s intervals

slides in a vertical position. Place the slides into the grooves in pairs back to back.

1. Immerse the slides in a Coplin jar containing enough xylene reagent to dissolve the paraffin; follow this step three times at 5 min intervals.

2. Place the slides into a Coplin jar containing a mixture of xylene-anhydrous ethanol 1:1 for 5 min, and then transfer the slides into a jar containing anhydrous ethanol.

3. Use the following descending concentrations of ethanol (95, 85, 70, and 50%) to immerse the slides before staining. The intervals take 5 min for each change (*see* **Note 23**).

4. Dissolve 1 g safranin O in 50 mL absolute ethanol, and then add 50 mL distilled water.

5. Dissolve 0.1 g Fast Green FCF in 75 mL absolute ethanol then add 25 mL clove oil.

6. Put the slides in a Coplin jar containing enough amount of safranin O in 50% ethanol for 6 h, then remove the slides from the safranin with forceps, and put in water for a few seconds to remove the excessive safranin.

7. A progressive transfer to anhydrous ethanol is then made through the indicated grades of ethyl ethanol; *see* Fig. 7. The intervals take a few seconds (*see* **Note 24**). Then the Fast Green FCF is applied.

8. Use clean and dry long cover glass for covering several rows of sections on a slide. Remove a slide from the last xylene jar and place with tissue upward on a black sheet.

9. Put a drop of mounting medium (Canada balsam or numerous synthetic resins) on tissues, lower a cover glass obliquely onto the resin, and leave at least 5 mm between the sections and the edge of the cover glass. Just to avoid drying of the tissue, it should be done quickly.

10. Place the preparations on a horizontal tray until dry in an oven at 53 °C for 1–2 days to harden the resin somewhat for safe handling (*see* **Note 25**). Observe under the microscope.

## 4  Notes

1. Somatic embryos are commonly produced from shoot tips, leaves, and immature inflorescence explants [5–7]. The source of sterile explant in vitro leaves of date palm is continuously available year-around.

2. Actually, the references [19, 20] mention dissolving 1 g gelatin in 100 mL distilled water at 30 °C and then adding 2 g phenol

in Haupt's adhesive. However, the same results are obtained using 1 g phenol thus reducing the consumption of phenol. However, it is even better to use 0.05 g thymol (2-isopropyl-5-methylphenol) instead of phenol as a preservative for the adhesive. Phenol changes the gelatin solution less transparent as compared to thymol. To accelerate dissolving gelatin, use a magnetic hot plate set at 35 °C.

3. Avoid use of large leaf explants because they have no regenerative ability and turn brown. The basal portion, 5–7 mm of the explant, is the most responsive on the induction medium since it contains the most undifferentiating vascular strands as compared to the middle and upper portions [21, 22]. Carefully excise the in vitro leaves with its basal portion to obtain the best embryogenic response. There will be no embryogenic response by excising above the basal portion.

4. This leaf sample at the time of culturing is crucial because it will be easier to identify any changes in the anatomical structure of the explants arising later by comparing with this sample.

5. Histological examination of samplings is scheduled according to the previous study [7, 15]. In fact, the matter of picking up a sample for the anatomical studies based on morphological changes in the explant is not enough to investigate all the ontogeny embryonic stages, because many of these stages occur inside the tissues and cannot be observed with the naked eye. It is necessary to take samples for the microscopic examination at these schedule periods even without any change in shape.

6. The histological analysis of the explant reveals the basal part contains numerous procambial cells *undifferentiated vascular tissues,* which can produce the somatic embryos directly. Two indications can prove this result: (1) no involvement of the mesophyll parenchyma or the epidermal cells in this process, and (2) no response is seen for the direct embryo formation in the more developed leaf parts (the middle and the upper parts) or when using large or older leaves as explants since they contain fully developed and mature vascular tissues.

7. Anatomical observations and measurements are carried out using a Leica light research microscope model DM 500 supplied with a digital camera LEICA ICC 50 HD with LAS E7 software version 2.1.0 2012. With this software, one can draw a straight line along the wide diameter of the globular embryo's microphotograph. The length of this line equals the diameter. Repeat this step 20 times with the observed globular embryos and calculate the average of these measurements. If this software for measurements is not available, use an eyepiece and stage micrometer [20].

8. By observing the stage of the bipolar embryo, the embryonic axis *shoot and root meristems* form the meristematic pole with small cells and dense contents, whereas the cotyledon forms the opposite one with more differentiated cells having large vacuoles.

9. Using 1 mg/L NAA and 1 g/L activated charcoal increased the thickness of plantlets, improved root formation, and stimulated secondary root formation.

10. Do not pull or press the in vitro leaves to avoid rupture of delicate tissues; such damage does not become visible until the finished slide is examined.

11. Pieces in which it is difficult to evacuate air will not be ready for the infiltration process.

12. Anhydrous ethanol is more expensive; therefore use the commercial ethanol (95%) to prepare the different concentrations of ethanol.

13. The extreme changes of ethanol concentrations can cause shrinkage of protoplasm and distortion of cells, and tissues become brittle. To soften the tissue, keep it for a longer time in low concentrations of dehydration fluid.

14. Xylene is a toxic reagent, so carry out processing either in a hood or in a laboratory with a large volume of air exchange.

15. Do not soak up much of the paraffin in the paper boat or in other commercially available molds and cassettes.

16. The heater of the gradient hot plate is fixed to one side of the apparatus. When switched on the surface of the plate will be hotter in one side more than the other end.

17. Never heat the bottom of the specimen vial because the tissues will become overheated and ruined.

18. The rapid cooling of paraffin is necessary to prevent crystallization, which appears as white spots and patches. Such paraffin is difficult to cut into sections by the microtome. In this case, cut the wax block and re-embed.

19. Avoid touching the plant tissue with the hot scalpel.

20. The length of the ribbon pieces is in accordance with the study and the cover length. Our histological studies need to observe series of sections; therefore, a long cover glass (24× 45 mm) is best. The length of ribbon pieces should be less than the cover length because the ribbon will expand.

21. The strength of the adhesive is reduced after 1 month of the preparation date.

22. Excessive heating of the slide will melt the paraffin ribbons and cause distortion in cellular arrangement and cellular details.

The less unexpanded ribbons do not adhere well; sections are easily separated from the slide during staining.

23. The transfers should be done quickly to avoid slides becoming dry.

24. Avoid extending the time over a few seconds because safranin will begin to fade during passage through ethanol.

25. The newly covered preparations must be handled carefully because the slide cover is not well fixed to the slide yet and can be easily moved causing specimen damage.

## References


1. Al Kaabi H, Rhiss A, Hassan M (2001) Effect of auxins and cytokinins on the in vitro production of date palm bud generative tissues and on the number of differentiated buds. In: Proceedings second international conference on date palm Al Ain, UAE, pp 47–86

2. Al-Khateeb A (2006) Role of cytokinin and auxin on the multiplication stage of date palm (*Phoenix dactylifera* L.) cv. Sukry. Biotech 5:349–352

3. Quiroz-Figueroa FR, Rojas-Herrera R, Galaz-Avalos RM, Loyola-Vargas VM (2006) Embryo production through somatic embryogenesis can be used to study cell differentiation in plants. Plant Cell Tissue Organ Cult 86 (3):285–301

4. Al-Khayri JM (2003) In vitro germination of somatic embryos in date palm: effect of auxin concentration and strength of MS salts. Curr Sci 84:680–683

5. Othmani A, Bayoudh C, Drira N, Marrakchi M, Trifi M (2009) Somatic embryogenesis and plant regeneration in date palm *Phoenix dactylifera* L., cv. Boufeggous is significantly improved by fine chopping and partial desiccation of embryogenic callus. Plant Cell Tissue Organ Cult 97(1):71–79

6. Tisserat B, DeMason D (1980) A histological study of development of adventive embryos in organ cultures of *Phoenix dactylifera* L. Ann Bot 46(4):465–472

7. Zayed ME, Abd Elbar OH (2015) Morphogenesis of immature female inflorescences of date palm in vitro. Ann Agric Sci 60 (1):113–120

8. Al-Khayri JM (2005) Date palm *Phoenix dactylifera* L. In: Jain SM, Gupta PK (eds) Protocol for somatic embryogenesis in woody plants. Springer, Dordrecht, pp 309–319

9. Sudhersan C, Abo El-Nil M, Al-Baiz A (1993) Occurrence of direct somatic embryogenesis on the sword leaf of in vitro plantlets of *Phoenix*

*dactylifera* L. cultivar Barhee. Curr Sci 65 (11):887–889

10. Fki L, Masmoudi R, Kriaa W, Mahjoub A, Sghaier B, Mzid R et al (2011) Date palm micropropagation via somatic embryogenesis. In: Jain SM, Al-Khayri JM, Johnson DV (eds) Date palm biotechnology. Springer, Dordrecht, pp 47–68

11. Abd El Bar OH, El Dawayati MM (2014) Histological changes on regeneration *in vitro* culture of date palm (*Phoenix dactylifera*) leaf explants. Aust J Crop Sci 8(6):848–855

12. Alkhateeb A (2008) Comparison effects of sucrose and date palm syrup on somatic embryogenesis of date palm (*Phoenix dactylifera* L.) Am J Biochem Biotech 4(1):19–23

13. Kunert K, Baaziz M, Cullis C (2003) Techniques for determination of true-to-type date palm (*Phoenix dactylifera* L.) plants: a literature review. Emir J Food Agric 15:1–16

14. Sané D, Aberlenc-Bertossi F, Gassama-Dia YK, Sagna M, Trouslot MF, Duval Y, Borgel A (2006) Histocytological analysis of callogenesis and somatic embryogenesis from cell suspensions of date palm (*Phoenix dactylifera*). Ann Bot 98:301–308

15. El Dawayati MM, Abd El Bar OH, Zaid ZE, El Din AFZ (2012) *In vitro* morpho-histological studies of newly developed embryos from abnormal malformed embryos of date palm cv. Gundila under desiccation effect of polyethelyne glycol treatments. Ann Agric Sci 57 (2):117–128

16. Kwaaitaal MA, de Vries SC (2007) The SERK1 gene is expressed in procambium and immature vascular cells. J Exp Bot 58(11):2887–2896

17. Nolan KE, Irwanto RR, Rose RJ (2003) Auxin up-regulates MtSERK1 expression in both *Medicago truncatula* root-forming and embryogenic cultures. Plant Physiol 133 (1):218–230

18. Murashige T, Skoog F (1962) A revised medium for rapid growth and bio assays with tobacco tissue cultures. Physiol Plant 15 (3):473–497

19. Berlyn GP, Miksche JP, Sass JE (1976) Botanical microtechnique and cytochemistry. Iowa State University Press, Ames, IA, pp 24–100

20. Sass JE (1951) Botanical microtechnique, 2nd edn. The Iowa State College Press, Ames, IA, pp 197–198

21. Caboni E, Tonelli M, Falasca G, Damiano C (1996) Factors affecting adventitious shoot regeneration in vitro in the apple rootstock 'Jork 9'. Adv Hortic Sci 10(3):146–150

22. Welander M (1988) Plant regeneration from leaf and stem segments of shoots raised *in vitro* from mature apple trees. J Plant Physiol 132(6):738–744

# Part III

Contamination, Hyperhydricity and Acclimatization

# Chapter 14

# Identifying and Controlling Contamination of Date Palm Tissue Cultures

## Abeer H.I. Abdel-Karim

## Abstract

Fungal and bacterial contaminations are major problems facing in vitro date palm (*Phoenix dactylifera* L.) proliferation. To overcome this problem, we must first identify the fungal (e.g., *Alternaria* sp., *Aspergillus niger*, *Penicillium* sp.) and bacterial (e.g., *Pseudomonas* sp.) spread in date palm in vitro cultures. Incorporating fungicides (e.g., copper oxychloride, Vitavax T, and Topsin M) or antibiotics (e.g., streptomycin, Banocin, and Bencid D) at 500 mg/L in medium significantly reduces the contamination rate during various stages of in vitro date palm culture. *Streptomyces chloramphenicol* (pharmacy) is highly effective in reducing the bacterial contamination of date palm cultures to below 10%, as well as enhancing growth vigor.

**Key words** Actinomycete, Antibiotics, Contamination, Fungicides, Tissue culture

## 1  Introduction

Preventing microbial contamination of date palm tissue cultures is critical for successful micropropagation. Microorganisms isolated from contaminated plant tissue cultures include fungi, yeast, bacteria, viruses, viroids, and micro-arthropods (mites, thrips) [1]. *Alternaria* sp., *Aspergillus niger*, *Penicillium* sp., and *Pseudomonas* sp. were isolated from *Phoenix dactylifera* cv. Zagloul at different stages of in vitro culture [2]. In addition, fungal species (*Alternaria alternate, Aspergillus niger, Penicillium* spp.) and bacteria genera (*Bacillus, Staphylococcus, Proteus*) were isolated and identified from different date palm tissue culture laboratories in Iraq [3]. Some bacteria are latent and invisible during plant growth in the medium through many subculture cycles. Contamination is not completely eliminated by surface sterilization treatment(s), as they are not exposed to the sterilant during explant disinfection. Most are introduced into plant tissue cultures through endogenously contaminated initial explants [4].

Fungicides (copper oxychloride, Vitavax T, Topsin M) and antibiotics (streptomycin, Banocin, Bencid D) eliminate contamination

Jameel M. Al-Khayri et al. (eds.), *Date Palm Biotechnology Protocols Volume 1: Tissue Culture Applications*,
Methods in Molecular Biology, vol. 1637, DOI 10.1007/978-1-4939-7156-5_14, © Springer Science+Business Media LLC 2017

by fungi and bacteria in date palm proliferation [2]. Antibiotics (gentamicin, ampicillin, streptomycin, rifampicin, tetracycline, cefotaxime, penicillin) inhibit mycelia growths of some bacterial species in oil palm micropropagation [5]. Some researchers use silver and its compounds as antimicrobial agents. The most common are silver sulfadiazine (Ag SD), silver metal, silver acetate, silver nitrate, and silver protein [6].

Actinomycete (streptomycin) selects resistant bacteria which are resistant to other antibiotics, and there is some measure of cross-resistance to other aminoglycosides [7]. The addition of 500 mg/L actinomycete (*Streptomyces chloramphenicol*) to the date palm culture medium resulted in increased number of contamination-free explants [8].

Initial steps for producing aseptic cultures of date palm require attention to indexing explants and cultures for contaminants, determination of the source of contaminants, as well as characterizing the contaminants and eliminating the contaminating organism with improved cultural practices and the use of fungicides, antibiotics, and actinomycete. This chapter addresses identification of fungi and bacteria types spread in date palm in vitro cultures and effective concentrations of fungicides (e.g., copper oxychloride, Vitavax T, and Topsin M) and antibiotics (e.g., streptomycin, Banocin, and Bencid D) to overcome date palm contamination problems.

## 2  Materials

### 2.1  Plant Material

1. Offshoots weighing 5–7 kg, inflorescence encased in its spathe, roots, and seeds from date palm Zagloul cv.

2. Bacteria-contaminated explants.

### 2.2  Solution and Culture Medium

1. Clorox disinfection solutions: 30% and 50% Clorox solutions (active ingredient: 5.25% w/v sodium hypochlorite) containing two drops Tween-20 per 100 mL solution.

2. MC disinfection solution: 1 g/L mercuric chloride ($HgCl_2$).

3. Potato dextrose agar (PDA): 4 g potato extract, dextrose 20 g, and agar 15 g at pH 5.6.

4. Actinomycete solution: 500 mg/L *Streptomycin chloramphenicol*.

5. Fungicide solution: 500 mg/L Topsin M.

6. Basal culture medium: Murashige and Skoog [9] medium (Table 1).

7. Growth regulator stock solutions: 10 mg/L 2,4-dichlorophenoxyacetic acid (2,4-D), 0.1 mg/L naphthaleneacetic acid (NAA), and 3 mg/L isopentenyladenine (2iP).

**Table 1**
**Components of Murashige and Skoog (1962) medium**

| Constituent | Concentration (mg/L) |
| --- | --- |
| Macronutrients | |
| $NH_4NO_3$ | 1650 |
| $KNO_3$ | 1900 |
| $CaCl_2 \cdot 2H_2O$ | 440 |
| $MgSO_4 \cdot 7H_2O$ | 370 |
| $KH_2PO_4$ | 170 |
| Micronutrients | |
| $MnSO_4 \cdot 4H_2O$ | 22.30 |
| $ZnSO_4 \cdot 4H_2O$ | 8.60 |
| $H_3BO_3$ | 6.20 |
| KI | 0.83 |
| $NaMoO_4 \cdot 2H_2O$ | 0.25 |
| $CuSO_4 \cdot 5H_2O$ | 0.025 |
| $CoCl_2 \cdot 6H_2O$ | 0.025 |
| Iron | |
| $Na_2EDTA$ | 37.25 |
| $FeSO_4 \cdot 7H_2O$ | 27.85 |
| Vitamins | |
| Nicotinic acid | 0.5 |
| Pyridoxine-HCl | 0.5 |
| Thiamine-HCl | 5 |
| Myoinositol | 100 |
| Biotin | 0.5 |
| Amino acid | |
| Glycine | 2 |
| Glutamine | 200 |
| Sodium and potassium | |
| $NaH_2PO_4$ | 170 |
| $KH_2PO_4$ | 120 |
| Sucrose | 30,000 |
| Agar | 7000 |

**Table 2**
**Hormonal, sucrose, and activated charcoal supplements to the culture medium used for date palm culture stages**

| Media additives (mg/l) | Culture stages | | | | |
|---|---|---|---|---|---|
| | Initiation | Callus formation | Somatic embryogenesis | Proliferation | Rooting |
| 2,4-Dichlorophenoxyacetic acid (2,4-D) | 10 | 10 | – | – | – |
| 2-Isopentenyladenine (2iP) | 3 | 3 | 0.2 | 3 | – |
| Naphthaleneacetic acid (NAA) | – | 5 | 0.1 | – | 0.1 |
| Adenine sulfate 2H$_2$O | 40 | – | – | – | – |
| Glutamine | 170 | – | – | – | – |
| Activated charcoal | 3000 | 1500 | 1500 | – | – |
| Sucrose | 30,000 | 30,000 | 30,000 | 30,000 | 30,000 |
| Agar | 6000 | 6000 | 6000 | 6000 | 6000 |

8. Medium additives for various culture stages are shown in Table 2.

### 2.3  Equipment

1. Glassware: Jars 150 mL, tubes, Petri dishes, and McCartney bottles.

2. Instruments: Laminar airflow hood, growth chamber, autoclave, hot plate with magnetic stirrer, and pH meter.

3. Surgical tools: Stainless steel sterile surgical blades, scalpel handle, and forceps.

## 3  Methods

### 3.1  Medium Preparation

1. MS medium: Prepare stock solutions using chemicals (Table 1) and demineralized water and then add growth regulators; after that add fungicides, antibiotics, or actinomycete, then add sugar, organic supplements, and agar, and finally adjust pH 5.6 ± 0.2 and the medium diluted to a final volume.

2. Potato dextrose agar (PDA): Prepare potato infusion by boiling 200 g sliced, unpeeled potatoes in 1 L distilled water for 30 min; filter through cheesecloth, saving effluent, which is potato infusion; after that mix with dextrose, agar, and water and boil to dissolve; finally adjust pH 5.6 ± 0.2.

3. Autoclave culture medium for 20–25 min at 121 °C (1.1 kg/cm$^2$) (*see* **Note 1**).

4. Autoclave culture vials with proper closure before pouring the medium for 30 min at 12 °C and 1.1 kg/cm$^2$, cool slightly, and then dispense in culture vessels.

### 3.2 Disinfection of Offshoot Explants

1. Select healthy and pathogen-free plants as explant sources (*see* **Note 2**).

2. Spray the donor date palm plants with fungicide (1–2 g/L Topsin M) and antibiotic (1 g/L Bencid) solutions (*see* **Note 3**).

3. Separate healthy offshoots carefully from adult date palm trees.

4. Remove outer leaves and fibrous tissue before transfer of explants to the laboratory.

5. Under aseptic conditions (*see* **Notes 4–6**), submerge explants in 70% ethanol for 30 s.

6. Immerse explants in 0.5 g/L mercuric chloride (HgCl$_2$) for 5 min and then wash them with sterilized distilled water three times.

7. Remove one or two outer primordial leaves from the sterilized explants.

8. Expose explants to double-surface disinfection; first in 50% Clorox for 25 min, thoroughly wash with sterilized distilled water, and second in 30% Clorox for 25 min, wash with sterilized distilled water three times.

9. Remove some surrounding primordial leaf carefully before culturing the explants [10].

### 3.3 Disinfection of Immature Inflorescence Explants

1. Remove spathes of adult female date palm trees by using a tapestry knife.

2. Rinse spathes under running water and liquid soap for ½ h.

3. Sterilize spathes for 20 min in 50% Clorox containing two drops of Tween-20.

4. Open sterilized female spathes under a laminar airflow hood with sterilized scalpel and forceps.

5. Isolate spikelets and sterilize by immersion in 0.2% mercuric chloride solution for 10 min.

6. Rinse sterilized spikelets three times with sterilized distilled water [11].

### 3.4 Disinfection of Root Explants

1. Select healthy offshoots producing new roots after planting for 2 months in the greenhouse.

2. Cut root segments, 1–10 cm in length including root tips.

3. Wash root segments under running water for 1–2 h.

4. Submerge explants in antioxidant solution containing 100 mg/l ascorbic acid and 150 mg/l citric acid for 30 min.

5. Dip explants in 70% ethanol for 1–5 s.

6. Transfer explants and dip in 20% calcium hypochlorite Ca (OCl)$_2$ for 5 min.

7. Rinse the explants three times with sterilized distilled water before culturing [12].

***3.5 Disinfection of Seed Explants***

1. Sterilize seed explants with 96% H$_2$SO$_4$ for 10 min.

2. Rinse seeds three times with sterilized distilled water.

3. Soak seed explants in sterilized distilled water for 24 h [13].

4. Culture disinfected explant and collect the contamination culture after 3 weeks during the culture initiation stage to identify the contamination.

***3.6 Identification of Contaminants***

1. Isolate microbial contaminants from the contaminated plant tissue culture by inoculating them on potato dextrose agar (PDA).

2. Incubate them for 6 days at 26 °C under 12 h photoperiod in the case of fungi, and on nutrient agar incubate for 3 days at 30 °C under 12 h photoperiod.

3. Collect pure isolates obtained by repeated subculturing of the isolates.

4. Place pure isolates in an agar slant in McCartney bottles and store at 4 °C in a refrigerator.

5. Identify the fungal isolates by using cultural characters and morphology and by comparison with standards [14].

6. In the case of bacteria, use the morphological characteristics, a number of biochemical (gram staining, spore staining, motility test, catalase production, oxidase test, indole production, citrate utilization, urease activity, hydrogen sulfide production, gelatin hydrolysis, starch hydrolysis, carbohydrate utilization) and physiological tests on the isolates [15, 16].

***3.7 Disinfecting Contaminated In Vitro Cultures***

1. Collect explants contaminated with bacteria from different stages of date palm somatic embryo formation protocol (initiation, callus, embryos, shoots, roots).

2. Clean contaminated tissues by washing in distilled sterilized water (*see* **Notes 7** and **8**).

3. Culture different explants from different stages in recommended MS medium described in Table 2 (*see* **Note 9**).

4. Add 500 mg/L *Streptomyces chloramphenicol* (pharmacy) to the culture medium during explant establishment, callus,

embryogenesis, germination, and rooting stages (*see* **Notes 10** and **11**).

5. Culture explants and incubate at 26 ± 2 °C. Subculture at 3-week intervals for 9 weeks (*see* **Note 12**).

6. Culture explants in medium without *Streptomyces chloramphenicol* to complete the culture stages until plantlet formation.

---

## 4   Notes

1. Media can become contaminated after autoclaving at ratio 2–5%, and *Bacillus* bacteria survive even after autoclaving at 110–120 °C [19].

2. Selection of explants from healthy parent plants coupled with an effective surface sterilization method is the goal in avoiding culture contamination [17].

3. Use fungicides and antibiotics separately or in combination to reduce external contaminants of cultivated plants prior to use for in vitro culture [18].

4. Operators must be well trained in aseptic techniques. Inadequate training of operators in aseptic techniques causes contamination by bacteria (*Staphylococcus* spp.) [20].

5. Sterilize instruments (scalpels, forceps, spatulas) carefully before and between culture, under aseptic conditions by dipping in 95% ethanol and flaming. Cool instruments before they are used to handle explants.

6. It is necessary to change alcohol regularly because there are some *Bacillus circulans* strains that can persist in alcohol for more than 1 week [21].

7. Many date palm in vitro cultures do not remain aseptic and become contaminated with fungi and bacteria after the sterilization process and culture incubation (Fig. 1). Bacterial and fungal contaminations often appear inside the culture tubes after 2–4 months from the start of culturing [22].

8. To ensure the best results of bacterial control using antibiotics, clean contaminated tissue by washing in distilled water, dip explants in antibiotic solution before culturing, sterilize culture manipulation tools such as scalpels and forceps, and use young date palm tissue in culture [23].

9. Add fungicide (Topsin M) at 500 mg/L to culture media. Endogenous fungi *Rhizoctonia*, *Alternaria*, and *Stilbella* are not eliminated by surface sterilization of plant material. Using fungicide (Dithane M-45) represses mycorrhizal fungi and does not have any inhibitory effect on the cultured tissue [24]. In order to reduce the fungal and bacterial contamination

**Fig. 1** Contamination by bacteria and fungi in date palm cultures in vitro. (**a**), (**b**), and (**c**) Initiation stages, (**d**) callus, (**e**) embryo germination, (**f**) and (**g**) elongation, (**h**) and (**i**) rooting

in in vitro date palm cultures, add copper oxychloride, Vitavax T, Topsin M, streptomycin, Banocin, or Bencid D at 500 or 1000 mg/L to the culture medium [2].

10. Addition of actinomycetes (*Streptomyces bobilii* or *S. chloramphenicol*) at the concentrations (50, 100, 250, 500, or 1000 mg/L) on the culture media of date palm bacterially contaminated explants significantly reduces contamination in establishment, callus, somatic embryo, shooting, and rooting stages. However, 500 mg/L *S. chloramphenicol* gave the highest number of contamination-free explants with the highest

number of explant survival and high growth vigor during growth and development stages [8].

11. When the bacterial contamination is high, increase the concentration of *S. chloramphenicol* to 1000 mg/L in the culture media for three subcultures.

12. Incubate callus and embryo cultures in the dark while incubating shooting and rooting cultures at 16-h photoperiod of cool-white florescent light (40 $\mu$mol/m$^2$/s).

## References


1. Cobrado JS, Fernandez AM (2016) Common fungi contamination affecting tissue-cultured abaca (*Musa textiles* nee) during initial stage of micropropagation. Asian Res J Agric 1(2):1–7

2. Abd-El Kareim AHE, Rashed MF, Sharabasy SF (2006) Impact of using some fungicides and antibiotics on controlling microbial contamination during all stages of date palm tissue culture protocol. J Agric Sci Mansoura Univ 31(5):2805–2814

3. Abas MA (2013) Microbial contaminants of date palm (*Phoenix dactylifera* L.) in Iraqi tissue culture laboratories. Emir J Food Agric 25 (11):875–882

4. Leary JV, Nelson N, Tisserat B, Allingham EA (1986) Isolation of pathogenic *Bacillus circulans* from callus culture and healthy offshoots of date palm (*Phoenix dactylifera*). Appl Environ Microbiol 52(5):1173–1176

5. Eziashi EI, Asemota O, Okwuagwu CO, Eke CR, Chidi NI, Dimaro EA (2014) Screening sterilizing agents and antibiotics for the elimination of bacterial contaminants from oil palm explants for plant tissue culture. Euro J Exp Biol 4(4):111–115

6. Herman EB (1996) Microbial contamination of plant tissue culture. Agritech Consultants Inc., Shrub Oak, NY. 84 pp

7. Leary JV, Chun WWC (1989) Pathogenicity of *Bacillus circulans* to seedlings of date palm (*Phoenix dactylifera*). Plant Dis 73(4):353–354

8. Abd-El Kareim AHE (2009) Using actinomycetes on controlling bacterial contamination of date palm during different stages in vitro. J Hortic Sci Ornam Plants 1(3):92–99

9. Murashige T, Skoog F (1962) A revised medium for rapid growth and bioassay with tobacco tissue cultures. Physiol Plant 15:473–497

10. Abd-El Kareim AHE, Hassan MM, Darwesh RSS, Hussien FA (2013) Comparative in vitro studies between 5-aminolevulinic acid (ALA) and different amino acids on growth and development of date palm cv. Khalas. J Biol Chem Environ Sci 8(4):319–341

11. Stino RG, El-Kosary S, Hassan MM, Kinawy AA (2015) Direct embryogenesis from inflorescences culture of Sewy date palm (*Phoenix dactylifera* L.) J Biol Chem Environ Sci 10 (1):173–186

12. Madboly EAR (2007) Biotechnological studies on date palm via tissue culture techniques. Ph. D. thesis, Department of Pomology, Faculty of Agriculture, Cairo University, Egypt, p 79

13. Sané D, Aberlenc-Bertossi F, Gassama-Dia YK, Sagna M, Trouslot MF, Duval Y, Borgel A (2006) Histocytological analysis of callogenesis and somatic embryogenesis from cell suspensions of date palm (*Phoenix dactylifera*). Ann Bot 98:301–308

14. Barnett HL, Hunter BB (1986) Illustrated genera of fungi, 4th edn. Millan Publishing Co., New York, p 212

15. Breed RS, Marfoy JGD, Hichnas AP (1974) Bergy's manual of determinative bacteriology. Williams and Wikins Co., Baltimore

16. Odutayo OI, Amusa NA, Okutade OO, Ogunsanwo YR (2007) Sources of microbial contamination in tissue culture laboratories in southwestern Nigeria. Afr J Agric Res 3:67–72

17. Webster SK, Seynour JA, Mitchell SA, Ahmad MH (2003) A novel surface sterilization method for reducing microbial contamination of field grown medicinal explants intended for in vitro culture. Biotechnology Centre, U.W.1, Mona, Kingston7, Jamaica, West Indies

18. George EF (1993) Plant propagation by tissue culture, Part 1. The Technology Exegetics Ltd., Edington

19. Leifert C, Morris CE, Waites WM (1994) Ecology of microbial saprophytes and pathogens in tissue culture and field-grown plants: reasons for contamination problems in vitro. Crit Rev Plant Sci 13:139–183

20. Leifert C, Cassells AC (2001) Microbial hazards in plant tissue and cell cultures. In Vitro Cell Dev Biol Plant 37:133–138

21. Leifert C, Ritchie JY, Waites WM (1991) Contaminants of plant-tissue and cell cultures. World J Microbiol Biotech 7:452–469

22. Oda ML, de Faria RT, Fonseca ICB, Silva GL (2003) Fungicide and germicide on contamination escaping in the in vitro propagation of *Oncidium varicosum* Lindl. Semin: Cien Agrár (Londrina) 24:273–276

23. Benjama A, Cherkaoui B, Al-Maii S (2001) Origin and detection of *Bacillus* contaminating date palm vitro-culture and importance of manipulations conditions. Al Awamia 104:73–74

24. Kritzinger EM, Vuuren RJV, Woodward B, Rong IH, Spreeth MH, Slabbert MM (1997) Elimination of external and internal contaminations in rhizomes of *Zantedeschia aethiopica* with commercial fungicides and antibiotics. In: Cassels AC (ed) Pathogen and microbial contamination management in micropropagation, vol 12. Kluwer Academic Publishers, Dordrecht, pp 161–167

# Chapter 15

# Controlling Hyperhydricity in Date Palm In Vitro Culture by Reduced Concentration of Nitrate Nutrients

## Maiada M. El-Dawayati and Zeinab E. Zayed

## Abstract

Hyperhydricity (or vitrification) is a fundamental physiological disorder in date palm micropropagation. Several factors have been ascribed as being responsible for hyperhydricity, which are related to the explant, medium, culture vessel, and environment. The optimization of inorganic nutrients in the culture medium improves in vitro growth and morphogenesis, in addition to controlling hyperhydricity. This chapter describes a protocol for controlling hyperhydricity during the embryogenic callus stage by optimizing the ratio of nitrogen salts of the Murashige and Skoog (MS) nutrient culture medium. The best results of differentiation from cured hyperhydric callus are obtained using modification at a ratio of $NH^{4+}/NO^{3-}$ at 10:15 (825:1425 mg/L) of the MS culture medium to remedy hyperhydric date palm callus and achieve the recovery of normal embryogenic callus and subsequent regeneration of plantlets. Based on the results of this study, nutrient medium composition has an important role in avoiding hyperhydricity problems during date palm micropropagation.

**Key words** Ammonium nitrate, Calcium compound, Ethylene, Hyperhydricity, In vitro culture, Oxidative stress, Tissue culture, Vitrification

## 1 Introduction

Although plant tissue culture strategies have proved indispensable for the propagation of date palm (*Phoenix dactylifera* L.), in vitro regeneration is still hampered by certain physiological disorders. Hyperhydricity (or vitrification) is a common physiological disorder encountered in date palm in vitro regeneration systems based on either somatic embryogenesis [1] or organogenesis [2, 3]. This waterlogged appearance is the result of the accumulation of water in the cultured explants. Several factors are associated with increasing the magnitude of hyperhydricity, such as deficiency in concentration of certain microelement such as $Ca^{++}$; imbalance in plant growth regulators; high ammonium concentration, using liquid medium; and accumulation of high levels of ethylene gas in the culture vessels [3–6]. These factors are known to trigger oxidative

Jameel M. Al-Khayri et al. (eds.), *Date Palm Biotechnology Protocols Volume 1: Tissue Culture Applications*, Methods in Molecular Biology, vol. 1637, DOI 10.1007/978-1-4939-7156-5_15, © Springer Science+Business Media LLC 2017

stress which is considered the main cause of hyperhydricity in plant tissue culture [7, 8].

In this context, Alkateeb [4] suggested medium modifications to reduce the incidence of hyperhydricity including increased agar concentration and decreased PGRs concentration. Zayed [9] found that modification of nitrogen sources and concentrations was an effective approach to reduce hyperhydricity. However, the addition of calcium pantothenate or silver nitrate to the culture medium of developed callus was found to reduce hyperhydricity [10]. Moreover, a partial desiccation procedure by adding polyethylene glycol (PEG) to the culture medium [11, 12] or by air dehydration [13] was found effective in countering the hyperhydricity problem in date palm somatic embryos.

This chapter describes an approach for controlling the occurrence of hyperhydricity during embryogenic callus proliferation stage by modifying the ratio of $NH^{4+}/NO^{3-}$ to 10:15 (825:1425 mg/L) of the MS culture medium to prevent hyperhydric date palm callus. In addition, the approach achieves the recovery of normal embryogenic callus and subsequent regeneration of plantlets, in order to realize a successful protocol for date palm micropropagation.

## 2  Materials

### 2.1  Plant Material

Hyperhydric embryogenic calli of the date palm cv. Siwy (*see* **Note 1**).

### 2.2  Culture Medium

1. Basal culture medium: The stock solutions of Murashige and Skoog (MS) [14] medium (Table 1).

2. Hormone stock solutions (1 mg/mL each): Benzyladenine (BA), naphthaleneacetic acid (NAA), indolebutyric acid (IBA), and paclobutrazol (PBZ).

3. Hyperhydricity-reducing medium (medium I): Modified MS salt solutions by reducing $NH_4NO_3$ salt to 825 mg/L and $KNO_3$ salt to 1425 mg/L, 30 g/L sucrose, 0.1 g/L activated charcoal, 40 mg/L adenine-sulfate, 0.05 mg/L BA, 0.1 mg/L NAA, and 8 g/L agar (Table 2).

4. Rooting medium (medium II): ½ strength basal salts of MS medium (Table 1) supplemented with 50 mg/L sucrose, 0.2 mg/L calcium pantothenate, 0.1 mg/L NAA, 0.1 mg/L IBA, 7 mg/L agar, and 0.5 g/L activated charcoal (Table 2).

5. Pre-acclimatization medium (medium III): ¼ strength basal salts of MS medium (Table 1) liquid medium supplement with 10 g/L sucrose and 0.4 mg/L PBZ (Table 2).

### 2.3  Acclimatization Requirements

1. Pots: Round plastic pots, 18.5 height × 5 cm diameter.

2. Soil mixture: Peat moss, vermiculite, and sand at 1:1:1 ratio.

**Table 1**

**Components of MS basal medium and additives used for date palm in vitro culture stages. Hormones, agar, and activated charcoal are added according to the culture stage as shown in Table 2**

| Medium composition | Stock concentration (mg/L) | Final concentration in culture medium (mg/L) |
|---|---|---|
| *Stock I: Major inorganic nutrients (20× stock) use 50 mL to prepare 1 L of medium* | | |
| $NH_4NO_3$ | 33,000 | 1650 |
| $KNO_3$ | 38,000 | 1900 |
| $CaCl_2 \cdot 2H_2O$ | 8800 | 440 |
| $MgSO_4 \cdot 2H_2O$ | 7400 | 370 |
| $KH_2PO_4$ | 3400 | 170 |
| $NaH_2PO_4 \cdot H_2O$ | 3400 | 170 |
| *Stock II: Minor inorganic nutrients (200× stock) use 5 mL to prepare 1 L of medium* | | |
| KI | 166 | 0.83 |
| $H_3BO_3$ | 1240 | 6.2 |
| $MnSO_4 \cdot 2H_2O$ | 4460 | 22.3 |
| $ZnSO_4 \cdot 7H_2O$ | 1720 | 8.6 |
| $Na_2.MoO_4 \cdot 2H_2O$ | 50 | 0.25 |
| $CuSO_4 \cdot 5H_2O$ | 5 | 0.025 |
| $CoCl_2 \cdot 6H_2O$ | 5 | 0.025 |
| *Stock III: Iron source (200× stock) use 5 mL to prepare 1 L of medium* | | |
| $FeSO_4 \cdot 7H_2O$ | 5560 | 27.8 |
| $Na_2EDTA \cdot 2H_2O$ | 7460 | 37.3 |
| *Stock IV: Vitamins (200× stock) use 5 mL to prepare 1 L of medium* | | |
| Myoinositol | 25,000 | 125 |
| Nicotinic acid | 200 | 1 |
| Pyridoxine·HCl | 200 | 1 |
| Thiamine·HCl | 200 | 1 |
| Glycine | 400 | 2 |
| Biotin | 200 | 1 |
| *Carbon source* | | |
| Sucrose | – | 30,000 |
| *Antioxidants* | | |
| Glutamine | – | 200 |
| Citric acid | – | 100 |
| Ascorbic acid | – | 100 |

**Table 2**
**Different culture stages and their corresponding hormones, activated charcoal, and sucrose additives supplemented to different strengths of MS medium (Table 1)**

| Culture stage and medium code | Hormones, agar, activated charcoal, and sucrose additives | | | | | | | |
| --- | --- | --- | --- | --- | --- | --- | --- | --- |
| | BA | NAA | IBA | PBZ | Agar | Activated charcoal | Sucrose | MS salt strength |
| Somatic embryo differentiation (medium I) | 0.05 mg/L | 0.1 mg/L | _ | _ | 8 g/L | 0.1 g/L | 30 g/L | Full MS with 825 mg/L $NH_4NO_3$ and 1425 mg/L $KNO_3$ |
| Rooting (medium II) | _ | 0.1 mg/L | 0.1 mg/L | _ | 7 g/L | 0.5 g/L | 50 g/L | ½ |
| Pre-acclimatization (medium III) | _ | 0.1 mg/L | _ | 0.4 mg/L | _ | _ | 10 g/L | ¼ |

3. Fungicide solution: Benlate, 0.5% w/v.

4. Greenhouse.

5. Transparent polyethylene bags, 30 × 20 cm.

*2.4  Equipment*

1. Glassware and culture vessels: Measuring cylinder, conical flask, pipettes and beakers, small culture jars (150 mL), and test tubes (2.5 × 25 cm) capped with polypropylene closures.

2. Instruments and tools: Magnetic stirrer, laminar airflow hood, autoclave and incubator, stainless scalpel, forceps, and spatula (*see* **Note 2**).

# 3  Methods

*3.1  Medium Preparation*

1. Prepare stock solutions of plant growth regulators by dissolving (1 mg/mL each) the (NAA), (IBA), and (PBA) in 95% ethanol or 1 N NaOH and (BA) using 1 N HCl, and make up the required volume by adding double-distilled water. Store in the refrigerator at 4 °C for 1 month.

2. Mix MS salts solutions with other components for each nutrient medium as listed in Tables 1 and 2.

3. Adjust pH to 5.7 ± 0.1 by HCl and NaOH solutions.

4. Make up the volume of nutrient medium with distilled water.

5. Add agar and heat the medium until agar is dissolved.

6. Distribute media in 150-mL culture jars. Each jar containing 30-mL culture nutrient medium for embryogenic callus

cultures and somatic embryo cultures and 45 mL for shoot cluster enlargement cultures. For rooting and pre-acclimatization, test tubes (2.5 × 25 cm) are filled with 25-mL nutrient medium.

7. Cap the culture jars or tubes immediately with polypropylene, and autoclave for 20 min at 121 °C and 1.1 $kg/cm^2$.

***3.2 Control Hyperhydricity in Embryogenic Callus***

1. Inoculate hyperhydric embryogenic callus on modified MS culture medium.

2. Incubate cultures for a 16-h photoperiod of cool-white florescent light, 40 $\mu mol/m^2/s$ and 24 ± 2 °C.

3. Transfer the hyperhydric embryogenic callus (*see* **Note 1**) on the same nutrient medium every 4 weeks for two re-cultures (*see* **Notes 3–5**).

4. Observe the new healthy differentiated somatic embryo appearance after two re-cultures (Fig. 1a).

5. Transfer all differentiated mature somatic embryos from hyperhydric embryogenic callus on the same composition of

**Fig. 1** Plantlets production from hyperhydric callus: (**a**) hyperhydric callus produced good somatic embryos after treated with $NH^{4+}/NO^{3-}$ ratio, (**b**) cluster of shoots after treatment with $NH^{4+}/NO^{3-}$ ratio, (**c**) elongation of shoots, (**d**) rooting of shoots, (e) and pre-acclimatization of plantlets

modified (MS) nutrient medium with the addition of 500 mg/L malt extract to enhance shoot formation (Fig. 1b) (*see* **Notes 6–8**).

6. Incubate cultures at $27 \pm 2\,°C$ and16-h photoperiod of cool-white florescent light ($100\ \mu mol/m^2/s$).

7. Continue subcultures to obtain healthy shoot clusters (Fig. 1c) for 3 subcultures at 8-week intervals to obtain shoot elongation (*see* **Note 9**).

### 3.3 Rooting Stage

1. Transfer shoots with at least 5 cm in length to nutrient medium for rooting stage for 4 subcultures at 8–10-week intervals.

2. Incubate cultures at $27 \pm 2\,°C$ and 16-h photoperiod of cool-white florescent light ($100\ \mu mol/m^2/s$) (Fig. 1d).

3. Transfer plantlets having 8–10 cm in length with 2 expanded leaves and an adequate root system (2–3 roots) to the same rooting medium composition for 2 subcultures at 4-week intervals.

4. Transfer plantlets for pre-acclimatization medium, and cover the tubes with aluminum foil caps.

5. Incubate cultures at $27 \pm 2\,°C$ and a 16-h photoperiod of cool-white florescent light ($200\ \mu mol//m^2/s$) for 4 weeks (Fig. 1e).

### 3.4 Acclimatization of Plantlets

1. Rinse the plantlets thoroughly with tap water to remove remaining agar from the root system, and then immerse in 0.5% (w/v) Benlate fungicide solution for 5 min (Fig. 2a).

2. Fill the plastic pots with prepared soil, and keep in the greenhouse under natural daylight and high relative humidity (90–95%) using a cover of white polyethylene bags for 1 month, and keep the plantlets under the tunnels for 4–5 months until new leaves emerge (Fig. 2b).

**Fig. 2** Stages of acclimatization of date palm: (**a**) plantlets immerse in 0.5% (w/v) Benlate fungicide solution for 5 min, (**b**) culture of plantlets in plastic pots, (c) and plantlets after 1 year from transfer to greenhouse

3. Water the plants with quarter strength MS inorganic salts, once a week; spray with the fungicide solution every 2 weeks (Fig. 2c).

## 4  Notes

1. Hyperhydric embryogenic callus has a waterlogged, glassy, yellowish appearance and exhibits undefined nodular shapes (Fig. 3b), where nonhyperhydric embryogenic callus exhibits defined nodular shapes with a creamy color appearance (Fig. 3a).

**Fig. 3** Different shapes of hyperhydric tissues: (**a**) nonhyperhydric embryogenic callus exhibiting defined nodular shapes with creamy color appearance, (**b**) hyperhydric callus exhibiting waterlogged, glassy, yellowish appearance and undefined nodular shapes, (**c**) aborted embryos (abnormalities), (**d**) cluster of hyperhydric embryos, and (**e**) cluster of hyperhydric shoots

2. Glassware: Glass goods should be of Corning or Pyrex or similar borosilicate, and all surgical tools should be of stainless steel.

3. There are various types of hyperhydric tissues which appear during the in vitro cycle of date palm micropropagation such as callus (Fig. 3a), embryo structures which are considered aborted embryos, incomplete maturity of somatic embryos (abnormalities) (Fig. 3c), and clusters of embryos or clusters of shoots which are brownish, swelling, glassy, and vitrified shoots with a curly leaf (Fig. 3d, e).

4. For other solutions for hyperhydricity problem occurrence, once hyperhydric tissue appears, swift intervention is needed by transferring the affected cultures to a culture medium supplemented with 0.2 mg/L AgNO$_3$ or PEG at 20 mg/L for 4 weeks, and then transfer them to a medium free of growth regulators for another 4 weeks.

5. Hyperhydric embryogenic callus or shoots of date palm can be treated in several ways; whereas, hyperhydric malformed embryos are not amenable to variable treatment.

6. Modifying the ratio of ammonia to nitrates in the MS salts composition with the addition of glutamine in order to increase the organic nitrogen presence in nutrient medium leads to improved somatic embryo differentiation of the injured hyperhydric callus.

7. Hyperhydricity may also be explained by the lack of a desiccation period and low endogenous ABA level. The positive effect of air desiccation or chemical desiccation by (PEG) to cure hyperhydric somatic embryo clusters has been demonstrated by the arising of new secondary embryos, with high potential to convert to normal plantlets. The osmotic effects have been attributed to increase in levels of endogenous (ABA) as determined by studies.

8. It is worth mentioning that, when hyperhydric shoot clusters cultured on MS medium which omitted $NH^{4+}/NO^{3-}$ salt for two to three subcultures, the best vegetative growth and vigorous shoots are achieved.

9. Most experimental designs do not always allow a clear distinction between the factors increasing hyperhydricity and those predisposing the culture to this condition.

## References

1. McCubbin MJ, Zaid A (2007) Would a combination of organogenesis and embryogenesis techniques in date palm micropropagation be the answer? Acta Hortic 736:255–259

2. Mazri MA, Meziani R (2013) An improved method for micropropagation and regeneration of date palm *Phoenix dactylifera* L. J Plant Biochem Biotechnol 22:176–184

3. Mazri MA (2015) Role of cytokinins and physical state of the culture medium to improve in vitro shoot multiplication, rooting and acclimatization of date palm *Phoenix dactylifera* L. cv. Boufeggous. J Plant Biochem Biotechnol 24:268–275

4. Al-Khateeb AA (2008) The problems facing the use of tissue culture technique in date palm *Phoenix dactylifera* L. Sci J King Faisal Univ 9:85–104

5. Oliveira Y, Pinto F, Silva AL, Guedes I, Biasi LA, Quoirin M (2010) An efficient protocol for micropropagation of *Melaleuca alternifolia* Cheel. In Vitro Cell Dev Biol Plant 46:192–197

6. Reed BM, Wada S, De Noma J, Niedz RP (2013) Mineral nutrition influences physiological responses of pear in vitro. In Vitro Cell Dev Biol Plant 49:699–709

7. Cassells A, Curry R (2001) Oxidative stress and physiological, epigenetic and genetic variability in plant tissue culture: implications for micropropagators and genetic engineers. Plant Cell Tiss Org Cult 64(2–3):145–157

8. Tian J, Cheng Y, Kong X, Liu M, Jiang F, Wu Z (2016) Induction of reactive oxygen species and the potential role of NADPH oxidase in hyperhydricity of garlic plantlets in vitro. Protoplasma 5:1–10

9. Zayed ZE, El-Dawayaty MM, Abdel-Gelil LM (2012) Hyperhydricity-phenomenon problem in embryogenic callus of date palm, solving by glutamine and $NH^{4+}$:$NO^{3-}$ ratio in basal nutrient medium. Egypt J Biotechnol 42:86–95

10. Zayed ZE, Sidky RA, Saber TY (2013) Prevention of hyperhydricity phenomenon and improving somatic embryogenesis in date palm *Phoenix dactylifera* L. Bull Fac Agric Cairo Univ 64:297–302

11. Sidky RA, Gadalla EG (2014) Somatic embryogenesis in *Phoenix dactylifera* maturation, germination and reduction of hyperhydricity during embryogenic cell suspension culture. In: Zaid A, Alhadrami GA (eds) Proceedings of the fifth international date palm conference, Khalifa International Date Palm Award, Abu Dhabi, UAE, pp 183–190

12. El - Dawayati MM, Abd El Bar OH, Zayed ZE, El Din AFM Z (2012) In vitro morpho-histological studies of new developed embryos from abnormal malformed embryos of date palm cv. Gundila under desiccation effect of polyethelyne glycol treatments. Ann Agr Sci 57 (2):95–106

13. El - Dawayati MM, Zayed ZE, Ibrahim IA (2013) In vitro morphogenesis changes and embryogenesis of recovered abnormal embryos of date palm under effect of air desiccation treatments. Arab J Biotech 16 (1):105–118

14. Murashige T, Skoog F (1962) A revised medium for rapid growth and bioassays with tobacco tissue cultures. Physiol Plant 15:473–497

# Chapter 16

# Improvement of In Vitro Date Palm Plantlet Acclimatization Rate with Kinetin and Hoagland Solution

## Mona M. Hassan

## Abstract

In vitro propagation of date palm *Phoenix dactylifera* L. is an ideal method to produce large numbers of healthy plants with specific characteristics and has the ability to transfer plantlets to ex vitro conditions at low cost and with a high survival rate. This chapter describes optimized acclimatization procedures for in vitro date palm plantlets. Primarily, the protocol presents the use of kinetin and Hoagland solution to enhance the growth of Barhee cv. plantlets in the greenhouse at two stages of acclimatization and the appropriate planting medium under shade and sunlight in the nursery. Foliar application of kinetin (20 mg/ L) is recommended at the first stage. A combination between soil and foliar application of 50% Hoagland solution is favorable to plant growth and developmental parameters including plant height, leaf width, stem base diameter, chlorophyll A and B, carotenoids, and indoles. The optimum values of vegetative growth parameters during the adaptation stage in a shaded nursery are achieved using planting medium containing peat moss/perlite 2:1 (v/v), while in a sunlight nursery, clay/perlite/compost at equal ratio is the best. This protocol is suitable for large-scale production of micropropagated date palm plantlets.

**Key words** Acclimatization, Hoagland nutrition solution, Foliar application, Kinetin, Micropropagation, Plant nutrition, Planting media

---

## 1 Introduction

Acclimatization is an important process for the adaptation of micropropagated plants to the greenhouse and field conditions. Normally, in vitro plantlets are exposed to high relative humidity and low light intensity, cultured on medium enriched with sugar as a carbon and energy source [1–3]. Under such conditions, the plantlets have less developed cuticle, epicuticular waxes, and functional stomatal apparatus causing high transpiration rates that cause transplantation shock during the first step of the acclimatization process [4–6].

One of the main important limiting factors of commercial date palm in vitro propagation is the acclimatization of regenerated plantlets. Rooting and subsequent acclimatization stages are crucial

Jameel M. Al-Khayri et al. (eds.), *Date Palm Biotechnology Protocols Volume 1: Tissue Culture Applications*, Methods in Molecular Biology, vol. 1637, DOI 10.1007/978-1-4939-7156-5_16, © Springer Science+Business Media LLC 2017

for commercial plant micropropagation [7]. Several techniques have been used to acclimatize date palm plantlets and improve their survival rate during establishment under greenhouse conditions. The composition of the planting medium affects growth and development of date palm plantlets, depending on the cultivar [8, 9]. Acclimatization of date palm plantlets derived from somatic embryos has been successful in many date palm cultivars [10–12].

This chapter focuses on the acclimatization stage of in vitro date palm plantlets derived from direct somatic embryogenesis using shoot-tip explants. It presents a protocol for improved acclimatization based on foliar applications of kinetin and a combination between soil and foliar applications of Hoagland solution at 50% after transplantation to the greenhouse using an optimized potting mixture.

## 2    Materials

### 2.1    Plant Materials

Date palm cv. Barhee in vitro plantlets having stem thickness of 5 mm and 10–12 cm in length with 2–3 leaves and 2–3 roots (Fig. 1a, b), regenerated from direct somatic embryogenesis, derived from shoot-tip explants.

### 2.2    Equipment

1. Glassware and culture vessels: beaker (1000 mL), graduated cylinders (100, 500, 1000 mL), and culture tubes (2.5 × 15, 2.5 × 25 cm).

2. Surgical tools: forceps and scalpels.

3. Instruments: sterilizer, laminar airflow hood, growth chamber, precision balance, magnetic stirrer, pH meter, and autoclave.

### 2.3    Stock Solutions

1. Basal culture medium: The stock solutions of Murashige and Skoog (MS) [13] medium and other additives as listed in Table 1.

2. Hormone stock solutions: Ancymidol (0.4 mg/L), indolebutyric acid (IBA, 1 mg/L), and naphthaleneacetic acid (NAA, 0.1 mg/L).

### 2.4    Culture Medium

1. Rooting medium (RM): ½ strength major salts and full strength minor salts of MS solid medium (Table 1) containing 1 mg/L IBA, 45 g/L sucrose, and 1 g/L activated charcoal (AC), dispensed in small test tube (2.5 × 15 cm).

2. Pre-acclimatization liquid medium (PLM): ½ strength major salts and full strength minor salts of MS liquid rooting medium supplemented with 0.1 mg/L naphthaleneacetic acid (NAA), 10 g/L sucrose, and 0.4 mg/L ancymidol (Table 1) dispensed into large test tubes (2.5 × 25 cm) at 15 mL.

**Fig. 1** Date palm cv. Barhee acclimatization stage: (**a**) Rooted plantlet in liquid rooting medium, (**b**) misted plantlets after washing in tap water, (**c**) plantlets immersed in fungicide solution, (**d**, **e**) plantlets cultivated in pots, (**f**) incubation of cultivated plantlet under transparent plastic bag inside the greenhouse

**Table 1**
**Composition of rooting and pre-acclimatization culture media consisting of Murashige and Skoog (MS) salts [13] and additives**

| Constituent | Stock concentration (g/L) | Final concentration in rooting culture medium (mg/L) | Final concentration in pre-acclimatization medium (mg/L) |
| --- | --- | --- | --- |
| Macronutrients: use 10 mL to prepare 1 L of the medium | | | |
| $NH_4NO_3$ | 82.5 | 850 | 850 |
| $KNO_3$ | 95 | 950 | 950 |
| $CaCl_2 \cdot 2H_2O$ | 22 | 220 | 220 |
| $MgSO_4 \cdot 7H_2O$ | 18.5 | 185 | 185 |
| $KH_2PO_4$ | 8.5 | 85 | 85 |
| Micronutrients: use 5 mL to prepare 1 L of the medium | | | |
| $MnSO_4 \cdot 4H_2O$ | 4.46 | 22.30 | 22.3 |
| $ZnSO_4 \cdot 4H_2O$ | 1.72 | 8.60 | 8.6 |
| $H_3BO_3$ | 1.24 | 6.20 | 6.2 |
| KI | 0.166 | 0.83 | 0.83 |
| $NaMoO_4 \cdot 2H_2O$ | 0.05 | 0.25 | 0.25 |
| $CuSO_4 \cdot 5H_2O$ | 0.005 | 0.025 | 0.025 |
| $CoCl_2 \cdot 6H_2O$ | 0.005 | 0.025 | 0.025 |
| Iron: use 5 mL to prepare 1 L of the medium | | | |
| $Na_2EDTA$ | 7.45 | 37.25 | 37.25 |
| $FeSO_4 \cdot 7H_2O$ | 5.57 | 27.85 | 27.8 |
| Vitamins: use 10 mL to prepare 1 L of the medium | | | |
| Nicotinic acid | 0.05 | 0.5 | – |
| Pyridoxine-HCl | 0.05 | 0.5 | – |
| Thiamine-HCl | 0.01 | 0.1 | – |
| Amino acid: use 10 mL to prepare 1 L of the medium | | | |
| Glycine | 0.2 | 2 | – |
| Growth regulators | | | |
| IBA | | 1 | - |
| Ancymidol | | – | 0.4 |
| NAA | | - | 0.1 |
| Sucrose | | 45 g/L | 10 g/L |
| Agar | | 6 g/L | – |
| Activated charcoal | | 1 g/L | – |

**Table 2**
**Composition of Hoagland and Arnon solution [14]**

| Chemical formulas | Stock solutions (g/L) | mL stock solution/1 L final solution |
| --- | --- | --- |
| Macronutrients | | |
| $KNO_3$ | 202 | 2.5 |
| $Ca(NO_3)_2 \cdot 4H_2O$ | 472 | 2.5 |
| $MgSO_4 \cdot 7H_2O$ | 493 | 1 |
| $NH_4NO_3$ | 80 | 1 |
| Micronutrients | | |
| $H_3BO_3$ | 2.86 | 1 |
| $MnCl_2 \cdot 4H_2O$ | 1.81 | 1 |
| $ZnSO_4 \cdot 7H_2O$ | 0.22 | 1 |
| $CuSO_4 \cdot 5H_2O$ | 0.051 | 1 |
| $Na_2MoO_4 \cdot 2H_2O$ | 0.12 | 1 |
| Iron | | |
| Fe-EDTA | 15 | 1.5 |
| Phosphate | | |
| $KH_2PO_4$ (pH to 6.0) | 136 | 0.5 |

***2.5 Greenhouse Material***

*2.5.1 Solutions*

1. Anti-transpiration agent (Stress Relief 35, active ingredient acrylic latex polymers).
2. Fungicide solution: 0.2% (w/v) Benlate.
3. Hoagland stock solutions (Table 2) [14].
4. Kinetin solution: 20 mg/L.

*2.5.2 Planting Media*

1. Clay soil (local substrates).
2. Compost.
3. Peat moss.
4. Perlite.
5. Vermiculite.
6. Washed sand.

*2.5.3 Acclimatization Vessels*

1. Pots 5 cm in diameter, 18 cm in height (P1).
2. Pots 12 cm in diameter, 8 cm in height (P2).
3. Pots 12 cm in diameter, 18 cm in height (P3).
4. Black plastic bags 20 cm in diameter, 30 cm in height (P4).
5. Pots 20 cm in diameter, 25 cm in height (P5).
6. Black mesh polypropylene cover, shade densities 73%.

## 3  Methods

### 3.1  Medium Preparation

1. Prepare MS medium stocks: Weigh the MS stocks major, minor, iron, and vitamin components individually. Dissolve the components of each stock separately in 500 mL distilled water with continuous stirring. Make up the volume to 1000 mL using graduated cylinders and store in a refrigerator at 4 °C until use.

2. Prepare growth regulator stocks (1 mg/mL): Weigh 100 mg from each growth regulator individually and dissolve in 0.1 N NaOH or 0.1 N HCl solutions as appropriate. Complete the solution to 100 mL using graduated cylinder, stir, and store it in a refrigerator.

3. Add MS stock solutions, growth regulators, and other components specified to rooting and pre-acclimatization media as described in Table 1 to 500 mL distilled water and then complete solution to 1000 mL.

4. Adjust the pH of the rooting solid medium to 5.7–5.8 and for pre-acclimatization liquid medium to 5–5.2.

5. Add 6 g/L agar and 1 g/L activated charcoal to solid rooting medium and heat to dissolve agar.

6. Dispense solid medium in small test tubes (2.5 × 15 cm) at 20 mL each, whereas pre-acclimatization liquid medium' in large test tubes (2.5 × 25 cm) at 15 mL each.

7. Cape culture tubes with polypropylene closures and autoclave for 20 min at 121 °C and 1.1 kg/cm$^2$.

### 3.2  First Stage Under In Vitro Conditions

1. Culture Barhee plantlets in test tubes containing 20 mL rooting medium (RM) and incubate cultures at 27 ± 2 °C and light intensity 50 μmol/m$^2$/s for 6 weeks (Table 1; *see* **Note 1**).

2. Transfer rooted plantlets to pre-acclimatization liquid medium (PLM) (Table 1) and incubate cultures at 27 ± 2 °C and light intensity 80 μmol//m$^2$/s for another 6 weeks [15] (Fig. 1a; *see* **Notes 2** and **3**).

3. One week before the acclimatization, loosen the test tube covers and finally open those to expose plantlets to lower humidity, while test tubes remain under growth-room conditions as a pre-acclimatization step (*see* **Note 4**).

### 3.3  Second Stage: Greenhouse Conditions

#### 3.3.1  Foliar Application of Anti-transpiration Agent

1. Carefully remove plantlets from the culture test tubes and wash the roots using tap water to remove residual sugar and nutrients (Fig. 1b).

2. Dip plantlets in 0.2% (w/v) fungicide solution for 5 min. It is important to keep plantlets wet until cultivation (Fig. 1c; *see* **Notes 5** and **6**).

3. Cultivate plantlets in pots (P1) filled with a mixture of peat moss/washed sand 3:1 (v/v).

4. Spray the plantlets twice with 0.5 mL/L anti-transpiration agent Stress Relief 35 (active ingredients acrylic latex polymers). The first one directly after planting and the second after 10 days to enhance survival percentage [16] (*see* **Note 7**).

5. Keep the plantlets growing in a transparent bag inside a Saran greenhouse in the shaded area for 6 weeks at $27 \pm 2\ ^\circ C$, relative humidity 80–90%, and natural daylight. Partially remove the plastic bags in order to reduce humidity for another 6 weeks (Fig. 1d; *see* **Note 8**).

*3.3.2 Foliar Application of Kinetin*

1. Cultivate plants, 18–20 cm length in plastic pots (P2) filled with peat/perlite (2:1 v/v).

2. Apply kinetin at 20 mg/L manual foliar solution (10 mL/plant) twice a week for 3 months (*see* **Note 9**).

3. Irrigate all plants in different treatments with 25% Hoagland solution (Table 2) as soil application every 3 days.

4. Record plant length (cm), leaf width (cm), stem base diameter (cm), and growth vigor after 3 months.

5. To observe the effectiveness of kinetin foliar application, spray control plants with water with similar conditions (*see* **Notes 10–16**).

*3.3.3 Soil and Foliar Application Combination of Hoagland Solution*

1. Cultivate 6-month-old plants, 25–30 cm length, having 4–5 leaves in larger pots (12 cm diameter and 18 cm height (P3) filled with planting medium peat/perlite 2:1 (v/v).

2. Irrigate plants with 50% Hoagland solution two times a week (200 mL/plant) as soil application for 6 months.

3. Add foliar application (15 mL/plant) of 50% Hoagland solution to plants two times a week for 6 months.

4. Estimate plant length (cm), leaf width (cm), growth vigor, and chemical compositions after 6 months (Fig. 2; *see* **Notes 17–20**).

**3.4  Third Stage at Shaded Nursery**

1. Cultivate 12-month-old plants, 25–30 cm length and 4–5 cm diameter, in black plastic bags (P4) filled with planting medium consisting of peat moss/perlite 2:1 (*see* **Note 21**).

2. Keep plants under shaded nursery conditions of black mesh polypropylene allowing 40–50% sunlight and temperature ranging from 18 $^\circ$C at night to 30–35 $^\circ$C at noon.

**Fig. 2** Effect of soil and foliar application of Hoagland solution on growth and development of cv. Barhee: (**a**) Soil application of Hoagland solution at 25% + foliar application at 0, 25, 50, and 100%; (**b**) soil application of Hoagland solution at 50% + foliar application at 0, 25, 50, and 100%; (**c**) soil application of Hoagland solution at 100% + foliar application at 0, 25, 50, and 100%

**Fig. 3** Effect of planting media on growth and development of date palm cv. Barhee under shaded nursery: (**a**) peat moss, (**b**) peat moss/perlite 2:1, (**c**) peat moss/vermiculite 1:1, (**d**) peat moss/compost 1:1 (v/v), (**e**) peat moss/vermiculite/perlite at equal ratio, (**f**) peat moss/compost /perlite 1:1:1, (**g**) peat moss/compost/vermiculite/perlite 1:1:1:1

**Fig. 4** Growth and development of Barhee cv. as affected by planting media during adaptation stage: (*1*) clay/sand 1:1 (v/v), (*2*) peat moss/perlite 2:1 (v/v) control treatment, (*3*) clay/sand/compost at equal ratio, (*4*) clay/compost/perlite at equal ratio

3. Add 50% Hoagland solution, as a soil fertilizer, at 3-day intervals and foliar application, at the same concentration, twice weekly to plants for 6 months.

4. Observe plant length (cm), stem base diameter (cm), growth vigor, plant fresh weight (g), and leaf number (normal and pinnate) after 6 months (Fig. 3; *see* **Note 22**)

### 3.5 Fourth Stage at Full Sun Nursery (Adaptation)

1. Cultivate 18-month-old plants, 40 cm height having 3–4 true leaves (pinnate leaves), in a black plastic pot (P5) which is filled with adaptation planting medium consisting of clay/compost/perlite in equal ratio for 6 months.

2. Fertilize plants with Hoagland solution 100% every 3 days and apply to the leaves at 50% twice per week for 6 months.

3. Measure plant fresh weight (kg), stem base diameter (cm), plant height (cm), true leaf numbers, and growth vigor after 6 months (Fig. 4; *see* **Notes 23–25**)

## 4 Notes

1. Activated charcoal in rooting medium adsorbs phenolic compounds, reduces light intensity in rooting area, and subsequently enhances growth and plantlet vigor.

2. By adding ancymidol, paclobutrazol, or abscisic acid (ABA) to the rooting medium, the stem length is shortened and roots thickened, and there were reduction in stomatal apertures,

increase in epicuticular wax and chlorophyll content, and subsequently reduction of wilting after transfer to soil [21].

3. The addition of growth retardants to liquid medium is more effective than in the solidified one.

4. To minimize the physiological stress in plantlets growing under greenhouse conditions, the epicuticular wax development increases on the upper leaf surfaces of the plantlets, therefore ensuring a better survival.

5. Washing the medium from the roots and dipping plantlets in fungicide solution are required to avoid bacterial and fungal growth that may cause plantlet rot and death.

6. Keep plantlets moistened during the process of soil transfer in order to avoid plantlet dehydration.

7. Using anti-transpiration agent (Stress Relief 35, active ingredients acrylic latex polymers) decreases stomata transpiration and subsequently control of water loss and enhances survival percentage up to 80% [1–4]. However, vigorous plantlets treated with PEG, ABA, or any growth retardants at recommended concentrations in rooting stage are able to survive in the greenhouse with higher survival percentage without the need to spray anti-transpiration.

8. Keep cultivated pots in a shaded area of the greenhouse to avoid direct sunlight. Under that condition, heat will increase under the cover and may kill the plants.

9. It is useful and effective to add Tween 20 to all foliar application as a detergent agent to increase benefit of foliar solution.

10. After 3 months, kinetin at 20 mg/L has significant stimulatory effects on vegetative growth parameters (Fig. 5).

**Fig. 5** Effect of kinetin foliar application on growth and development of date palm cv. Barhee under greenhouse conditions: (*1*) zero, (*2*) 5 mg/L, (*3*) 10 mg/L, (*4*) 20 mg/L, (*5*) 40 mg/L

11. Application of kinetin at 5, 10, or 40 mg/L also enhances vegetative growth parameters compared with untreated plants (Fig. 5). However, values of all parameters are less than 20 mg/L.

12. Cytokinins are important plant hormones that regulate various processes of plant growth and development; play an important role in regulation of cell division in apical meristems and cambium; enhance leaf expansion, nutrient mobilization, and accumulation of greater photosynthetic pigments; and delay senescence subsequently leading to better growth parameters [17, 18].

13. Foliar application of gibberellic acid (GA₃) at 5, 10, or 20 mg/L can be used instead of kinetin.

14. GA₃ at 5 mg/L enhances leaf width and stem base diameter, while increasing concentration to 10 and 20 mg/L is favorable with growth vigor and plant length, respectively (Fig. 6).

15. GA₃ has the capability of modifying the growth pattern of treated plants by affecting the DNA and RNA levels, cell division and expansion, and biosynthesis of enzymes, protein, carbohydrates, and photosynthetic pigments [19].

16. Foliar application of kinetin or GA₃ enhanced chemical contents of acclimatized plants (chlorophyll A, chlorophyll B, and indoles) compared with control. However, phenols were reduced significantly to lowest values with these regulators (Figs. 7 and 8).

**Fig. 6** Effect of GA₃ foliar application on growth and development of date palm cv. Barhee under greenhouse conditions: (*1*) zero, (*2*) 5 mg/L, (*3*) 10 mg/L, (*4*) 20 mg/L, (*5*) 40 mg/L

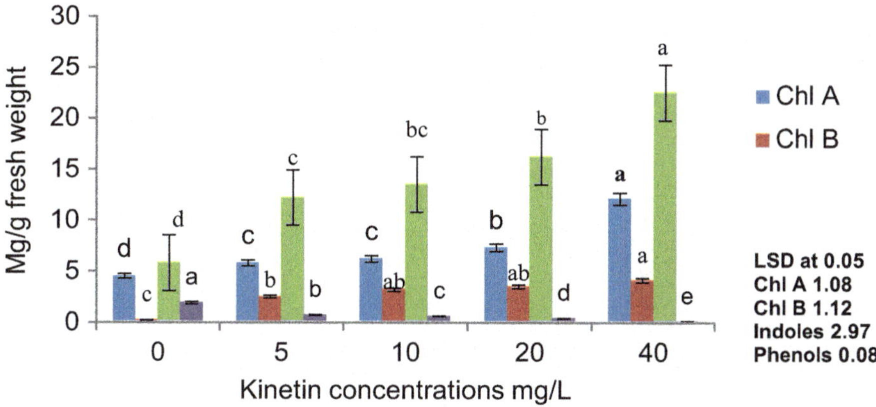

Fig. 7 Effect of foliar application of kinetin on chemical compositions of date palm cv. Barhee, 3 months after transfer to the greenhouse

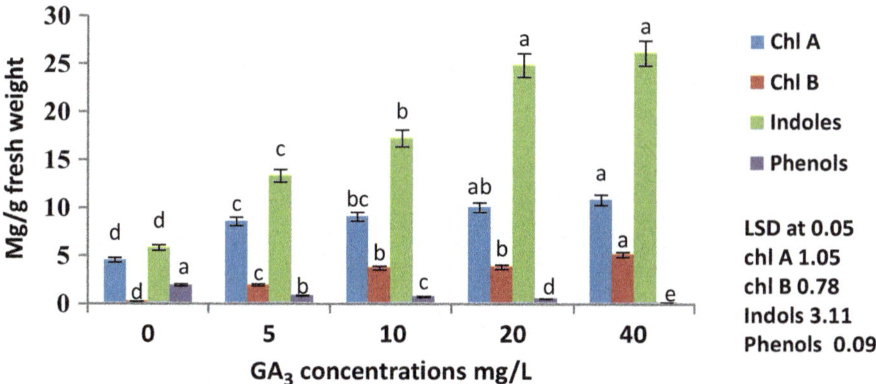

Fig. 8 Effect of foliar application of kinetin on chemical compositions of date palm cv. Barhee, 3 months after transfer to the greenhouse

17. Combination of soil and foliar application of Hoagland solution at 50% has a significant effect on plant length (cm), leaf width (cm), and growth vigor under greenhouse conditions (Fig. 2).

18. Generally, different combinations of soil and foliar application of Hoagland solution at 25, 50, or 100% are effective for date palm acclimatization. Moreover, when soil application is used at 25%, foliar application must be increased to (100%) obtain highest significant values of vegetative parameters (Fig. 2).

19. Combination of Hoagland solution in both foliar and soil application at 50% increases chemical contents (chlorophyll A, B, carotenoids, and indoles), while reducing phenol concentrations (Fig. 9).

20. The improvement in vegetative and extractable measures resulting from soil and foliar applications may be attributed

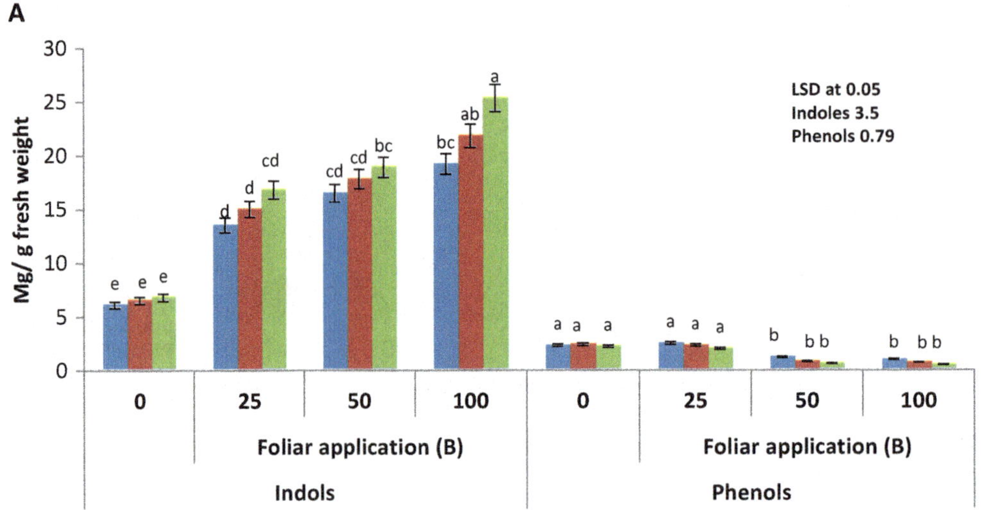


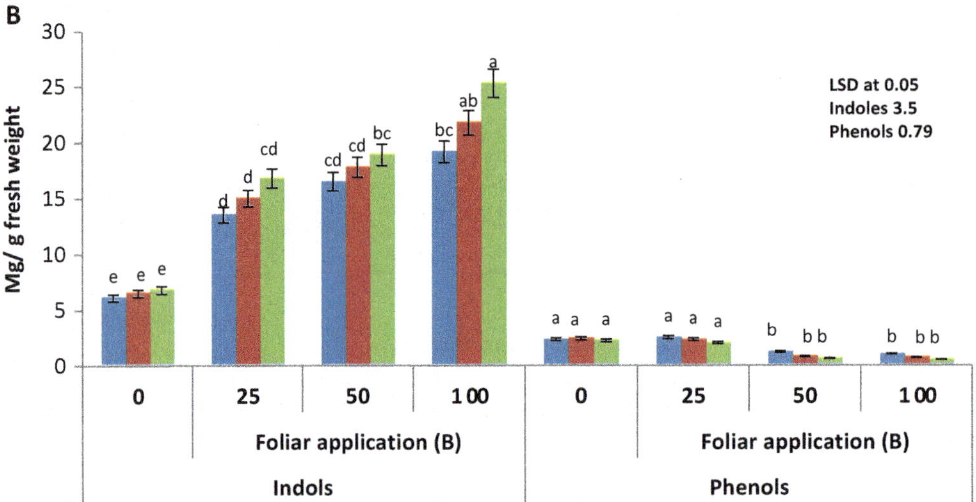


**Fig. 9** Effect of soil and foliar nutrient applications of Hoagland solution on: (**a**) chlorophyll A, B contents; (**b**) indole and phenol contents of date palm cv. Barhee (6 months old) after 6 months in the greenhouse

to the photosynthesis process which certainly reflects positively on both vegetative growth and extractable measures of the leaf [20].

21. Small punctures must be made at the bottom of poly bags using a paper punch to allow aeration and drainage of excessive water.

22. Cultivation of plants in peat moss alone, peat moss/perlite/ vermiculite/compost at equal ratio, and peat moss/vermiculite (1:1) also promotes plant growth and increases number of pinnate leaves (true leaves) in this stage, while using peat moss/compost 1:1 (v/v) or peat moss/vermiculite/perlite at equal ratio is not recommended at this stage (Fig. 3).

23. Cultivation of plants on clay/compost/perlite at equal ratio enhances significantly the plant fresh weight (kg), stem base diameter (cm), plant height (cm), and growth vigor (Fig. 4).

24. Using planting medium consisting of peat moss/perlite at 2:1 (v/v) in this stage is suitable with higher cost than the previous one.

25. Avoid cultivating plants in clay/sand 1:1 (v/v). Plants under such conditions showed restricted growth in all tested parameters (Fig. 4).

## References


1. Preece JE, Sutter EJ (1991) Acclimatization of micropropagated plants to the greenhouse and field. In: Debergh PC, Zimmerman RH (eds) Micropropagation technology and application. Kluwer Academic, London, pp 71–93

2. Sciutti R, Morini S (1993) Effect of relative humidity in *in vitro* culture on some growth characteristics of a plum rootstock during shoot proliferation and rooting and on plantlet survival. Adv Hortic Sci 7:153–156

3. Pospisilova J, Ticha I, Kadlecek S, Haisel D, Pizakova S (1999) Acclimatization of micropropagated plants in *ex vitro* conditions. Biol Plant 42:481–497

4. Hazarika BN (2006) Morpho-physiological disorders in *in vitro* culture of plants. Sci Hortic 108:105–120

5. Chandra S, Bandopadhyay R, Kuma V, Chandra R (2010) Acclimatization of tissue cultured plantlets: from laboratory to land. Biotechnol Lett 32:1199–1205

6. Kumar K, Rao IU (2012) Morphophysiological problems in acclimatization of micropropagated plants in *ex vitro* conditions – a review. J Ornam Hort Plant 2(4):271–283

7. Hassan MM, Gadalla EG, Abd El-Kareim AH (2008) Effect of sucrose and abscisic acid on *in vitro* growth and development of date palm rooting stage. Arab J Biotechnol 11(2):281–292

8. Burasheed RK, El-Wakeel HM, Desouky IM (2006) Some factors affecting in vitro propagation of Barhee and Khalas date palm cultivars. Ann Agric Sci 51(1):191–201

9. Khirallah HS, Badr SM (2007) Micropropagation of date palm (*Phoenix dactylifera* L.) var Maktoom through direct organogenesis. Acta Hortic 736:213–223

10. Al-Khayri JM (2010) Somatic embryogenesis of date palm (*Phoenix dactylifera* L.) improved by coconut water. Biotechn 9:477–484

11. Othmani A, Bayoudh C, Drira N, Marrakchi M, Trifi M (2009a) Somatic embryogenesis and plant regeneration in date palm (*Phoenix dactylifera* L.) cv. Boufeggous is significantly improved by fine chopping and partial desiccation of embryogenic callus. Plant Cell Tissue Organ Cult 97:71–77

12. Othmani A, Bayoudh C, Drira N, Marrakchi M, Trifi M (2009b) Regeneration and molecular analysis of date palm (*Phoenix dactylifera* L.) plantlets using RAPD markers. Afr J Biotech 8:813–820

13. Murashige T, Skoog F (1962) A revised medium for rapid growth and bioassays with tobacco tissue cultures. Physiol Plant 15:473–497

14. Hoagland D, Arnon DI (1950) The water-culture method for growing plants without soil Circ 347. University of California Agricultural Experiment Station, Berkeley, CA

15. Ibrahim AI, Hassan MM, Taha RA (2011) Morphological studies on date palm micropropagation as a response to growth retardants application. The third international conference of genetic engineering and its applications. Sharm El-Sheikh, South Sinai Governorate, 5–8 Oct. 2011, pp 291–304

16. Mohamed MH, El-Wakeel H, Abd El-Hamid A, El-Bana A, Hassan MM (2011) Effect of some *in vitro* and *ex vitro* treatments on rooting and acclimatization of dry date palm cvs, Sakkouty and Bartamuda plants. J Biol Chem Environ Sci 6(4):583–606

17. El-Keltawi NE, Croteau R (1987) Influence of foliar applied cytokinin in growth and essential oil content of several members of the Lamiacea. Phytochemistry 26:891–895

18. Hassan MM (2012) Stimulatory effect of kinetin, gibberellic acid, nutrient foliar applications and planting media on growth and chemical composition of date palm cv Barhee during

acclimatization process. J Biol Chem Environ Sci 7(3):21–42

19. Davies PJ (1995) Plant hormones: physiology, biochemistry and molecular biology. Kluwer, Alphen aan den Rijn

20. Dahmardeh M, Mehravaran L, Naderi S (2011) Eucalyptus plantlet growth in relation to foliar application with complete fertilizers in Southeast of Iran. Afr J Biotechnol 10 (66):14812–14815

21. Hazarika BN, Parthasarathy VA, Nagarju V (2001) Influence in vitro preconditioning of citrus microshoots with paclobutrazol on ex vitro survival. Acta Botanica Croatiica 60:25–29

# Part IV

# Cell Suspension, Protoplast and Bioreactors

# Chapter 17

# Plant Regeneration from Somatic Embryogenic Suspension Cultures of Date Palm

## Mansour A. Abohatem, Yazid Bakil, and Mohmmed Baaziz

## Abstract


Somatic embryogenesis is one of the most important technologies for plant regeneration of elite date palm cultivars. Recently, considerable progress has been made in the development and optimization of this technique from embryogenic cell suspension cultures. This chapter describes a procedure for the rapid development of a large number of somatic embryos from embryogenic cell suspension cultures. An efficient plant regeneration protocol via somatic embryogenesis from cell suspension cultures starting with shoot-tip explants to plantlet acclimatization also is fully described. Low concentrations of 6-benzylaminopurine (BAP) to 0.3 mg/L and high rate of subcultures each 7 days lead to improve the establishment and multiplication of somatic embryos in suspension cultures by limiting oxidative browning, associated with high total phenols and peroxidase activities. The detailed morphological observations have revealed the cells destined to become somatic embryos. Activated charcoal (AC) at 0.15 g/L has a positive effect on growth rate of somatic embryos by reducing tissue and medium browning, phenolics, and peroxidase activity.

**Key words** Activated charcoal, BAP, Peroxidase, Phenolics, Somatic embryogenesis, Suspension culture


## 1  Introduction

Normally solid culture media are routinely used to produce somatic embryos from embryogenic calli and are helpful to monitor different developmental stages of somatic embryogenesis in date palm [1]. However, solid media cannot be used to ensure large-scale propagation. For this reason, utmost efforts have been made to establish embryogenic cell suspension cultures with high morphogenetic potential [2, 3] and successfully plant regeneration [2–6]. These studies demonstrated that cell suspension cultures are the most suitable source of somatic embryos for mass propagation of several date palm cultivars. Additionally, suspension culture proved useful in conducting biochemical studies involving the role of protein, peroxidase activities, and phenolics accumulation in date palm somatic embryogenesis [6–8].


Jameel M. Al-Khayri et al. (eds.), *Date Palm Biotechnology Protocols Volume 1: Tissue Culture Applications*,
Methods in Molecular Biology, vol. 1637, DOI 10.1007/978-1-4939-7156-5_17, © Springer Science+Business Media LLC 2017

The brown polymers of phenols hamper tissue culture of many plants, such as Scots pine [9], apple [10], and date palm [11]. Peroxidase, a phenol-oxidizing enzyme, highly represented in date palm [12] and polyphenol oxidase is involved in oxidative browning in date palm [11]. Attempts have been made to prevent the browning of callus by supplementing the initial basal medium supplemented with different additives to limit the production of phenolics or chelate-forming substances [10], which can simultaneously alter the metabolism of plant tissue. Therefore, frequent subcultures of callus are carried to the fresh media as an alternative solution to reduce phenolic compounds and soluble fractions of peroxidase and polyphenol oxidase [11]. Activated charcoal (AC) is added in the MS culture medium at every developmental stage of somatic embryogenesis: 0.15 g/L for callogenesis and embryogenesis and 0.25 g/L for embryo maturation and germination [3]. However, addition of 0.3 g/L AC in the liquid culture medium resulted in the differentiation of large number of somatic embryos [2].

This chapter describes a procedure for the induction and proliferation of date palm cv. Bouskri callus and an efficient plant regeneration protocol via somatic embryogenesis from cell suspension cultures. It also provides the methodology for extracting and analyzing phenolics and peroxidase contents.

## 2    Materials

### 2.1  Plant Material and Sterilization

1. Date palm cv. Bouskri offshoots used as a source of explants (*see* **Note 1**).

2. Chilled antioxidant solution: 150 mg/L ascorbic acid, 150 mg/L citric acid.

3. Disinfectant solution: 1.6% (w/v) sodium hypochlorite solution (30% v/v Clorox, commercial bleach) with 300 mg/L potassium permanganate.

### 2.2  Culture Medium

1. The stock solutions of Murashige and Skoog (MS) [13] medium (MS stock I, II, III, and Fossard vitamins stock IV [14]) (Table 1) (*see* **Note 2**).

2. Hormone stock solutions: 2,4-dichlorophenoxyacetic acid (2,4-D) (1 mg/mL), 6-benzylaminopurine (BAP) (1 mg/mL) (*see* **Note 3**).

3. Solutions to adjust pH: 0.1 and 1 N NaOH and 0.1 and 1 N HCl.

### 2.3  Culture Medium Additives Used for Various Culture Stages

1. Culture initiation medium: MS medium (Table 1) supplemented with 5 mg/L 2,4-D, 5 mg/L BAP, 2 mg/L riboflavin, and 150 mg/L activated charcoal with 7 g/L agar (Table 2, *see* **Note 4**).

**Table 1**
**Composition of MS salts [13] and De Fossard vitamins [14] used for date palm tissue culture**

| Constituent | Formula | Concentration in the stock solution (mg/L) | Concentration in the culture medium (mg/L) |
|---|---|---|---|
| *Macronutrients solution (stock I – 10×) use 100 mL/L medium* | | | |
| Potassium nitrate | $KNO_3$ | 19000 | 1.900 |
| Ammonium nitrate | $NH_4NO_3$ | 16500 | 1.650 |
| Calcium chloride | $CaCl_2\ 2H_2O$ | 4400 | 440 |
| Magnesium sulfate | $MgSO_4.7H_2O$ | 3700 | 370 |
| Potassium phosphate | $KH_2\ PO_4$ | 1700 | 170 |
| *Micronutrients solution (stock II – 100×) use 10 mL to prepare 1 L of medium* | | | |
| Boric acid | $H_3BO_3$ | 620 | 6.2 |
| Manganese sulfate | $MnSO_4.4H_2O$ | 2230 | 22.3 |
| Zinc sulfate | $ZnSO_4.7H_2O$ | 860 | 8.6 |
| Sodium molybdate | $Na_2MoO_4.2H_2O$ | 35 | 0.35 |
| Copper sulfate | $CuSO_45H_2O$ | 25 | 0.025 |
| Cobalt chloride | $CoCl_2.6H_2O$ | 25 | 0.025 |
| Potassium iodide | $KI$ | 83 | 0.83 |
| *Iron chelate solution (stock III – 100×) use 10 mL to prepare 1 L of medium* | | | |
| Sodium EDTA | Na2EDTA | 3730 | 37.30 |
| Ferrous sulfate | $Fe\ SO_4.7H_2O$ | 2780 | 27.80 |
| *De Fossard vitamins (stock IV – 100×) use 4 mL to prepare 1 L of medium* | | | |
| Glycine | | 370 | 0.37 |
| Nicotinic acid | | 460 | 0.46 |
| Pyridoxine–HCl | | 620 | 0.62 |
| Thiamine–HCl | | 670 | 0.67 |
| *Myo*-inositol | | 5400 | 5.4 |
| Biotin | | 48 | 0.048 |
| Ascorbic acid | | 170 | 0.17 |
| L. cysteine | | 460 | 0.46 |
| Choline chloride | | 140 | 0.14 |
| Ca pantothenate D | | 480 | 0.48 |

2. Callus proliferation medium: MS medium (Table 1) containing 0.5 mg/L 2,4-D, 0.1 mg/L BAP, and 150 mg/L activated charcoal with 7 g/L agar (Table 2) (*see* **Note 5**).

3. Cell suspension culture medium: MS medium (Table 1) amended with 0.3 mg/L BAP and 0.1 mg/L 2,4-D (Table 2).

4. Somatic embryo maturation medium: half-strength MS liquid medium (Table 1) without plant growth regulator and containing 0.15 g/L activated charcoal (Table 2).

5. Somatic embryo germination medium: MS medium (Table 2) without plant growth regulator supplemented with 7 g/L agar.

6. Elongation medium: MS medium (Table 2) containing 1 mg/L gibberellic acid ($GA_3$) (Table 2).

7. Rooting medium: MS medium (Table 1) amended with 0.1 mg/L NAA and 7 g/L agar (Table 2).

### 2.4 Acclimatization Stage

1. Potting mixture: 1:1:1 peat moss, Coco peat, and sand in 5 cm polyethylene nursery pots (Fig. 5c).

2. Fungicide solution: 2 g Tolex 50 WP fungicide in 1 L distilled water.

### 2.5 Extraction and Analysis of Peroxidase

1. Tris-maleate 0.1 M, pH 6.5.

2. Tris-maleate-Gaicol-$CaCl_2$ 0.1 M, PH 6.5.

### 2.6 Equipment

1. Pestle and mortar.

2. Water bath.

3. Centrifuge machine.

4. Orbital shaker.

5. Spectrophotometer.

## 3 Methods

### 3.1 Preparation of Explant

1. Take off offshoots from adult palm trees using sharp tools. Remove the outer leaves exposing the shoot-tip region (about 8 cm long) and immediately dip in chilled antioxidant solution to prevent oxidation-induced browning.

2. Wash the shoot-tips with distilled water and soak for 20 min in the disinfection solution.

3. Rinse the shoot-tip tissue in sterilized distilled water 3–5 times, each for 5 min. Remove the tissue surrounding the shoot-tip terminal until it is 1 cm long. Excise the tip and section longitudinally into 6–10 small sections (0.8 cm) inside the laminar flow hood.

**Table 2**

Composition of different culture mediums used during the culture initiation, callus proliferation, cell suspension, embryo maturation, embryo germination, elongation, and rooting phases

| Composition | Culture initiation medium | Callus proliferation medium | Cell suspension medium | Somatic embryo maturation medium | Somatic embryo germination medium | Elongation medium | Rooting medium |
|---|---|---|---|---|---|---|---|
| Basal medium | MS | MS | MS/2 | MS/2 | MS | MS | MS |
| Vitamins | DeFossard | DeFossard | DeFossard/2 | DeFossard/2 | DeFossard | DeFossard | DeFossard |
| 2,4-D (mg/L) | 5 | 0.5 | 0.1 | – | – | – | – |
| BAP (mg/L) | 5 | 0.1 | 0.3 | – | – | – | – |
| NAA (mg/L) | – | – | – | – | – | – | 0.1 |
| $GA_3$ (mg/L) | – | – | – | – | – | 1 | – |
| Activated charcoal (g/L) | 0.15 | 0.15 | 0.15 | 0.15 | 0.15 | 0.15 | 0.15 |
| Sucrose (g/L) | 30 | 30 | 30 | 30 | 30 | 30 | 30 |
| Agar (g/L) | 7 | 7 | – | – | 7 | 7 | 7 |

**3.2  Culture Medium Preparation**

1. Soak glassware in liquid detergent for 1 h and thoroughly wash with hot water. Rinse the glassware with double-distilled water and air-dry.

2. MS stocks: weigh MS stocks I, II, and III components individually and dissolve each of them separately in 800 mL distilled water by stirring with magnetic stirrer; raise the volume up to 1000 mL by adding distilled water.

3. Transfer the stock solutions into reagent bottles and store at 4 °C until use.

4. Adjust pH 5.8 of the medium by adding 0.1 mL either 1 N NaOH or 1 N HCl after adding the hormones and activated charcoal in the culture medium (Table 2).

**3.3  Induction and Proliferation of Embryogenic Callus**

1. For callus culture initiation, place shoot tips on the surface of solid culture initiation medium (Table 2) (*see* **Note 4**) and maintain them at 25 ± 2 °C for 12–16 weeks in the dark.

2. Subculture the callus cultures to a fresh medium at each 4–5 week interval.

3. After 8–10 weeks of culture, callus initiates from the shoot-tip explants.

4. Transfer callus to the proliferation medium and maintain for an additional 12 weeks to obtain sufficient growth for subsequent experiments (Table 2).

**3.4  Establishment of Cell Suspension**

1. Collect actively growing friable callus from the solid cultures (Fig. 1a) and inoculate 500 mg fresh callus biomass per 250 mL conical flask, containing 50 mL liquid medium (Table 2, *see* **Note 6**).

2. Pass the content of each flask through a stainless steel sieve, pore size 500 μ, and grow them shaking on a rotary shaker (100 rpm), at 25 ± 2 °C under a 16 photoperiod of 28 μmol/ m$^2$/s photon flux.

3. Transfer the cell suspension cultures to fresh culture medium weekly.

**3.5  Development of Somatic Embryos in Cell Suspension Culture**

1. The development and division of embryogenic cells from the 15-day-old cell suspension cultures is observed under the microscope (Fig. 1).

2. Morphological characteristics of somatic embryogenesis at different developmental stages including globular, elongation, and cotyledonary are recorded from the cell suspension cultures (Fig. 1, *see* **Note 7**).

3. Add 0.3 mg/L BAP to cell suspension medium for studying the effect of BAP on the establishment and development of somatic embryos (Fig. 2, *see* **Note 8**).

**Fig. 1** The morphology of date palm cell suspension culture during the induction of somatic embryogenesis. (a) Embryogenic callus, (b) cell suspension after 7 days of culture, (c) small clumps of cells during the first week under ×100 enlargement with light microscope, (d) large clumps of cells during the second week under ×100 enlargement with light microscope, (e) embryogenic cell in globular stage after 14 days of culture, (f) conversion of globular stage to elongated embryos after 21 days of culture, (g) cotyledonary embryos after 32 days of culture, (h) somatic embryo after 40 days of culture. Scale bar: 0.1 mm

4. Subculture embryogenic suspension cultures to fresh liquid medium at 7-day intervals (Fig. 3, *see* **Note 9**).

5. Add 0.15 g/L AC to the suspension culture medium and the control is without AC for comparison is desired (Fig. 4, *see* **Note 10**).

*3.6  Maturation and Germination of Somatic Embryos*

1. For maturation of somatic embryos, transfer the somatic embryos to half-strength MS liquid medium without plant growth regulator and incubate for 2 weeks.

2. Transfer the mature somatic embryos to full strength MS solid medium devoid of plant growth regulators for the germination (Fig. 5a).

*3.7  Elongation and Rooting of Shoots*

1. After the germination of embryos, transfer the embryos to elongation medium for 4 weeks for the elongation of shoots.

**Fig. 2** Effect of BAP on the establishment and multiplication of date palm somatic embryos in suspension culture. (**a**) Embryogenic suspension with 0.5 mg/L BAP, (**b**) embryogenic suspension with 0.3 mg/L BAP. Scale bar: 10 mm

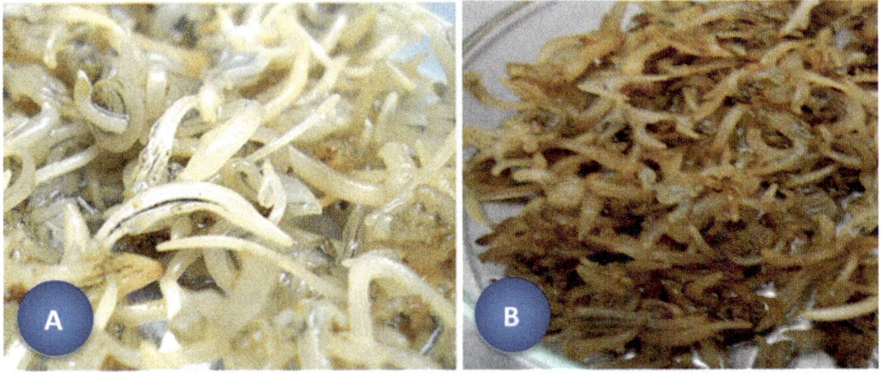

**Fig. 3** Effect of subculture rates on the establishment and multiplication of date palm somatic embryos in suspension cultures including 0.3 mg/L BAP and derived from BSK cultivar. (**a**) Somatic embryos subcultured every 7 days, (**b**) somatic embryos subcultured every 20 days. Scale bar: 10 mm

2. Transfer the shoots on rooting medium until root induction and shoot elongation (12–20 weeks) leading to complete plant formation (Fig. 5b).

### 3.8 Plant Acclimatization

1. Select all well-developed plantlets forming two leaves or more and roots from the cultures on the rooting medium; gently rinse them in water to remove residual medium from the root region.

2. Soak the roots into the fungicide solution for 15 min and transfer to 5 cm polyethylene nursery pots with the potting mixture (Fig. 5c). Water the plantlets with 100 mg/L N-P-K fertilizer (20-20-20) and subsequently as needed.

3. Cover the plantlets with polythene bag to maintain the humidity preventing the plants from dehydration. Maintain the plants in the greenhouse under natural sunlight at $30 \pm 2\ °C$ and 80%

**Fig. 4** Effect of activated charcoal on phenolics and tissue browning of date palm somatic embryos in suspension cultures. (**a**) Somatic embryos with activated charcoal, (**b**) somatic embryos without activated charcoal

**Fig. 5** Stages of date palm in vitro regeneration. (**a**) Somatic embryos germination, (**b**) plantlets on elongation and rooting medium for shoot and root development, (**c**) plantlets in polyethylene nursery pots with the potting mixture during the acclimatization stage, (**d**) the plantlets in larger polyethylene nursery bags in the greenhouse, (**e**) acclimatized plants grown in a shade house in preparation for field transfer

relative humidity. Remove gradually the polythene bags over a period of 3 weeks.

4. Transfer the plantlets to a larger polyethylene bags (Fig. 5d). After 6 months of culture, transfer the plants to a shade house (Fig. 5e) and maintain for 10–18 months before transfer them to the field for further growth.

**3.9 Analysis of Biochemical Parameters**

*3.9.1 Extraction and Analysis of Phenolics*

1. Grind fresh 250 mg somatic embryogenic tissue in 2 mL 80% methanol, at 4 °C.

2. Centrifuge the homogenate three times at $7000 \times g$ for 3 min.

3. Take 50 μL phenolic extract and adjust to 2 mL volume with distilled water.

4. Add 0.25 mL Folin-Ciocalteu to the mixture, shake vigorously, and add 0.5 mL 20% sodium carbonate.

5. Incubate the homogenate at 40 °C for 30 min. Read the absorbance at 760 nm and express the results as mg-equivalent of catechin/g FW.

*3.9.2 Extraction and Analysis of Peroxidase*

1. To prepare Tris-maleate 0.1 M, pH 6.5, add 100 mL water to a glass beaker. Weigh 2.422 g Tris–HCl and transfer to a glass beaker. Add water to a volume of 200 mL. Mix and adjust pH with HCl and store at 4 °C.

2. To prepare Tris-maleate-Gaicol-$CaCl_2$ 0.1 M, PH 6.5, add125 mL Tris-maleate 0.1 M, pH, and 0.25 mL Gaicol. Weigh and mix 70 mg $CaCl_2$. Adjust pH with HCl and store at 4 °C.

3. Grind fresh 250 mg somatic embryogenic tissue in 2 mL cold 0.1 M Tris-maleate buffer, pH 6.5.

4. Centrifuge the homogenate for 10 min at $9000 \times g$. Collect the supernatant that corresponds to enzyme extract.

5. Assay peroxidase activity by measuring the oxidation of guaiacol at 470 nm.

6. Take variable volumes of enzyme extract (20 μL) and add 1960 μL reaction mixture containing 0.1 M Tris-maleate buffer (pH 6.5) and 25 mM guaiacol.

7. Incubate the sample in a water bath at 30 °C for 30 min. Add 20 μL 10% $H_2O_2$ and read the absorbance at 470 nm after 1, 2, and 3 min.

# 4 Notes

1. After cutting greater part of offshoot leaves and discarding the soil coated rooted part, the appropriate size of offshoots is 40–50 cm long, 12–15 cm wide, and weighing 3 kg.

2. Prepare the MS medium (Table 1) stock solutions of major, minor inorganic nutrients, iron source, and vitamins (organic supplements) separately. Keep the stock solutions in the refrigerator at 4 °C until use for a maximum of 2–3 months to avoid crystallization and loss of activity.

3. Prepare the stock solutions of plant growth regulators by dissolving 2,4-D in 95% ethanol or 1 N NaOH and 2iP in 1 N

HCl and make the required volume by adding double-distilled water. Store in the refrigerator at 4 °C for up to 1 month.

4. This protocol indicates that the embryogenic cultures can be induced on a medium containing lower concentrations of 2,4-D (5 mg/L) as compared to the commonly used 100 mg/L in most date palm tissue culture protocols. Both abnormal somatic embryo differentiation and somaclonal variation are associated with the reducing 2,4-D concentration. The number of abnormal somatic embryos and somaclones are significantly minimized [15].

5. Clean all glassware with a liquid detergent and wash thoroughly with tap water. Rinse the glassware with double-distilled water and air-dry before use. Sterilize all media and the potting mixture using autoclave at 121 °C for 15 min and 1.1 kg/cm$^2$ atmospheric pressure.

6. An optimized protocol for the establishment of cell suspension from embryogenic suspension cultures of date palm has been described by our group [4, 6, 16].

7. The morphological observations have revealed that the cells destined to become somatic embryos divided into spherical proembryos (globular stage) within 7–15 days, with subsequent conversion of globular stage to elongation stage after 17 days and cotyledonary stage after 27 days of establishment of cell suspension cultures.

8. Addition of 0.3 mg/L BAP promotes growth and stimulation of somatic embryos in cell suspension cultures of date palm cultivars Boufeggouss (BFG) and Bouskri (BSK). These results show improvement of date palm somatic embryogenesis (Fig. 2).

9. The transfer of cultures on fresh culture medium at each 7-day interval results in a substantial reduction of tissue/cell oxidative browning, which is due to the reduction of phenolic compounds and decrease in peroxidase activities promoting proliferation of embryogenic cells. These results suggest that oxidative browning is mainly caused by peroxidase in date palm cell suspension cultures. The negative correlation between the number of subcultures and growth of somatic embryos vs. the intensity of browning and the levels of phenolics and peroxidase activity suggests the enzymatic oxidation of phenolic compounds by peroxidases and polyphenol oxidases [11].

10. Addition of 0.15% activated charcoal (AC) improved growth rate of somatic embryos, reduced tissue and medium browning, and significantly reduced phenolics and peroxidase activity. Activated charcoal is highly effective to prevent browning during the first few months of suspension culture establishment [1, 5].

## References


1. El Hadrami I, Cheikh R, Baaziz M (1995) Somatic embryogenesis and plant regeneration from shoot-tip explants in *Phoenix dactylifera* L. Biol Plantarum 37:205–211

2. Fki L, Masmoudi R, Drira N, Rival A (2003) An optimised protocol for plant regeneration from embryogenic suspension cultures of date palm, *Phoenix dactylifera* L., cv. Deglet nour. Plant Cell Rep 21:517–524

3. Zouine J, El Bellaj M, Meddich A, Verdeil J, El Hadrami I (2005) Proliferation and germination of somatic embryos from embryogenic suspension culture in *Phoenix dactylifera* L. Plant Cell Tissue Organ Cult 82:83–92

4. Zouine J, El Hadrami I (2007) Effect of 2,4-D, glutamine and BAP on embryogenic suspension culture of date palm (*Phoenix dactylifera* L.) Sci Hortic 112:221–226

5. Othmani A, Bayoudh C, Drira N, Marrakchi M, Trifi M (2009) Somatic embryogenesis and plant regeneration in date palm *Phoenix dactylifera* L. cv. Boufeggous is significantly improved by fine chopping and partial desiccation of embryogenic callus. Plant Cell Tissue Organ Cult 97:71–79

6. Abohatem M, Zouine J, El Hadrami I (2011) Low concentrations of BAP and high rate of subcultures improve the establishment and multiplication of somatic embryos in date palm suspension cultures by limiting oxidative browning associated with high levels of total phenols and peroxidase activities. Sci Hortic 130:344–348

7. Zouine J, El Hadrami I (2004) Somatic embryogenesis in *Phoenix dactylifera* L.: effect of exogenous supply of sucrose on proteins, sugars, phenolics and peroxidases activities during the embryogenic cell suspension culture. Biotechnology 3:114–118

8. Abohatem M, Baaziz M (2014) Multiplication and germination of somatic embryos obtained from cell suspensions of date palm (*Phoenix dactylifera*). In: Zaid A, Alhadrami GA (eds) Proceedings of the fifth international date palm conference, Abu Dhabi, United Arab Emirates 16–18 March, 2014, pp. 229–236

9. Laukkanen H, Rautiainen L, Taulavuori E, Hohtola A (2000) Changes in cellular structures and enzymatic activities during browning of scots pine callus derived from mature buds. Tree Physiol 20:467–475

10. Dobranszki J, Teixeira da Silva J (2010) Micropropagation of apple - a review. Biotechnol Adv 28:462–488

11. El Hadrami I (1995) L'embryogeneèse somatique chez *Phoenix dactylifera* L.: quelques facteurs limitants et marqueurs biochimiques. Thèse de Doctorat d'Etat. Université Cadi Ayyad, Faculté des Sciences-Semlalia, Marrakech, Morocco

12. Baaziz M, Aissam F, Brakez Z, Bendiab K et al (1994) Electrophoretic patterns of acid soluble proteins and active isoforms of peroxidase and polyphenoloxidase typifying calli and somatic embryos of two reputed date palm cultivars in Morocco. Euphytica 76:159–168

13. Murashige T, Skoog F (1962) A revised medium for rapid growth and bioassays with tobacco tissue cultures. Physiol Plant 15:473–497

14. De Fossard RA, Myint A, Lee ECM (1974) A broad spectrum tissue culture experiment with tobacco (*Nicotiana tabacum*) pith tissue culture. Physiol Plant 30:125–130

15. Fki L, Masmoudi R, Kriaâ W, Mahjoub A, Sghaier B, Mzid R, et al (2011) Date palm micropropagation via somatic embryogenesis In: Jain SM, Al-Khayri JM, Johnson DV (eds) Date palm biotechnology. Springer, Dordrecht, pp 47–68

16. Abohatem M, Baaziz M (2015) In vitro date palm somatic embryo from cell suspension culture. Lambert Academic Publishing, Berlin, pp 65– 79

<h1>Chapter 18</h1>

# Synchronization of Somatic Embryogenesis in Date Palm Suspension Culture Using Abscisic Acid

**Hussain A. Alwael, Poornananda M. Naik, and Jameel M. Al-Khayri**

## Abstract


Somatic embryogenesis is considered the most effective method for commercial propagation of date palm. However, the limitation of obtaining synchronized development of somatic embryos remains an impediment. The synchronization of somatic embryo development is ideal for the applications to produce artificial seeds. Abscisic acid (ABA) is associated with stress response and influences in vitro growth and development. This chapter describes an effective method to achieve synchronized development of somatic embryos in date palm cell suspension culture. Among the ABA concentrations tested (0, 1, 10, 50, 100 μM), the best synchronized growth was obtained in response to 50–100 μM. Here we provide a comprehensive protocol for in vitro plant regeneration of date palm starting with shoot-tip explant, callus initiation and growth, cell suspension establishment, embryogenesis synchronization with ABA treatment, somatic embryo germination, and rooting as well as acclimatized plantlet establishment.

**Key words** Abscisic acid, In vitro, Micropropagation, Somatic embryo development, Somatic embryogenesis, Synchronization


## 1  Introduction

Somatic embryogenesis is the most efficient in vitro propagation technique with great economical value for commercial propagation [1]. Synchronized somatic embryos grown in cell suspension cultures can be used to optimize reliable clonal production by synthetic seeds through restraint development of somatic embryos [2, 3].

Abscisic acid (ABA) plant hormone is related to stress response and plays a significant role in plant growth and development. It has been found to influence in vitro growth and differentiation in a number of plant species, like rapeseed (*Brassica napus* L.) [4] and coconut (*Cocos nucifera* L.) [5]. However, the influence on growth and differentiation is related to the concentration of ABA within plants, which is balanced between the biosynthesis and catabolism in plants [6]. ABA has an important role in the formation of pre-


Jameel M. Al-Khayri et al. (eds.), *Date Palm Biotechnology Protocols Volume 1: Tissue Culture Applications*,
Methods in Molecular Biology, vol. 1637, DOI 10.1007/978-1-4939-7156-5_18, © Springer Science+Business Media LLC 2017

globular embryonic structures, increases the number of somatic embryos [5], and enhances embryo tolerance to desiccation [4]. In relation to stress, increasing stress agents reduced germination rate in cotton [7], limited callus and embryo growth in date palm [8, 9], increased antioxidative enzyme activities in grapevine [10], and inhibited shoot and root fresh mass in rice [11].

A few studies have been conducted to evaluate the effect of ABA on date palm in vitro suspension cultures. Adding ABA to culture medium has increased the production and maturity of somatic embryos [12] and enhanced the accumulation of sugars and stockpiling of proteins in date palm somatic embryos. Moreover, it has increased the proliferation rate and protein content [13]. Increasing ABA can reduce different growth parameters, organic compounds, and physiological responses of date palm [14]. Al-Khayri et al. [9] studied the effect of different ABA concentrations (0–100 μM) on somatic embryos of date palm and found that 1 μM ABA added to the liquid media can suppress the growth and development of date palm somatic embryos. However, a high concentration of ABA restrains the elongation of somatic embryos at the small globular stage and thus induces synchronization in embryo size. Hassan et al. [15] reported a significant increase in root formation and growth and reduction in shoot length in response to ABA.

This chapter describes the methodology of establishing cell suspension culture and determining the effects of ABA on the synchronization of somatic embryo development of date palm.

## 2 Materials

### 2.1 Plant Material and Explant Sterilization

1. Offshoots of 3–4-year-old date palm (cv. Nabout Saif) to be used as a source of explants (*see* **Note 1**).

2. Ethanol, 70%.

3. Chilled antioxidant solution: 150 mg/L ascorbic acid, 150 mg/L citric acid.

4. Disinfectant solution: 30% v/v Clorox (1.6% w/v sodium hypochlorite) with 0.1 mL Tween 20 per 100 mL disinfectant solution.

### 2.2 Composition of Culture Medium

1. Murashige and Skoog (MS) [16] medium: stock solutions and final concentrations of the components of the culture medium are listed in Table 1 (*see* **Note 2**).

2. Hormonal stocks: 50 mg/mL 2,4-dichlorophenoxyacetic acid (2,4-D), 1 mg/mL naphthaleneacetic acid (NAA), 1 mg/mL abscisic acid (ABA), and 1 mg/mL 2-isopentenyladenine (2iP) (*see* **Note 3**).

3. pH adjustment solutions: NaOH (0.1 and 1 N) and HCl (0.1 and 1 N).

**Table 1**
**Chemical composition of the MS medium used for date palm tissue culture**

| Chemical combination | Concentration (mg/L) | Final concentration (mg/L) |
| --- | --- | --- |
| *Stock I: major inorganic nutrients (20×) use 50 mL for preparing 1 L of medium* | | |
| $NH_4NO_3$ | 33,000 | 1650 |
| $KNO_3$ | 38,000 | 1900 |
| $CaCl_2 \cdot 2H_2O$ | 8800 | 440 |
| $MgSO_4 \cdot 2H_2O$ | 7400 | 370 |
| $KH_2PO_4$ | 3400 | 170 |
| $NaH_2PO_4.H_2O$ | 3400 | 170 |
| *Stock II: minor inorganic nutrients (200×) use 5 mL for preparing 1 L of medium* | | |
| KI | 166 | 0.83 |
| $H_3BO_3$ | 1240 | 6.2 |
| $MnSO_4 \cdot 2H_2O$ | 4460 | 22.3 |
| $ZnSO_4 \cdot 7H_2O$ | 1720 | 8.6 |
| $Na_2.MoO_4 \cdot 2H_2O$ | 50 | 0.25 |
| $CuSO_4 \cdot 5H_2O$ | 5 | 0.025 |
| $CoCl_2 \cdot 6H_2O$ | 5 | 0.025 |
| *Stock III: iron source (200×) use 5 mL for preparing 1 L of medium* | | |
| $FeSO_4 \cdot 7H_2O$ | 5560 | 27.8 |
| $Na_2EDTA \cdot 2H_2O$ | 7460 | 37.3 |
| *Stock IV: organic supplements (200×) use 5 mL for preparing 1 L of medium* | | |
| *myo*-inositol | 25,000 | 125 |
| Nicotinic acid | 200 | 1 |
| Pyridoxine·HCl | 200 | 1 |
| Thiamine·HCl | 1000 | 5 |
| Glycine | 400 | 2 |
| Calcium pantothenate | 200 | 1 |
| Biotin | 200 | 1 |
| *Other additives: hormones and activated charcoal* | | |
| Hormones | According to phase as specified in Table 2 | |
| Activated charcoal | According to phase as specified in Table 2 | |
| pH | 5.7 | |

**2.3 Medium Additives for Different Culture Phases**

The basal culture medium described in Table 1 is supplemented with additives according to the culture phase as specified in Table 2. The media needed for each of the culture phases are as following:

1. Culture initiation (CT) medium: MS medium (Table 1) supplemented with 100 mg/L 2,4-D, 3 mg/L 2iP (453 μM 2,4-D, 15 μM 2iP), and 1.5 g/L activated charcoal with 7 g/L agar.

2. Callus induction (CD) medium: MS medium (Table 1) supplemented with 3 mg/L of 2iP and 10 mg/L NAA (15 μM 2iP, 54 μM NAA) and 1.5 g/L activated charcoal with 7 g/L agar.

3. Callus proliferation (CP) medium: MS medium (Table 1) supplemented with 6 mg/L 2iP, 10 mg/L NAA (30 μM 2iP, 54 μM NAA), and 1.5 g/L activated charcoal with 7 g/L agar.

4. Callus maintenance (CM) medium: MS medium (Table 1) supplemented with 1.5 mg/L 2iP and 10 mg/L NAA (7.5 μM 2iP, 54 μM NAA) and 7 g/L agar.

5. Somatic embryogenesis (SE) medium: MS liquid medium (Table 1) supplemented with 0, 0.26, 2.63, 13.15, and 26.31 mg/L (0, 1, 10, 50, 100 μM) of ABA (*see* **Note 4**).

6. Somatic embryo germination (EG) medium: hormone-free MS medium (Table 1) supplemented with 7 g/L agar.

7. Rooting (RT) medium: MS medium (Table 1) supplemented with 0.2 mg/L NAA (1 μM) and 2.5 g/L Gelrite (*see* **Note 5**).

**2.4 Acclimatization Stage**

1. Potting mixture: 1:1:1 peat moss, vermiculite, and sand in 9 cm polyethylene nursery pots (*see* **Note 5**).

2. Fungicide solution: 1 g Captan 50% fungicide in 1 L distilled water.

---

## 3 Methods

**3.1 Preparation of Culture Medium**

1. Combine an adequate volume of each stock solution of MS medium as described in Table 1 in a flask containing half the final volume of distilled water.

2. Seven culture media differing in hormones and activated charcoal content are needed for various culture stages. Add these additives as described in Subheading 2.3 (Table 2) and adjust the medium to the final volume.

3. Adjust the medium to pH 5.7 after adding hormones and activated charcoal using NaOH (0.1 and 1 N) and HCl (0.1 and 1 N).

**Table 2**

Different hormonal combination and other additives used for culture initiation, callus induction, callus maintenance, establishment of cell suspension, embryo development, and rooting phases

| Media additives | Culture phase | | | | | | |
| --- | --- | --- | --- | --- | --- | --- | --- |
| | Culture initiation (CT) | Callus induction (CD) | Callus proliferation (CP) | Callus maintenance (CM) | Cell suspension (CS) | Somatic embryo germination (EG) | Rooting (RT) |
| 2,4-Dichlorophenoxyacetic acid (2,4-D) | 100 mg/L | – | – | – | – | – | – |
| 2-Isopentenyladenine (2iP) | 3 mg/L | 30 mg/L | 6 mg/L | 1.5 mg/L | – | – | – |
| Naphthaleneacetic acid (NAA) | – | 10 mg/L | 10 mg/L | 10 mg/L | – | – | 0.2 mg/L |
| Abscisic acid (ABA) | – | – | – | – | 0, 0.26, 2.63, 13.15, 26.31 mg/L | – | – |
| Sucrose | 30 g/L | 30 g/L | 30 g/L | 30 g/L | 30 g/L | 30 g/L | 30 g/L |
| Activated charcoal | 1.5 g/L | 1.5 g/L | 1.5 g/L | – | – | – | – |
| Agar | 7 g/L | 7 g/L | 7 g/L | 7 g/L | – | 7 g/L | – |
| Gelrite | – | – | – | – | – | – | 2.5 g/L |

4. Add agar or Gelrite as specified in Subheading 2.3 (Table 2) and heat the medium until the gelling agent is completely dissolved.

5. Dispense the medium in 125 mL tissue culture jar with screw cap (40 mL per jar), 125 mL flasks for cell suspension medium (25 mL per flask) capped using aluminum foil and rubber bands, and in 150 × 25 mm test tubes (15 mL per tube) for the rooting phase. Sterilize all the media using an autoclave for 15 min at 121 °C and 1.1 kg/cm$^2$ atmospheric pressure.

### 3.2 Explant Preparation

1. Isolate 3–4-year-old offshoots of date palm from the mother trees, trim and remove the outer leaves to expose the shoot-tip region, and then immediately dip in chilled antioxidant solution (*see* **Note 6**).

2. Under the laminar flow, surface-sterilize about 8-cm-long shoot tips using the 70% ethanol for 1 min followed by 15 min in disinfection solution.

3. Rinse the tissue four times in sterile distilled water for 5 min each.

4. Carefully remove the tissue surrounding the shoot tip until you reach the shoot, and excise into 6–12 small sections of about 1-cm-long shoot tips.

### 3.3 Induction and Maintenance of Callus

1. Place the explants on the surface of the semisolid CT medium (Fig. 1a) and incubate the explants for 12 weeks. Callus will develop from the shoot-tip explants at 8–10 weeks of culture (*see* **Note 7**).

2. Transfer and maintain the resultant callus with the original explant for 3 weeks on CD medium (Fig. 1b) (*see* **Note 7**).

3. Separate and transfer the callus to the CP medium and culture for 9 weeks (*see* **Note 7**).

4. Transfer and maintain the obtained embryogenic callus to the CM medium for subsequent use (*see* **Note 7**).

5. Transfer 750 mg embryogenic callus, maintained on CM medium, in the liquid SE medium (Fig. 1c) (*see* **Notes 7** and **8**).

### 3.4 Somatic Embryo Development and Synchronization

*3.4.1 Determination of Fresh Weight*

1. After 6 weeks of culture in liquid SE medium, filter all of the cells using Whatman Grade 1 filter paper fitted into a funnel.

2. Record the fresh weight using a weighing balance.

3. Determine the fresh weight rate in relation with the ABA concentrations (*see* **Note 9**).

*3.4.2 Determination of Total Somatic Embryo Number and Size Distribution*

1. After an additional 8 weeks of culture, filter all of the cells from the prior liquid SE medium using Whatman Grade 1 filter paper fitted into a funnel and collect in the petri dishes.

**Fig. 1** Stages of date palm in vitro regeneration. (**a**) Date palm explant on culture initiation medium, (**b**) callus formation on callus induction medium, (**c**) somatic embryos formation in suspension culture medium, (**d**) plantlets on rooting medium for shoot and root development, (**e**) plantlets in polyethylene nursery pots with the potting mixture for acclimatization stage

2. Collect and place the cells in the petri dish on an illuminated colony counter.

3. Count the number of embryos in relation with ABA concentrations (*see* **Note 9**).

4. After determining total embryo number, measure embryos size using a ruler and classify them by each class size (Fig. 2a) (*see* **Note 10**).

*3.5 Rooting and Acclimatization*

1. Transfer the resultant embryos to EG medium for 12 weeks for the induction and elongation of shoots (*see* **Note 7**).

2. After the appearance of mature shoots from EG medium, transfer them to RT medium for 12 weeks to stimulate root induction and shoot elongation for the complete plant formation (Fig. 1d) (*see* **Note 7**).

3. Collect the plantlets of nearly 8–10 cm in length from RT medium and gently rinse under slow stream of tap water to remove residual medium from the rooting region.

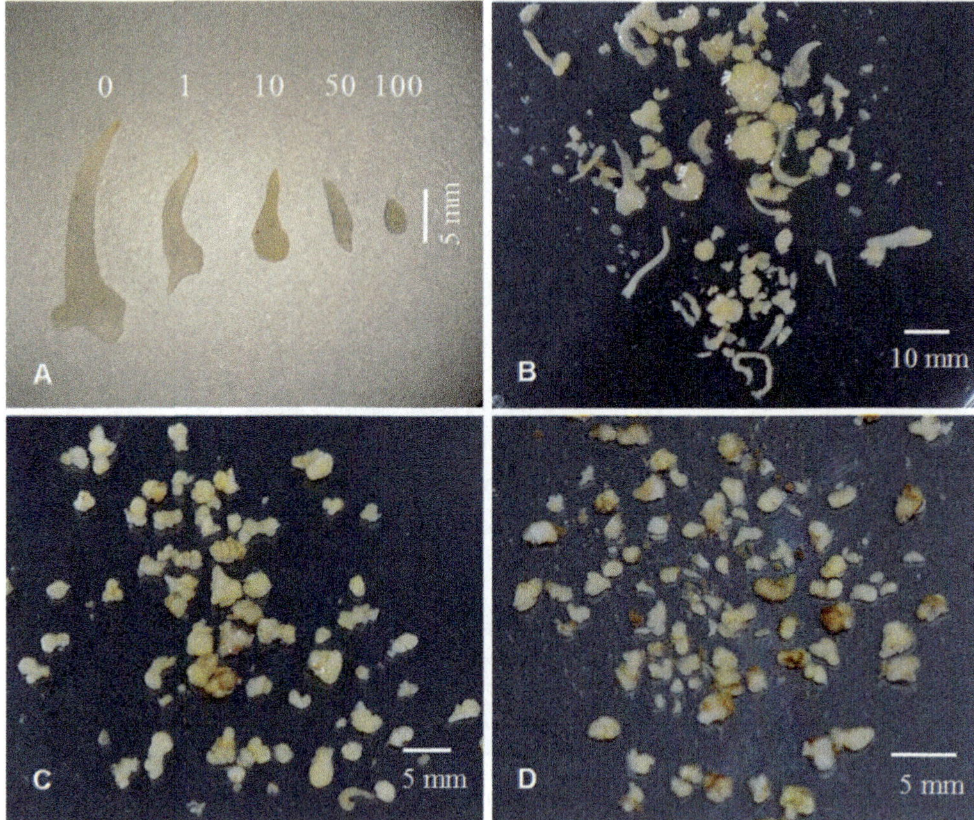

**Fig. 2** Date palm somatic embryo size distribution in response to ABA concentrations. (**a**) Comparison of somatic embryo sizes in response to ABA concentrations under the microscope with 10× magnification (numbers represent ABA concentration in μM), (**b**) embryo size in control conditions, (**c**) embryo size in 50 μM ABA, (**d**) embryo size in 100 μM ABA

4. Dip the roots into the fungicide solution and transfer to 9 cm polyethylene nursery pots with the potting mixture (Fig. 1e). Water the plantlets with 100 mg/L N-P-K fertilizer (20-20-20) and subsequently as needed. Mist the plantlets with water during the process of soil transfer to prevent desiccation.

5. Cover the plantlets with polyethylene bag to maintain the humidity preventing the plants from dehydration. Maintain the plants in a greenhouse under natural sunlight at $27 \pm 2\,°C$ and 65% relative humidity. Perforate the polyethylene bags after 1 week to adjust airflow and gradually remove them over a period of 3 weeks (*see* **Note 11**).

6. Transfer the plantlets to a larger polyethylene pots. After 3–6 months of culture, transfer the plants to a shade house and maintain for 12–24 months after which transfer to the permanent field for further growth.

## 4  Notes

1. It is preferable to collect offshoots during cool season due to the reduced phenolic compound accumulation, compared to the hot seasons that stimulate plants to produce high levels of phenolic compounds. The increased accumulation of phenolics can lead to tissue browning which decreases vitality of the tissue and may cause eventual death of the explant tissue [17].

2. Prepare MS medium stock solutions (stocks I, II, III, and IV) (Table 1) separately. Add the prepared stock solution separately while preparing the medium to avoid precipitation of the media. Store the stock solutions in the refrigerator at 4 °C and the vitamins at −20 °C to avoid precipitation. Use freshly prepared stock solutions for a maximum of 2–3 months to avoid crystallization and loss of activity.

3. To prepare hormonal stock solutions, dissolve 2,4-D, NAA, and ABA in 95% ethanol whereas 2iP in 1 N HCl. Make the required volume by adding double distilled water. Keep the stock solutions in the refrigerator at 4 °C until use for a maximum of 2–3 months to avoid crystallization and loss of activity and subsequently prepare it freshly.

4. Add different concentrations of ABA (0, 1, 10, 50, and 100 μM) in the liquid SE medium for testing their effects on synchronization of somatic embryos.

5. Clean all glasswares with a liquid detergent and wash thoroughly with tap water. Rinse the glasswares with double distilled water and air-dry before use. Sterilize all medium and the potting mixture using autoclave at 121 °C for 15 min and 1.1 kg/cm$^2$ atmospheric pressure.

6. Prevent the fast oxidation and induction of phenolics of the explant followed by browning of the tissue that leads to death of tissue. Young offshoots (1-year-old or younger) show less in vitro browning. However, they only supply various explants per offshoot and vice versa [18].

7. During CT and CM phases, incubate cultures in total darkness at 24 ± 3 °C and incubate the rest of cultures under cool-white florescent light at 16 h photoperiods (50 μmol/m$^2$/s) in 24 ± 3 °C growth chamber. Transfer the whole explant into a fresh medium at 3-week intervals during the initial phases prior to callus formation; thereafter, subculture callus by dividing and transfer to fresh culture medium every 3 weeks.

8. Embryos will develop in liquid SE culture medium. Maintain cultures on rotary shaker at 100 rpm. Use a pipette to replace half of the liquid media every 2 weeks. Allow the suspension to settle down at the bottom of the flask prior replacing it.

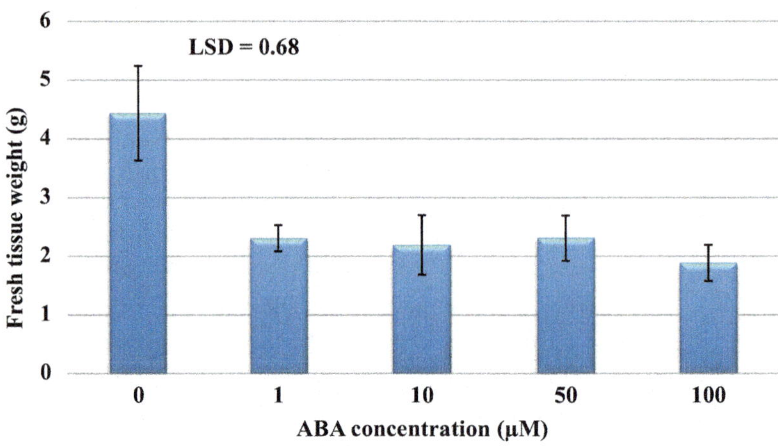

**Fig. 3** Tissue fresh weight of date palm suspension cultures with respect to different concentrations of ABA. Produced based on data from [5]

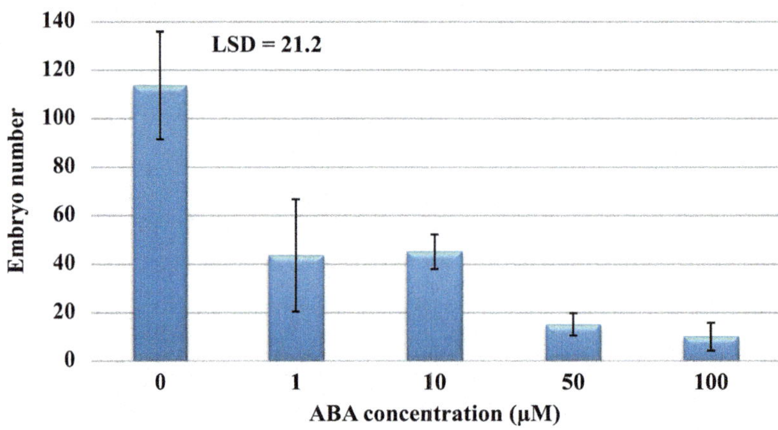

**Fig. 4** Effect of ABA concentrations on the number of embryos in date palm suspension cultures. Produced based on data from [5]

9. The ABA with 1 μM significantly reduces tissue fresh weight, whereas at 10 μM concentration further growth is inhibited, which is critical for physiological studies related to stress (Fig. 3). You can also notice the inhibition impact of ABA on embryo number (Fig. 4).

10. Observe different sizes of somatic embryos are synchronized to one class size in the presence of 50–100 μM ABA treatments (Fig. 2c–d) as compared to the control (Fig. 2b). Fig. 5 shows the majority of the embryos were of small size (>3 mm) whenever 50–100 μM ABA. These results represent Nabout Saif cultivar and may differ depending on the cultivar [8].

11. When plantlets show any sign of water stress during acclimatization, mist with water immediately and seal the polyethylene bags. After a few days, remove plastic bags to expose the ex vitro plantlets to normal relative humidity.

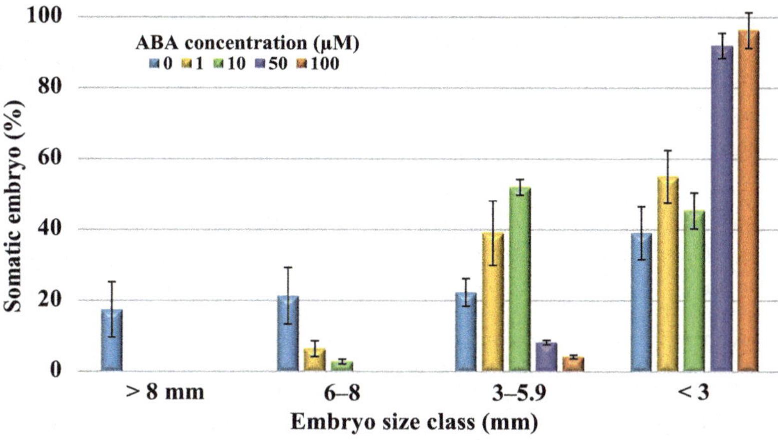


**Fig. 5** Response of date palm somatic embryos to ABA concentrations classified in each class size, error bars indicate standard deviation. Produced based on data from [5]

# Acknowledgment

The authors wish to express gratitude to the Deanship of Scientific Research at King Faisal University, Saudi Arabia, for the financial support of this research project (Grant No. 120070).

# References

1. Corredoira E, Vieitez AM, San José MC, Vieitez FJ, Ballester A (2016) Advances in somatic embryogenesis and genetic transformation of European chestnut: development of transgenic resistance to ink and blight disease. In: Park YS, Bonga JM, Moon HK (eds) Vegetative propagation of forest trees. National Institute of Forest Science (NIFoS), Seoul, Korea, pp 279–301

2. Ravi D, Anand P (2012) Production and applications of artificial seeds: a review. Int Res J Biol Sci 1:74–78

3. Colova VM, Bordallo PN, Phills BR, Bausher M (2015) Synchronized somatic embryo development in embryogenic suspensions of grapevine *Muscadinia rotundifolia* (Michx.) small. Vitis 46(1):15–18

4. Angoshtari R, Afshari RT, Kalantari S, Omidi M (2009) Effects of abscisic acid on somatic embryogenesis and induction of desiccation tolerance in *Brassica napus*. Asian J Plant Sci 8 (4):276–284

5. Fernando SC, Weerakoon LK, Gunathilake TR (2004) Micropropagation of coconut through plumule culture. Cocos 16:1–10

6. Nambara E, Marion-Poll A (2005) Abscisic acid biosynthesis and catabolism. Annu Rev Plant Biol 56:165–185

7. Du L, Cai C, Wu S, Zhang F, Hou S, Gou W (2016) Evaluation and exploration of favorable qtl alleles for salt stress related traits in cotton cultivars (*G. hirsutum* L.) PLoS One 11(3): e0151076

8. Al-Khayri JM, Al-Bahrany AM (2004) Genotype-dependent in vitro response of date palm (*Phoenix dactylifera* L.) cultivars to silver nitrate. Sci Hortic 99:153–162

9. Al-Khayri JM, Al-Bahrany AM (2012) Effect of abscisic acid and polyethylene glycol on the synchronization of somatic embryo development in date palm (*Phoenix dactylifera* L.) Biotechnology 11(6):318–325

10. Galbiati M, Matus JT, Francia P, Rusconi F, Cañón P, Medina C et al (2011) The grapevine guard cell-related VvMYB60 transcription factor is involved in the regulation of stomatal activity and is differentially expressed in response to ABA and osmotic stress. BMC Plant Biol 11:142

11. Mishra P, Bhoomika K, Dubey RS (2013) Differential responses of antioxidative defense system to prolonged salinity stress in salt-tolerant and salt-sensitive Indica rice (*Oryza sativa* L.) seedlings. Protoplasma 250:3–19

12. Zouine J, El Bellaj M, Meddich A, Verdeil JL, El Hadrami I (2005) Proliferation and germination of somatic embryos from embryogenic suspension cultures in *Phoenix dactylifera*. Plant Cell Tissue Organ Cult 82(1):83–92

13. Sghaier-Hammami B, Jorrin-Novo JV, Gargouri-Bouzid R, Drira N (2010) Abscisic acid and sucrose increase the protein content in date palm somatic embryos, causing changes in 2-DE profile. Phytochemistry 71:1223–1236

14. Al-Khayri JM (2002) Growth, proline accumulation, and ion content in sodium chloride-stressed callus of date palm. In Vitro Cell Dev Biol Plant 38:79–82

15. Hassan MM, Gadalla EG, Abd-El Kareim AH (2008) Effect of sucrose and abscisic acid on in vitro growth and development of date palm during rooting stage. Arab J Biotech 11:281–292

16. Murashige T, Skoog F (1962) A revised medium for rapid growth and bioassays with tobacco tissue cultures. Physiol Plant 15:473–497

17. Al-Khayri JM (2013) Factors affecting somatic embryogenesis in date palm (*Phoenix dactylifera* L.) In: Aslam J, Srivastava PS, Sharma MP (eds) Somatic embryogenesis and genetic transformation in plants. Narosa Publishing House, New Delhi, pp 15–38

18. Al-Khayri JM (2005) Date palm *Phoenix dactylifera* L. In: Jain SM, Gupta PK (eds) Protocol for somatic embryogenesis in woody plants. Springer, Netherlands, pp 309–319

# Microcalli Induction in Protoplasts Isolated from Embryogenic Callus of Date Palm



## Khayreddine Titouh, Nazim Boufis, and Lakhdar Khelifi



## Abstract



Date palm (*Phoenix dactylifera* L.) production is severely hampered due to several pests and diseases. Biotechnological tools such as protoplast fusion appear as an alternative to ensure rapid genetic improvement and multiplication of this species. However, establishment of an effective system of plant regeneration from protoplasts culture is a prerequisite for date palm somatic hybridization. In this chapter, we describe an effective protocol to induce microcalli in protoplasts isolated from nodular callus of important Algerian date palm cultivars. In this protocol, the main factors influencing the isolation (i.e., enzymatic solution, mannitol concentration, duration, and mode of maceration) of protoplasts from the calli of Algerian date palm cultivars were optimized. Purified protoplasts were cultured on a semisolid medium supplemented with a hormonal balance of auxin and cytokinin to obtain microcalli formation.

**Key words** Cellulase, Embryogenic callus, Microcalli, Pectinase, Protoplast



## 1 Introduction

The date palm (*Phoenix dactylifera* L.) is a perennial dioecious monocot, with a diploid genome ($2n = 2x = 36$) and belongs to the Palmae or Arecaceae family. Date palm is usually propagated vegetatively by offshoots, which maintains the genetic homogeneity of cultivars [1]. However, it remains limited by the number of offshoots produced, 10–15 per tree [2]. Furthermore, two of the main date palm cultivars, Deglet Noor and Medjool, are severely affected by bayoud disease, caused by a telluric fungus, *Fusarium oxysporum* f. sp. *albedinis* [3]. Several chemical and biological methods have been advocated for controlling this fungal disease. An alternative approach is to develop resistant cultivars, which seems to be an appropriate method. There are a few available resistant date palm cultivars able to withstand this fungus, namely, Akerbouch and Takerboucht [4]. However, these cultivars produce fruits of poor quality. In addition, classical genetic improvement programs require a long time (15–30 years) to produce resistant plant



Jameel M. Al-Khayri et al. (eds.), *Date Palm Biotechnology Protocols Volume 1: Tissue Culture Applications*, Methods in Molecular Biology, vol. 1637, DOI 10.1007/978-1-4939-7156-5_19, © Springer Science+Business Media LLC 2017

materials with good fruit quality [5, 6]. Biotechnological tools are an alternative to ensure rapid genetic improvement and multiplication of this species.

Somatic hybridization by protoplast fusion may be an appropriate approach for developing bayoud resistant date palm cultivars [7]. This technique has been successfully applied to the genetic improvement of many other species [1]. Protoplast fusion allows combining the nuclear and cytoplasmic genomes of related species or cultivars, entirely or partially, at the interspecific and intergeneric levels, overcoming barriers of natural sexual incompatibility [8]. However, this technique has not been applied to date palm improvement [1]. In many dicot species, leaves are the common plant material for protoplast isolation [1, 9], while in monocots, including date palm, callus and embryogenic cell suspension cultures are the preferred source for protoplast isolation and culture [3, 6, 10, 11]. Protoplast isolation is highly dependent on the genotype, material source, isolation and culture conditions, and type and concentration of plant growth regulators in the culture medium [1, 12].

In date palm, there are few reports of protoplast isolation and culture [5, 10, 13]. Therefore, the development of an effective protocol of regeneration from protoplasts of date palm is a necessary first step toward application of somatic hybridization to develop new cultivars, resistant to bayoud with acceptable fruit quality. Although this is yet to be realized, this chapter describes a protocol for protoplast isolation from nodular callus of date palm, followed by cell division and microcallus formation.

## 2 Materials

1. Plant material and sterilization solution.

2. Apical parts excised from shoot tips of offshoots of date palm cultivars Deglet Noor, Degla Beida, and Akerbouch.

3. Sterilization solution: 150 mg/L $HgCl_2$ (*see* **Note 1**).

### 2.1 Culture Medium and Maceration Enzymes

1. Basal culture medium: stock solutions for inorganic salts and organic elements of Murashige and Skoog medium (MS [14]) (*see* **Note 2**, Table 1).

2. Hormone stock solutions: 2,4-dichlorophenoxyacetic acid (2,4-D; 1 mg/mL) and 6-benzylaminopurine (BAP; 1 mg/mL) (*see* **Note 3**).

3. pH adjustment solutions: 1 N HCl and NaOH each.

4. Enzymes: cellulase from *Trichoderma longibrachiatum* (0.61 Units/mg) and pectinase (Macerozyme R10) from *Rhizopus* sp. (0.54 units/mg).

**Table 1**
**Murashige and Skoog medium [14] formulation used for date palm tissue culture**

| Elements | Composition of MS medium (mg/L) | Concentration of the stock solution | Final concentration in the stock solution (g/L) | Preparation of MS media (ml/L) |
|---|---|---|---|---|
| *Macroelements* | | | | |
| $NH_4NO_3$ | 1650 | ×20 | 33 | 50 |
| $KNO_3$ | 1900 | | 38 | |
| $MgSO_4.7H_2O$ | 440 | | 8.8 | |
| $CaCl_2.2H_2O$ | 370 | | 7.4 | |
| $KH_2PO_4$ | 170 | | 3.4 | |
| *Microelements* | | | | |
| $H_3BO_3$ | 6.20 | ×100 | 0.62 | 10 |
| $MnSO_4.4H_2O$ | 22.30 | | 2.23 | |
| $ZnSO_4.4H_2O$ | 8.60 | | 0.86 | |
| KI | 0.83 | | 0.083 | |
| $Na_22Mo\ O_4.2H_2O$ | 0.25 | | 0.025 | |
| $CuSO_4.5H_2O$ | 0.025 | | 0.0025 | |
| $CoCl_2.6\ H_2O$ | 0.025 | | 0.0025 | |
| *Iron (Fe-EDTA)* | | | | |
| $FeSO_4.7H_2O$ | 27.85 | ×100 | 2.785 | 10 |
| $Na_2$-EDTA | 37.25 | | 3.725 | |
| *Vitamins* | | | | |
| Glycine | 2.00 | ×1000 | 2.00 | 1 |
| Nicotinic Acid | 0.50 | | 0.50 | |
| Pyridoxine HCl | 0.50 | | 0.50 | |
| Thiamine HCl | 0.10 | | 0.10 | |
| *Others* | | | | |
| Myo-inositol | | | 100 mg/L | |

**2.2 Culture Medium and Protoplast Isolation and Purification Solutions**

1. Initiation and maintenance of embryogenic calli (CM) medium: MS medium (Table 1) supplemented with 100 mg/L L-glutamine, 40 mg/L adenine, 300 mg/L activated charcoal, 10 mg/L 2,4-D, 1 mg/L BAP, 45 g/L sucrose, and 7 g/L agar (Table 2).

2. Protoplast isolation solution (IS): MS medium (Table 1) containing 9.1 g/L mannitol, 15.2 g/L KCl, 7.4 g/L $CaCl_2$, 1.5% w/v cellulase (0.15 g in 10 mL of IS), and 1% w/v macerozyme (0.1 g in 10 mL of IS) (Table 2).

3. Protoplast purification solution A (PSA): MS medium (Table 1) containing 9.1 g/L mannitol, 15.2 g/L KCl, and 7.4 g/L $CaCl_2$ (Table 2).

**Table 2**
**Initiation and maintenance medium of embryogenic calli and protoplast isolation, purification, and culture medium composition**

| Medium components | Initiation and maintenance of embryogenic calli (CM) | Isolation solution (IS) | Purification solutions (PS) | | 25% Sucrose purification solution | Protoplast culture medium (PCM) |
|---|---|---|---|---|---|---|
| | | | A | B | | |
| Basal medium | MS | MS | MS | MS | MS | MS |
| Mannitol[a] | – | 9.1 | 9.1 | 9.1 | – | – |
| Chemical products[a] | | | | | | |
| KCl | – | 15.2 | 15.2 | – | – | – |
| $CaCl_2$ | – | 7.4 | 7.4 | 7.4 | – | – |
| Carbohydrates source[a] | | | | | | |
| Sucrose (g/L) | 45 | – | – | – | 250 | 40 |
| Glucose (mg/L) | – | – | – | – | – | 72 |
| Enzymes[a] (X g of powder product For 10 mL MS medium) | | | | | | |
| Cellulase | – | 0.15 g | – | – | – | – |
| Macerozyme R10 | – | 0.1 g | – | – | – | – |
| Hormones[a] (mg/L) | | | | | | |
| 2,4-D | 10 | – | – | – | – | 2 |
| BAP | 1 | – | – | – | – | 0.5 |
| Other components[a] (mg/L) | | | | | | |
| L-Glutamine | 100 | – | – | – | – | – |
| Adenine | 40 | – | – | – | – | – |
| Activated charcoal | 300 | – | – | – | – | – |
| Solidifying agent[a] (g/L) | | | | | | |
| Agar | 7 | – | – | – | – | – |
| Agarose | – | – | – | – | – | 3 |

[a]These quantities are added directly to the medium except the enzymes, activated charcoal, and solidifying agent which are added after adjusting pH

4. Protoplast purification solution B (PSB): MS medium (Table 1) containing 9.1 g/L mannitol and 7.4 g/L CaCl$_2$ (Table 2).

5. Sucrose purification solution: MS medium (Table 1) containing 250 g/L sucrose (Table 2).

6. Protoplast culture medium (PCM): MS medium (Table 1) containing 2 mg/L 2,4-D, 0.5 mg/L BAP, 40 g/L sucrose, 72 mg/L glucose, and 3 g/L agarose (Table 2).

***2.3  Equipment***

1. Tools and supplies: stainless steel scalpel and forceps, stainless steel·grid with pore size of 350 μm, 0.2 μm sterile microfilters, cotton, aluminum foil, and Parafilm.

2. Instruments: balance, magnetic stirrer, pH meter, hot plate, refrigerator, autoclave, oven, laminar air flow cabinet, orbital shaker, centrifuge with swinging bucket rotor, inverted microscope, ocular micrometer, pipet pump (micropipette), Malassez Cell (h × w × d) (2.5 × 2 × 0.2 mm), and growth chamber.

3. Glassware: beakers (100, 500 mL), Erlenmeyer flasks (50, 100, 250, 500, 1000 mL), glass funnel, pipettes and graduated cylinders (25, 50, 100 mL), 13 mL screw-capped glass centrifuge tubes, Pasteur pipet, 55 and 90 mm sterile Petri dishes, and sterile syringes (10 mL).

---

# 3  Methods

***3.1  Preparation of Basal Culture Medium***

1. Prepare the MS stock solutions of inorganic and organic elements according to the formulation given in Table 1 by dissolving the weighed quantities in 1000 mL distilled water (*see* **Note 2**, Table 1).

2. Place distilled water (1/3 of the final volume) in 1000 mL Erlenmeyer flask.

3. While stirring, add required amounts of the MS stock solutions and other components as specified in Table 1.

4. Adjust medium to pH 5.7 with 1 N NaOH or 1 N HCl and sterilize by autoclaving for 20 min at 121 °C and 1.4 kg/cm$^2$.

***3.2  Plant Material Disinfection***

1. Remove outer leaves of the offshoot until the last two leaves protecting the offshoot core including apical part containing the meristematic area.

2. Rinse the extracted offshoot core with tap water.

3. Immerse the offshoot core in the sterilizing solution for 1 h inside the laminar airflow hood.

4. Remove the last two apparent leaves with sterile forceps and scalpel.

5. Immerse the offshoot core in the sterilizing solution a second time for 1 h.

6. Rinse three consecutive times with sterile distilled water.

7. Remove necrotic tissues damaged by the $HgCl_2$ with forceps and scalpel.

8. Divide the sterilized apical part of offshoot core in small explants of 0.5 $cm^3$ size.

### 3.3 Embryogenic Calli Induction

1. Culture the shoot tip explants in test tubes containing 20 mL solid medium (CM) (*see* Table 2).

2. Incubate cultures in the dark at 25 °C ± 2 for 16 weeks. Subculture every 8 weeks and eliminate brown or necrotic parts of explants during subculturing.

3. Excise embryogenic callus and maintain on CM medium until protoplast isolation (*see* **Note 4**, Fig. 1a, b).

### 3.4 Preparation of Enzymatic Solution

1. Prepare 10 mL of protoplast isolation solution (IS) (Table 2) in a 50 mL Erlenmeyer flask (*see* Subheading 2.3, **item 2**).

2. Stir the solution until complete dissolution of enzymes (*see* **Note 5**, Fig. 1c).

3. Under laminar airflow hood, sterilize the enzymatic solution using a sterile 10 mL syringe, fitted with a sterile 0.22 μm-microfilter (Fig. 1d, e).

### 3.5 Protoplast Isolation

1. Under a laminar flow hood, weigh 0.6 g actively growing embryogenic nodular callus in a 90 mm sterile Petri dish.

2. Cut the callus into small pieces using stainless steel scalpel and forceps (*see* **Note 6**).

3. Add 10 mL enzymatic solution (IS) and seal the Petri dish with Parafilm and incubate the in the dark at 27 °C on an orbital shaker set at 50 rpm (*see* **Note 7**).

4. After 14–16 h of incubation (*see* **Note 8**), filter the mixture through a sterile stainless steel grid placed on a sterile glass funnel, and recover the protoplast suspension in 13 mL screw-cap glass tube.

5. Centrifuge the suspension in a swinging bucket rotor for 5 min at 65 × *g*.

6. Discard the supernatant and dilute the pellet containing the protoplasts in 1 mL protoplast purification solution A (PSA) (Table 2).

7. Wash the protoplasts twice by centrifuging at 65 × *g* for 5 min and resuspend the pellet each time in 1 mL PSA (*see* **Note 9**).

8. Wash the protoplasts once with the purification solution B (PSB) (*see* **Note 10**) by centrifuging at 65 × *g* for 5 min.

**Fig. 1** Date palm protoplasts isolation, purification, and culture. (**a–b**) Date palm embryogenic callus of Degla Beida and Deglet Noor (G = ×10 and ×45), (**c**) preparation of the enzymatic solution (bar = 2.8 cm), (**d**) 0.22 μm sterile filters (bar = 0.9 cm), (**e**) sterilization of the enzymatic solution (bar = 2 cm), (**f**) purification of protoplasts on a layer of 25% of sucrose (bar = 0.9 cm, *arrow* indicates the thin layer of the purified protoplasts), (**g–h**) Deglet Noor and Degla Beida isolated protoplasts (Bar = 30.5 μm), (**i–j**) first cell divisions of Akerbouch and Deglet Noor protoplasts after 3 days of culture (bar = 9.5 μm; *arrows* indicate separation of the cells after division and thickening of the protoplast periphery due to regeneration of a new cell wall), (**k**) microcalli visible to the naked eye after 10 weeks of culture (bar = 1.1 cm; *arrows* indicate microcalli derived from protoplast culture), (**l–m**) detail of the microcalli and microcolony of Deglet Noor and Akerbouch obtained after 10 weeks of culture (bar = 117.4 and 35.3 μm). *Source:* Photos b, g, h, i, j, k, l and m are taken from Titouh et al. [7]

9. Suspend the pellet in 1 mL PSB and gently transfer to 2–3 mL 25% sucrose solution contained in a new glass tube.

10. Centrifuge during 5 min at 65 × *g* and recover carefully the protoplasts from interphase of the two solutions of different density using 1 mL pipet pump or a sterile Pasteur pipet with the tip broken (*see* **Note 11**, Fig. 1f).

11. Adjust the protoplasts suspension volume to 1 mL by adding the solution PSB.

12. With a sterile Pasteur pipet, take one drop from the protoplasts suspension in a plate.

13. Add one drop of 0.1% methylene blue solution (*see* **Note 12**), mix well, and let react a few seconds.

14. Take one drop and count the number of protoplasts in a Malassez cell counter (hemocytometer) (*see* **Notes 13** and **14**). Live protoplasts appear in yellow fluorescent color and the dead protoplasts appear in blue color.

### 3.6 Protoplasts Culture and Microcallus Formation

1. Prepare 1 L PCM semisolid medium in Erlenmeyer flask (*see* **Note 15**).

2. Plug Erlenmeyer flasks with cotton plugs and cover with aluminum foil and autoclave at 121 °C and 1.4 kg/cm$^2$ for 20 min. Dispense the autoclaved medium in 55 mm sterile Petri dishes and keep inside the laminar airflow hood until complete solidification.

3. Before protoplast culture, adjust their density in MS medium to $1 \times 10^5$ protoplasts mL$^{-1}$ (*see* **Note 16**, Fig. 1g, h).

4. Spread uniformly 1 mL protoplast suspension on the entire surface of protoplast culture medium (PCM) (*see* **Note 17**) and seal Petri dishes with Parafilm.

5. Incubate the cultures in the dark at 27 °C and observe cell wall regeneration and mitotic activity using an inverted microscope (*see* **Note 18**).

6. After 3 weeks of culture, add 1 mL liquid PCM medium containing 4.5 g/L mannitol (instead sucrose and glucose), 2 mg/L 2,4-D, and 0.5 mg/L BAP (*see* **Note 19**).

7. After 4 weeks of culture, add 1 mL liquid PCM medium containing 2.25 g/L mannitol, 2 mg/L 2,4-D, and 0.5 mg/L BAP (*see* **Note 19**).

8. After 6 weeks of culture, add 1 mL liquid PCM medium containing 1.13 g/L mannitol, 2 mg/L 2,4-D, and 0.5 mg/L BAP (*see* **Note 19**).

9. After 10–12 weeks of culture, transfer the visible microcalli (Fig. 1k, l) to a new PCM semisolid medium for further growth and subsequent use in plant regeneration experiments.

## 4  Notes

1. As HgCl$_2$ is highly toxic to human health, while handling take all appropriate precautions: glasses, mask, and gloves.

2. Stock solutions are stored at 4 °C and renewed periodically to avoid sedimentation and contaminations.

3. Hormonal solutions are prepared by dissolving 100 mg 2,4-D in 100 mL distilled water and 100 mg BAP in few drops of 1 N NaOH and then add distilled water while stirring until reaching the final volume.

4. Nodular callus is a potentially embryogenic callus. Its texture is more or less friable and its color is white without browning.

5. An effective enzymatic formula allows achieving good protoplast yield while excessive enzyme concentration decreases protoplast viability [9], perhaps by enzymatic toxicity [15] or due to a strong and rapid enzymatic activity often causing membrane damage [16].

6. To increase the enzymatic solution efficiency, use calli with small nodules easily separable or disintegrated tissue in small pieces [1] and avoid too friable calli which contain a high number of non-meristematic cells [17].

7. A gentle agitation (<50 rpm) is required to liberate protoplasts from digested calli. Nevertheless, agitation is particularly effective at the end of the incubation when protoplasts are released from digested tissues [18]. Therefore, good protoplasts liberation may be achieved after stationary incubation followed with a gentle agitation for 15–30 min.

8. The yield of viable protoplasts increases with low concentrations of enzymes and overnight incubation duration.

9. After centrifugation, remove the maximum of the supernatant, avoiding disturbance of the pellet containing the protoplasts in order to reduce the bursting risk and protoplasts loss between purification steps.

10. Washing protoplasts without KCl allows increasing the viability of protoplasts and prevents their bursting.

11. Purification of the protoplasts by flotation on sucrose solution with a swinging bucket rotor allows obtaining a marked interphase easily recoverable.

12. 0.1% methylene blue solution is prepared by dissolving 0.1 g methylene blue in 100 mL PSB solution.

13. Perfect spherical protoplasts are viable as well as showing yellow fluorescent cytoplasm, while dead protoplasts turn blue with a nonspherical shape.

14. Malassez cell counter contains in total 100 rectangles of four kinds, each one with dimensions (h × w × d) $0.2 \times 0.25 \times 0.2$ mm giving a volume of $0.01$ mm$^3$ or $0.01$ µL. In addition, average diameter of date palm protoplasts varied with genotype, tissue source, osmotic agent, as well as

enzymatic mixture. Therefore, take care of the rectangle depth by varying the focus to count all the protoplasts within this volume.

15. Protoplasts are cultured in the liquid medium. However, culture on a solidified medium is more advantageous due to easy manipulation [8].

16. An optimal preparation can yield around $4$–$5 \times 10^5$ protoplasts per gram callus fresh weight [7].

17. To spread homogeneously the protoplasts suspension on culture medium incline the Petri dish to drain the suspension on the entire surface of the culture medium.

18. Spherical protoplasts are elongated or take nonspherical shape with a regular periphery without interruption while cell wall forms. Generally, a cell wall forms after 48 h of culture. Equally, mitotic activity begins just after a complete reformation of a new cell wall. Microcalli will be visible to the naked eye after culturing for 2 months (Fig. 1i–m).

19. Protoplasts are highly sensitive to osmotic shock and other physical disturbances during the culture period, and that can have a negative impact of their viability [19]. Mannitol must be reduced gradually after the cell wall regeneration and cellular division.

## References


1. Assani A, Chabane D, Shittu H, Bouguedoura N (2011) Date palm cell and protoplast culture. In: Jain SM, Al-Khayri JM, Johnson DV (eds) Date palm biotechnology. Springer, Netherland, pp 605–629

2. Bouguedoura N, Michaux-Ferrière N, Bompar J-L (1990) Comportement in vitro de bourgeons axillaires de type indéterminé du palmier dattier (*Phoenix dactylifera*). Can J Bot 68 (9):2004–2009

3. Chabane D (2007) Amélioration du palmier dattier *Phoenix dactylifera* L. par fusion de protoplastes de deux cultivars Deglet Noor sensible et Takerboucht résistant au Bayoud. Dissertation, University of Science and Technology Houari Boumediene, Algiers, Algeria

4. El Hadrami A, El Idrissi-Tourane A, El Hassni M, Daayf F, El Hadrami I (2005) Toxin-based in-vitro selection and its potential application to date palm for resistance to the bayoud Fusarium wilt. C R Biol 328(8):732–744

5. Chabane D, Assani A, Bouguedoura N, Haicour R, Ducreux G (2007) Induction of callus formation from difficile date palm protoplasts by means of nurse culture. C R Biol 330 (5):392–401

6. Boufis N, Khelifi-Slaoui M, Djillali Z, Zaoui D, Morsli A, Bernards MA, Khelifi L (2014) Effects of growth regulators and types of culture media on somatic embryogenesis in date palm (*Phoenix dactylifera* L. cv. Degla Beida). Sci Hortic 172(9):135–142

7. Titouh K, Khelifi L, Slaoui M, Boufis N, Morsli A, Titouh-Hadj Moussa K et al (2015) A simplified protocol to induce callogenesis in protoplasts of date palm (*Phoenix dactylifera* L.) cultivars. Iran J Biotechnol 13(1):26–35

8. Davey MR, Anthony P, Power JB, Lowe KC (2005) Plant protoplasts: status and biotechnological perspectives. Biotechnol Adv 23 (2):131–171

9. Sun Y, Zhang X, Huang C, Nie Y, Guo X (2005) Plant regeneration via somatic embryogenesis from protoplasts of six explants in Coker (*Gossypium hirsutum*). Plant Cell Tissue Organ Cult 201(82):309–315

10. Rizkalla AA, Badr-Elden AM, Nower AA (2007) Protoplast isolation, salt stress and

callus formation of two date palm genotypes. J Appl Sci Res 3(10):1186–1194

11. Assani A, Haïcour R, Wenzel G, Foroughi-Wehr B, Bakry F, Côte F-X et al (2002) Influence of donor material and genotype on protoplast regeneration in banana and plantain cultivars (*Musa* spp.) Plant Sci 162 (3):355–362

12. Borgato L, Pisani F, Furini A (2007) Plant regeneration from leaf protoplasts of *Solanum virginianum* L. (Solanaceae). Plant Cell Tissue Organ Cult 88(3):247–252

13. Gabr MF, Tisserat B (1984) Parameters involved in the isolation, culture, cell wall regeneration and callus formation from palm and carrot protoplasts. Date Palm J 3 (2):359–365

14. Murashige T, Skoog F (1962) A revised medium for rapid growth and bioassays with tobacco tissue cultures. Physiol Plant 15:473–497

15. Zhang J, Shen W, Yan P, Li X, Zhou P (2011) Factors that influence the yield and viability of protoplasts from *Carica papaya* L. Afr J Biotechnol 10(26):5137–5142

16. Monteiro M, Appezzato-Da-Glória B, Valarini MJ, De Oliveira CA, Vieira MLC (2003) Plant regeneration from protoplasts of alfalfa (*Medicago sativa*) via somatic embryogenesis. Sci Agric 60:683–689

17. Sané D, Aberlenc-Bertossi F, Gassama-Dia YK, Sagna M, Trousslot MF, Duval Y et al (2006) Histological analysis of callogenesis and somatic embryogenesis from cell suspensions of date palm (*Phoenix dactylifera*). Ann Bot 98:301–308

18. Sinha A, Wetten AC, Caligari PDS (2003) Effect of biotic factors on the isolation of *Lupinus albus* protoplasts. Aust J Bot 51 (1):103–109

19. Horine RK, Ruesink AW (1972) Cell wall regeneration around protoplasts isolated from *Convolvulus* tissue culture. Plant Physiol 50 (4):438–445

# Chapter 20

# Temporary Immersion System for Date Palm Micropropagation

Ahmed Othmani, Chokri Bayoudh, Amel Sellemi, and Noureddine Drira

## Abstract

The temporary immersion system (TIS) is being used with tremendous success for automation of micropropagation of many plant species. TIS usually consists of a culture vessel comprising two compartments, an upper one with the plant material and a lower one with the liquid culture medium and an automated air pump. The latter enables contact between all parts of the explants and the liquid medium by setting overpressure to the lower part of the container. These systems are providing the most satisfactory conditions for date palm regeneration via shoot organogenesis and allow a significant increase of multiplication rate (5.5-fold in comparison with that regenerated on agar-solidified medium) and plant material quality, thereby reducing production cost.

**Key words** In vitro, Micropropagation, Multiplication rate, Shoot organogenesis, Somatic embryogenesis, TIS

## 1  Introduction

Many reports have described the use of in vitro tissue culture techniques, with the potential to multiply date palm genotypes of superior value on a massive scale in a shorter period than the conventional methods. These techniques have used agar-solidified medium and are difficult to automatize and entail high production cost and are unsuitable for mass clonal propagation [1]. Thus, scaling up of the production methods is indispensable to develop efficient automated systems.

A bioreactor is a self-contained, sterile environment that capitalizes on liquid nutrient or liquid/air inflow and outflow systems and is designed to provide the most favorable growth conditions by enabling a high degree of control over chemical and/or physical factors. These factors include pH, dissolved oxygen and carbon dioxide concentrations, ethylene, illumination regime, aeration rate, and temperature, which are compatible with the automation of micropropagation procedures along with a

Jameel M. Al-Khayri et al. (eds.), *Date Palm Biotechnology Protocols Volume 1: Tissue Culture Applications*,
Methods in Molecular Biology, vol. 1637, DOI 10.1007/978-1-4939-7156-5_20, © Springer Science+Business Media LLC 2017

reduction of production costs [2]. Currently, the temporary immersion system (TIS) plays a crucial role in scaling up the production for commercialization of somatic embryogenesis and multiplication of clusters of meristem and bud-based plant micropropagation [3].

The present chapter describes procedures used for regeneration of date palm, cv. Deglet Bey from juvenile inflorescences through somatic embryogenesis and shoot organogenesis processes. It also describes the establishment of a protocol for large-scale automation of date palm micropropagation using the temporary immersion system.

## 2   Materials

### 2.1   Plant Material

Immature inflorescences (*see* **Note 1**) excised from mature field-grown date palm cv. Deglet Bey (Fig. 1a) growing in Deguèche, southern Tunisia (*see* **Note 2**).

### 2.2   Surface Sterilization and Browning Prevention

1. Disinfectant solutions: 70% ethanol, 0.01% mercuric chloride ($HgCl_2$), Tween 20 (3 per 100 mL solution).

2. Antioxidant solution: Ascorbic acid and citric acid, 150 mg/L each.

### 2.3   Culture Medium

1. Basal culture medium: The stock solutions of Murashige and Skoog (MS) [4] medium stock solutions I, II, III, and IV (Table 1; *see* **Note 3**).

2. Medium additives used for various culture stages: MS basal medium supplemented with various additives according to the culture stage (*see* Table 2), including culture initiation (CI), shoot multiplication (SM), embryogenic callus multiplication and somatic embryos differentiation (EMD) temporary immersion system liquid medium (SML) (*see* **Note 4**), and shoot and somatic embryo rooting medium (RM).

3. pH adjustment solutions: 0.1 and 1 N NaOH and HCl.

### 2.4   Equipment

1. Culture room: Laminar airflow hood, bead sterilizer.

2. Surgical tools: stainless steel forceps, sharp blade, surgical scalpels, removable sterile surgical blades, sterile craft paper, and cotton.

3. Medium preparation: Weighing balances, magnetic stirrer, hot plate, culture tube racks, pH meter, microwave, micropipettes, magnets, peristaltic pump, refrigerator, and autoclave.

4. Glassware: Conical flasks (100, 250 mL capacity) and beakers (250, 500 mL capacity).

**Fig. 1** Stages of shoot cluster formation from juvenile date palm inflorescences. (**a**) Date palm source of explants, (**b**) spathes, (**c**) opened spathes, (**d**) explants on culture initiation medium, (**e**) swelling florets after 3 months of culture, (**f**) shoots initiation after 6 months of culture, (**g**) initiation of shoot cluster, (**h**) shoot cluster on multiplication medium

**Table 1**
**Components of MS basal medium [4] and additives used for date palm in vitro culture stages**

| Compound | Formula | Concentration of the final medium (mg/L) | Quantity for stock solution (mg/L) | Volume of stock solution (mL) | Volume of stock solution required for 1 L of medium |
|---|---|---|---|---|---|
| *Stock I: Major inorganic nutrients* | | | | 1000 | 50 |
| Ammonium nitrate | $NH_4NO_3$ | 1650 | 33 | | |
| Potassium nitrate | $KNO_3$ | 1900 | 38 | | |
| Magnesium sulfate | $KH_2PO_4$ | 370 | 7.4 | | |
| Potassium dihydrogen phosphate | $MgSO_4 \cdot 7H_2O$ | 170 | 3.4 | | |
| Calcium chloride | $CaCl_2 \cdot 2H_2O$ | 440 | 8.8 | | |
| *Stock II: Minor inorganic nutrients* | | | | 200 | 10 |
| Manganese sulfate | $MnSO_4 \cdot 4H_2O$ | 22.3 | 0.446 | | |
| Zinc sulfate | $ZnSO_4 \cdot 7H_2O$ | 8.6 | 0.172 | | |
| Boric acid | $H_3BO_3$ | 6.2 | 0.124 | | |
| Potassium iodide | KI | 0.83 | 0.0166 | | |
| Sodium molybdate | $Na_2MoO_4 \, 2H_2O$ | 0.25 | 0.005 | | |
| Copper sulfate | $CuSO_4 \cdot 5H_2O$ | 0.025 | 0.0005 | | |
| Cobalt chloride | $CoCl_2 \cdot 6H_2O$ | 0.025 | 0.0005 | | |
| *Stock III: Iron source* | | | | 200 | 10 |
| Ferrous sulfate | $FeSO_4 \cdot 7H_2O$ | 27.8 | 0.556 | | |
| EDTA disodium | $Na_2EDTA \cdot 2H_2O$ | 37.2 | 0.744 | | |
| *Stock IV: Vitamins* | | | | 20 | 1 |
| Nicotinic acid | | 0.5 | 0.01 | | |
| Pyridoxine HCl | | 0.5 | 0.01 | | |
| Thiamine HCl | | 0.1 | 0.002 | | |
| Glycine | | 2 | 0.04 | | |
| Myoinositol | | 100 | 2 | 200 | 10 |
| *Carbon source* | | | | | |
| Sucrose | | 50,000 | | | |

**Table 2**
**Specific additives to the culture medium used for various culture stages**

| Media additives | Culture stage | | | | |
| --- | --- | --- | --- | --- | --- |
| | Culture initiation (CI) | Shoot multiplication on gelified medium (SM) | Shoot multiplication in liquid medium (SML) | Embryogenic callus multiplication and somatic embryos differentiation (EMD) | Rooting medium (RM) |
| 2,4-Dichloro-phenoxyacetic acid (2,4-D) | 10 mg/L | – | – | – | – |
| Benzylaminopurine (BAP) | – | 0.2 mg/L | 0.2 mg/L | – | – |
| Indolebutyric acid (IBA) | – | – | – | – | 3 mg/L |
| Kinetin (Kin) | – | 0.02 mg/L | 0.02 mg/L | – | – |
| Naphthaleneacetic acid (NAA) | – | 0.04 mg/L | 0.04 mg/L | – | – |
| Activated charcoal | 3 g/L | – | – | – | – |
| Agar | 8 | – | – | 8 | – |
| Gellan gum | – | 2 | – | – | – |

5. Culture vessels: Reagent bottles, conical flasks (250, 1000 mL capacity), beakers, measuring cylinders, and culture tubes (25 × 150 mm) with polypropylene caps (*see* **Note 5**).

# 3   Methods

## 3.1   Preparation of Culture Media

### 3.1.1   Glassware Cleaning

1. Scrub glassware with liquid detergent solution for 1 h and thoroughly wash with hot water.

2. Rinse glassware with double-distilled water and dry in a hot air oven.

### 3.1.2   MS Stocks Preparation

1. Prepare separate MS stock I, II, III, and IV solutions for macronutrients, micronutrients, iron, and vitamins (Table 1).

2. Weigh and dissolve the components of each stock in 500 mL double-distilled water by using magnetic stirrer and make up the final volume to 1000 mL by adding distilled water.

*3.1.3  Protocol to Prepare 1 L MS Medium*

1. Pour 800 mL of distilled water in a 1 L flask with a magnetic stirring bar, add 30 g sucrose, and dissolve.

2. Add required stock solutions according to the order in Table 1.

3. Add aliquot hormones and activated charcoal according to the culture stages (*see* Table 2).

4. Adjust the pH to 5.8 with pH meter by adding 0.1 or 1 N NaOH or HCl, with constant stirring.

5. Add 8 g agar and make up the volume with double-distilled water 1 L by using 1 L measuring cylinder.

6. Heat the culture medium until it starts boiling, dispense 12 mL medium per culture tube using peristaltic pump, and cap the tubes.

7. Autoclave at 121 °C with a pressure of 15 psi (1.06 kg/cm$^2$), for 20 min.

8. Take out the culture medium from the autoclave and allow it to cool and solidify at room temperature (*see* **Note 6**).

### 3.2  Preparation of Explants

1. After cutting off the date palm plant, remove all mature leaves gradually from the outer ring toward the center by using a hatchet and a serrated knife. Collect immature female inflorescences with their protective sheaths intact (spathe) (Fig. 1b), place in a clean plastic cover, and take it directly to the laboratory (*see* **Note 7**).

2. Disinfect the surface of the laminar flow with 75% alcohol after UV light exposure for 20 min.

3. Surface sterilize the spathes in 70% ethanol for 1 min, gently open spathes with sterilized scalpel in laminar airflow hood (Fig. 1c), and then immediately soak spikelets in disinfectant solution (0.01% HgCl$_2$ containing three drops of Tween 20 per 100 mL) for 1 h.

4. Rinse the tissue three times with sterilized distilled water.

5. Excise each spikelet into 1–2 cm long pieces carrying 2–5 florets, inside the laminar flow hood; dip them in chilled antioxidant solution for 10 min then rinse three times with sterilized distilled water to prevent browning.

### 3.3  Culture Initiation

1. In order to induce shoots and embryogenic callus formation, place culture inflorescence explants (2–4 pieces) on CI medium (Fig. 1d).

2. Incubate cultures at 27 ± 2 °C in darkness and subculture at a 4-week interval for a 6–8-month culture period.

3. Explants start to expand in the initial 3 months of culture (Fig. 1e). After 6 months of culture, shoots and embryogenic callus initiate from florets (Figs. 1f and 2a).

**3.4  Shoot Multiplication and Elongation**

1. For stimulating multiplication and elongation of the shoots, transfer all expanding explants (Fig. 1g) on SM medium supplemented with 2 g/L Gellan gum instead of agar, and incubate at 27 ± 2 °C and 16-h photoperiods (80 μmol/m$^2$/s) provided by fluorescent lamps.

2. Maintain for an additional 2 months to obtain sufficient shoot cluster growth for subsequent experiments.

**3.5  Embryogenic Callus Multiplication and Somatic Embryo Differentiation**

1. For stimulating multiplication of embryogenic callus and differentiation of somatic embryos, transfer embryogenic callus (Fig. 2a) on EMD medium, and incubate at 27 ± 2 °C and 16-h photoperiods (80 μmol/m$^2$/s) provided by fluorescent lamps.

2. After approximatively 5 months, embryogenic callus gives rise to well-structured somatic embryos (Fig. 2b–d).

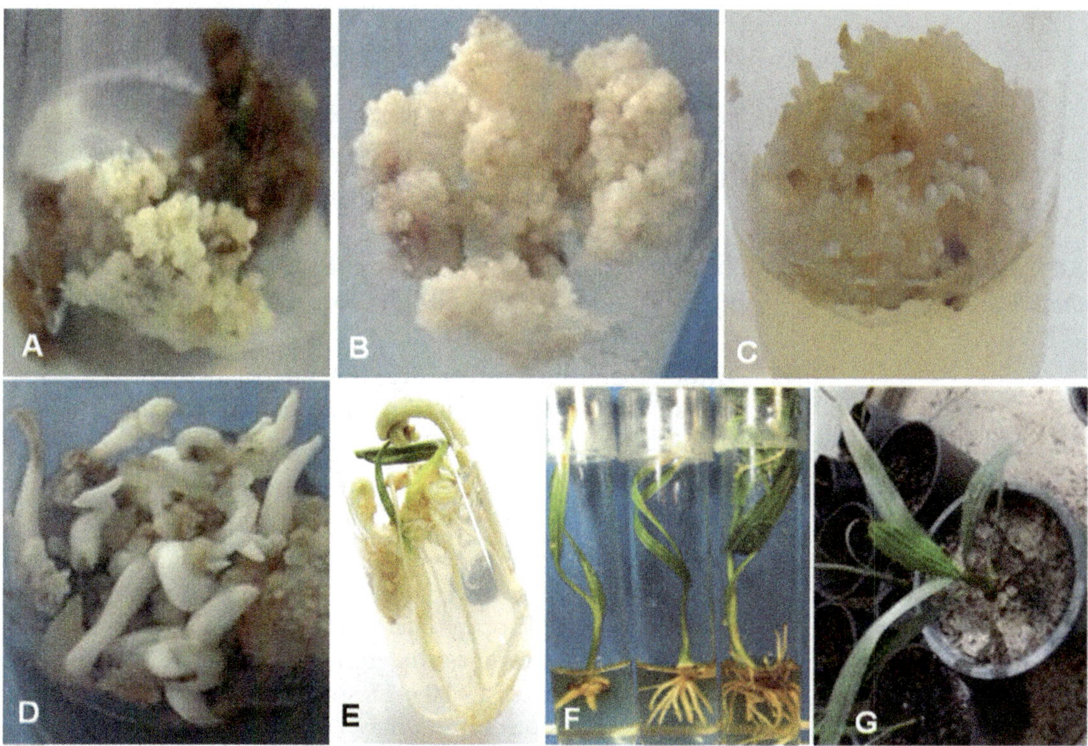


**Fig. 2** Stages of somatic embryos formation from juvenile date palm inflorescences. (**a**) Embryogenic callus initiation after 8 months of culture, (**b**) callus multiplication, (**c**) initiation of somatic embryos differentiation, (**d**) mature somatic embryo explants on culture initiation medium, (**e**) germination of somatic embryos, (**f**) conversion of somatic embryos into vigorous vitroplants, (**g**) acclimated vitroplant after 1 year of transfer to the greenhouse

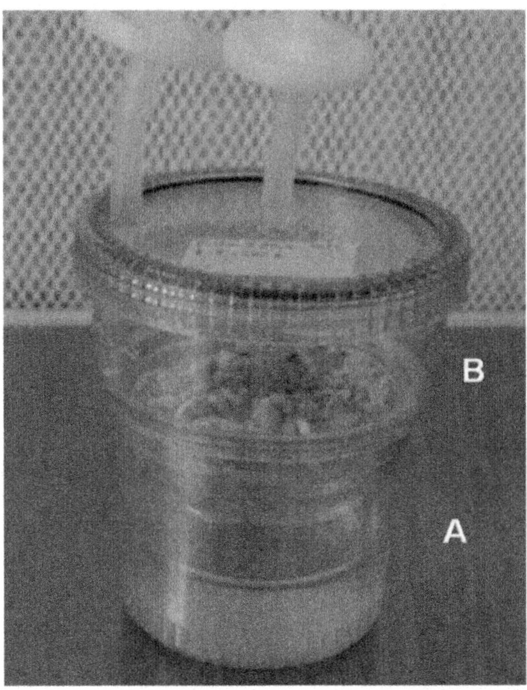

**Fig. 3** Temporary immersion bioreactor container (RITA™, Cirad, France) consisting of two compartments: (**a**) lower part and (**b**) upper part of the bioreactor

**3.6 Establishing Culture in TIS (RITA®)**

1. Pour 200 mL MS liquid medium (SML) containing 0.2 mg/L BAP, 0.02 mg/L Kin, and 0.04 mg/L NAA in the lower part of RITA bioreactor [5] (Fig. 3).

2. Divide 10 g shoot clusters (Fig. 1h), into small parts (Fig. 4a, b) with at least two shoots, and then place onto the upper part of RITA® bioreactor (Fig. 4c).

3. Use an immersion frequency of 5 min every 8 h for a 6-week culturing period.

4. Connect the container to an automated air pump which sets overpressure to the lower part of the container pushing the medium to the upper part through the filter (Fig. 5). The overpressure escapes through an air vent in the lid of the container (*see* **Notes 8–10**). Despite many promising results, TIS-growing shoot clusters may face hyperhydricity (vitrification) and contamination risks (*see* **Notes 10** and **11**). These difficulties must be considered before using bioreactors for commercial production of shoot clusters.

**3.7 Shoot and Somatic Embryo Rooting**

1. To promote rooting of shoots, transfer well-structured somatic embryos (Fig. 2e) and shoots derived from RITA system (Fig. 4d) after culturing for 6 weeks separately onto RM.

**Fig. 4** Developmental stages of shoot clusters after transfer in the TIB system. (**a**) Shoot cluster, (**b**) parts of divided shoot cluster, (**c**) small parts of shoot clusters placed on the upper part of RITA bioreactor, (**d**) well-structured shoots after a week culture period in RITA bioreactor, (**e**) rooting of two shoots after 1 month of culture on rooting medium, and (**f**) potted plants 18 months after transfer to the greenhouse. Source: [3]

2. Under these conditions, rooting initiates after 2 weeks of culture, and the whole plantlets are obtained within 2 months (Figs. 2f and 4e) with a 96% success rate (*see* **Note 12**).

***3.8 Hardening and Acclimatization***

1. Remove individually regenerated plantlets with well-developed shoots and roots from the vessels, and wash carefully away the agar sticking to roots, under running tap water.

2. Place plantlets in a potting mixture made of peat moss and sand (2:1 ratio) and cultivate it in the greenhouse under natural sunlight with 27 ± 2 °C and 50–60% relative humidity (Figs. 2g and 4f).

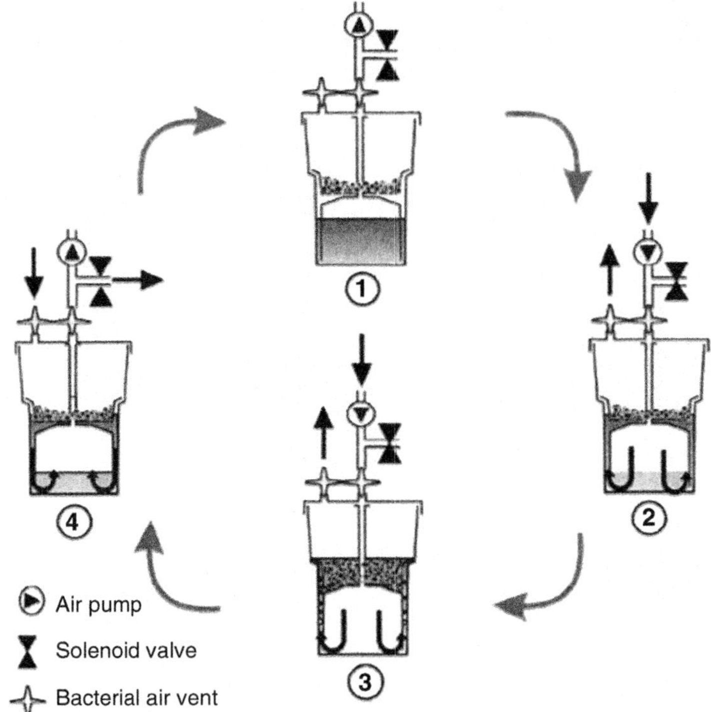



**Fig. 5** Operation procedure and design of the RITA® bioreactor. Source: http:/www.setis-systems.be/tis-technology/

## 4    Notes

1. In Tunisian oases, the sampling of date palm inflorescences used for in vitro plant regeneration is carried out in the beginning of February.

2. Store macro, iron, and hormones at 4 °C and micro, vitamins, and myoinositol at −20 °C.

3. Transfer the stocks into reagent bottles and keep the solutions in a refrigerator at 4 °C, and use within 3 months.

4. RITA™ contains 15 cm high cylindrical vessels that consist of two compartments (Fig. 3); the upper part holds the plant material on a polyurethane filter (1 cm) and the lower part contains the liquid culture medium.

5. Autoclave glass tubes, jars, and RITA bioreactor parts prior to use to avoid any contamination during the usage period.

6. Immediately after the autoclaving, soak the basal ends of culture tubes containing the culture medium in 4 °C water bath to avoid precipitation of activated charcoal.

7. For some date palm varieties, especially cv. Deglet Bey, explants must be cultured on the same day of collect; otherwise, necrosis

starts and reduces drastically the success of in vitro plant regeneration.

8. TIS technology has the potential for producing large numbers of plants economically and efficiently, which is due to the direct contact of the medium with the plant material, renewal of the culture growing environment on each immersion, and the reduction of asphyxia, necrosis, and tissue hyperhydration phenomena [5].

9. The TIS-grown shoot clusters of date palm, cv. Deglet Bey, clearly improve 5.5-fold yield of regenerated shoots versus agar-solidified medium [3].

10. Hyperhydricity increases with the immersion frequency of cultures [6]. Furthermore 15 min immersions, 2–6 times per day, result in 64 and 90% hyperhydric embryos, respectively. Thereby, adjust the immersion time of cultures in the liquid nutrient medium in order to significantly prevent hyperhydricity [7].

11. Date palm micropropagation in the TIB system is faced with enormous proliferation of endophytic bacterial growth that immediately affects shoot growth and their rooting ability on the rooting medium [8]. Disease can be controlled using healthy shoot clusters, completely free from endophytic bacteria.

12. In date palm cv. Deglet Bey, rooting efficiency is very poor (10%) when kept in the TIS container. However, after their transfer to MS agar-solidified rooting medium, 96% elongated shoots exhibited prolific adventitious rooting after 1 month of culture.

## References


1. Etienne H, Berthouly M (2002) Temporary immersion systems in plant micropropagation. Plant Cell Tissue Organ Cult 69:215–231

2. Georgiev V, Schumann A, Pavlov A, Bley T (2014) Temporary immersion systems in plant biotechnology. Eng Life Sci 14:607–621

3. Othmani A, Mzid R, Bayoudh C, Drira N (2011) Bioreactors and automation in date palm micropropagation. In: Jain SM, Al-Khayri JM, Johnson DV (eds) Date palm biotechnology. Springer, Dordrecht, pp 119–136

4. Murashige T, Skoog F (1962) A revised medium for rapid growth and bioassays with tobacco tissue cultures. Physiol Plant 15:473–497

5. Alvard D, Côte F, Teisson C (1993) Comparison of methods of liquid medium culture for banana micropropagation. Effects of temporary immersion of explants. Plant Cell Tissue Organ Cult 32:55–60

6. Marbun CLM, Toruan-Mathius N, Utomo C, Liwang T (2015) Micropropagation of embryogenic callus of oil palm (*Elaeis guineensis* Jacq.) using temporary immersion system. Procedia Chem 14:122–129

7. Fki L, Bouaziza N, Kriaa W, Benjemaa-Masmoudi R, Gargouri-Bouzidb R, Rival A, Drira N (2011) Multiple bud cultures of 'Barhee' date palm (*Phoenix dactylifera* L.) and physiological status of regenerated plants. J Plant Physiol 168:1694–1700

8. Albarran J, Bertrand B, Lartaud M, Etienne H (2005) Cycle characteristics in a temporary immersion bioreactor affect regeneration, morphology, water and mineral status of coffee (*Coffea arabica* L.) somatic embryos. Plant Cell Tissue Organ Cult 81:27–36

**Chapter 21**

# Plantform Bioreactor for Mass Micropropagation of Date Palm


**Abdulminam H.A. Almusawi, Abdullah J. Sayegh, Ansam M.S. Alshanaw, and John L. Griffis Jr**


## Abstract


A novel protocol for the commercial production of date palm through micropropagation is presented. This protocol includes the use of a semisolid medium alternation or in combination with a temporary immersion system (TIS, Plantform bioreactor) in date palm micropropagation. The use of the Plantform bioreactor for date palm results in an improved multiplication rate, reduced micropropagation time, and improved weaning success. It also reduces the cost of saleable units and thus improves economic return for commercial micropropagation. The use of the Plantform bioreactor successfully addresses other hindrances that can occur during the scale-up of date palm micropropagation, including asynchrony of somatic embryos, limited maturation of somatic embryos, and highly variable germination frequencies of embryos.

**Key words** Bioreactor, Commercial micropropagation, Partial desiccation, Plantform, Temporary immersion system


## 1  Introduction

Date palm (*Phoenix dactylifera* L.) is considered to be one of the oldest fruit trees domesticated by man since the dawn of civilization. Currently, however, date palm cultivation is challenged by various biotic and abiotic factors [1]. Commercial micropropagation is an alternative method for large-scale propagation of disease-free date palm varieties. Micropropagated date palm plants are produced in many countries around the world. The list of available cultivars is expanding as the result of market demands.

Somatic embryogenesis and organogenesis are the two pathways of choice for rapid and large-scale propagation of date palm. They have been successfully used for the micropropagation of elite genotypes using various explants including shoot tips, lateral buds, and inflorescence tissues [2–4]. Several different methods for in vitro propagation of date palm have been developed, but all of these protocols have used semisolid gelled media as a component


Jameel M. Al-Khayri et al. (eds.), *Date Palm Biotechnology Protocols Volume 1: Tissue Culture Applications*,
Methods in Molecular Biology, vol. 1637, DOI 10.1007/978-1-4939-7156-5_21, © Springer Science+Business Media LLC 2017

of the system. However, these methods (using only the semisolid gels) are labor intensive and slow, thus increasing production costs [5].

The temporary immersion system (TIS), also known as a RITA bioreactor, is an ideal method for scaling up the micropropagation process. The Plantform bioreactor is a recently developed TIS system. It has higher plant headspace, uses low-pressure fish tank air pumps and digital timers, and allows active aeration to cultures. The Plantform bioreactor can be sterilized with a medium using either an autoclave or a kitchen microwave. The sterile filters are connected to the body afterward, and two timers and two air pumps are used to regulate gas exchange and immersion cycles via silicon tubes. Its modularity and shape allow space savings [6]. This is especially true in the case of somatic embryogenesis, as the bioreactors offer large containers, good aeration for the cultures, and a high degree of control over other culture conditions such as undesirable ethylene buildup and desirable increases in carbon dioxide concentrations [7]. Ibraheem et al. [8] used the TIS for scaling up the production of date palm somatic embryos.

This chapter describes a novel date palm micropropagation (somatic embryogenesis and organogenesis) mixed system (semisolid and liquid medium) using the Plantform bioreactor. Currently, this system is used in Fadak Date Palm Production Laboratory, Basra, Iraq. This mixed system achieved reduction in costs and reduced the time frame for plantlet production to less than 3 years.

## 2  Materials

**2.1  Plant Materials and Sterilization**

Explant source: Apical shoot tips of date palms taken from 3-year-old offshoots of cultivar Barhee.

**2.2  Culture Medium**

1. Basal culture medium: Murashige and Skoog (MS) salts [9] (Table 1).

2. Hormone stock solutions: Auxins including 2,4-dichloro-phenoxyacetic acid (2,4-D, 1 mg/mL), naphthaleneacetic acid (NAA, 1 mg/mL), β-naphthoxyacetic acid (NOA, 1 mg/mL), indole-3-acetic acid (IAA, 1 mg/mL), and indole-3-butyric acid (IBA, 1 mg/mL). Cytokinins including benzylaminopurine (BAP, 1 mg/mL), kinetin (KN, 1 mg/mL), 6-($\gamma,\gamma$-dimethylallylamino) purine (2iP, 1 mg/mL), and thidiazuron (TDZ, 1 mg/mL).

3. Solutions to adjust pH: 0.1 M and 1 M NaOH and 0.1 M and 1 M HCl.

4. Other chemicals: Polyvinylpyrrolidone (PVP, MW 10,000) and activated charcoal (AC).

**Table 1**
Composition of modified MS medium [9] used for date palm tissue. Hormones, agar, PVP, and activated charcoal are added according to the culture stage as shown in Tables 2 and 3

| Composition | Concentration in the stock solution (mg/L) | Concentration in MS culture medium (mg/L) |
|---|---|---|
| *Stock I: Macronutrients (10× stock) use 100 mL to prepare 1 L of medium* | | |
| $NH_4NO_3$ | 16,500 | 1650 |
| $KNO_3$ | 19,000 | 1900 |
| $CaCl_2 \cdot 2H_2O$ | 4400 | 440 |
| $MgSO_4 \cdot 2H_2O$ | 3700 | 370 |
| $KH_2PO_4$ | 1700 | 170 |
| $NaH_2PO_4 \cdot H_2O$ | 1700 | 170 |
| *Stock II: Micronutrients (100× stock) use 10 mL to prepare 1 L of medium* | | |
| KI | 83 | 0.83 |
| $H_3BO_3$ | 620 | 6.2 |
| $MnSO_4 \cdot 2H_2O$ | 2230 | 22.3 |
| $ZnSO_4 \cdot 7H_2O$ | 860 | 8.6 |
| $Na_2.MoO_4 \cdot 2H_2O$ | 25 | 0.25 |
| $CuSO_4 \cdot 5H_2O$ | 2.5 | 0.025 |
| $CoCl_2 \cdot 6H_2O$ | 2.5 | 0.025 |
| *Stock III: Iron source* | | |
| Fe EDTA Na salt | Added fresh | 40 |
| Myoinositol | Added fresh | 100 |
| Ca-pantothenate (20 mg/L) | Added fresh | *20* |
| *Stock IV: Vitamins (200× stock) use 5 mL to prepare 1 L of medium* | | |
| Nicotinic acid | 200 | 1 |
| Pyridoxine·HCl | 200 | 1 |
| Thiamine·HCl | 200 | 1 |
| Glycine | 400 | 2 |
| Biotin | 200 | 1 |
| *Carbon source* | | |
| Sucrose | – | 30,000 |
| *Antioxidants* | | |
| Glutamine | – | 200 |
| Other additives | | |
| Sodium dihydrogen phosphate dihydrate | Added fresh | 187.5 |
| Calcium nitrate | Added fresh | 1500 |

1. Callus induction medium (CIM): Modified MS medium enriched with 100 mg/L 2,4-D medium, 3 mg/L 2iP, sucrose 30 g/L, and 2 g/L activated charcoal.

2. Callus proliferation and maintenance medium (CPM): Modified MS medium supplemented with 30 mg/L NAA, 30 mg/L sucrose, and 2 g/L activated charcoal.

3. Somatic embryo germination medium (SEM1): Modified MS medium fortified with 100 mg/L myoinositol, 0.4 mg/L thiamine-HCL 1 mg/L, biotin, 0.5 mg/L pyridoxine, 0.5 mg/L nicotinic acid, 200 mg/L glutamine, 500 mg/L PVP, and 30 g/L sucrose.

4. Scaling-up medium (SUM): Modified MS liquid medium supplemented with 1.5 mg/L NOA, 1 mg/L NAA, 1 mg/IAA L, and 1 mg/L 2iP.

5. Direct somatic embryogenesis induction medium (DSIM): Modified MS medium supplemented with 50 mg/L NAA, 5 mg/L 2,4-D, and 3 mg/L 2iP; or 50 mg/L NAA, 10 mg/L 2,4-D, and 3 mg/L 2iP; or 50 mg/L NAA, 5 mg/L 2,4-D, and 0.1 mg/L TDZ; or 50 mg/L NAA, 10 mg/L 2,4-D, and 0.1 mg/L TDZ.

6. Somatic embryo germination (SEM2): Modified MS medium fortified with 0.1 mg/L NAA, 4 mg/L BAP, 4 mg/L KN, and 2 g/L activated charcoal.

7. Large-scale embryogenesis medium (LEM): Modified MS medium supplemented with 0.1 mg/L BAP, 0.1 mg/L 2iP, and 0.1 mg/L KN and enriched with a higher level of sucrose (60–90 g/L).

8. Direct organogenesis (DOM): Modified MS medium containing 6 mg/L NAA, 3 mg/L NOA, 0.1 mg/L BAP, 0.2 mg/L 2iP, 500 mg/L PVP, and 6 g/L agar. Sucrose concentration must increase to 50 g/L.

9. Bud growth and proliferation medium (BGPM): Modified MS medium with 0.1 mg/L BAP, 0.1 mg/L Kinetin, and 0.1 mg/L 2iP. Sucrose concentration must increase to 50 g/L.

10. Elongation medium (EM): Modified MS medium supplemented with 80 mg/L adenine sulfate and 1.5 g/L $Ca(NO_3)_2 \cdot 4H_2O$ and 1 mg/L each of NAA, 2iP, BAP, and KN as well as 0.5 g/L activated charcoal and 0.5 g/L PVP (*see* **Note 2**). Sucrose increased to 80 g/L.

11. Rooting medium (RM1): Modified MS solid medium supplemented with 1 mg/L NAA and 40 g/L sucrose.

12. Rooting medium (RM2): Modified MS liquid medium containing 0.01 mg/L IBA and 40 g/L sucrose.

13. Pre-weaning medium (PWM): Sugar-free MS liquid medium containing 0.01 mg/L IBA and 1 mg/L KN.

**Table 2**

**Different culture stages for somatic embryogenesis and the corresponding hormonal, PVP, and activated charcoal additives supplemented to the MS medium**

| Culture stage (medium code) | Plant growth regulators, agar, PVP, and activated charcoal additives | | | | | | | | | | |
| --- | --- | --- | --- | --- | --- | --- | --- | --- | --- | --- | --- |
| | 2,4-D, mg/L | 2iP, mg/L | NAA, mg/L | NOA, mg/L | IAA, mg/L | BAP, mg/L | Kinetin, mg/L | Adenine sulfate, mg/L | Agar, g/L | Activated charcoal, g/L | PVP, mg/L |
| Callus induction medium (CIM) | 100 | 3 | – | – | – | – | – | 40 | 7 | 2 | – |
| Callus proliferation and maintenance medium (CPM) | – | 3 | 30 | – | – | – | – | 40 | 7 | 1 | – |
| Somatic embryo germination medium (SEM1) | – | – | – | – | – | – | – | 40 | 7 | – | – |
| Scaling-up medium (SUM), liquid medium | – | 1 | 1 | 1.5 | 1 | – | – | 40 | – | | 500 |
| Direct somatic embryogenesis induction medium (DSIM) | 10 | 3 | 50 | – | – | – | – | 40 | 7 | 2 | – |
| Somatic embryo germination (SEM2) | – | – | 0.1 | – | – | 4 | 4 | 80 | 7 | 2 | – |
| Large-scale embryogenesis medium (LEM), liquid medium | – | 0.1 | – | – | – | 0.1 | 0.1 | 40 | – | – | 500 |

**Table 3**
**Different culture stages for organogenesis and the corresponding hormonal, PVP, and activated charcoal additives supplemented to the MS medium**

| Culture Stage | Plant growth regulators, agar, PVP, and activated charcoal additives | | | | | | | | | | | |
|---|---|---|---|---|---|---|---|---|---|---|---|---|
| | 2,4-D, mg/L | 2iP, mg/L | NAA, mg/L | NOA, mg/L | IAA, mg/L | BAP, mg/L | IBA, mg/L | Kinetin, mg/L | Adenine sulfate, mg/L | Agar, g/L | Activated charcoal, mg/L | PVP, mg/L |
| Callus induction medium (CIM) direct organogenesis medium (initiation stage) (DOM) | – | 0.2 | 6 | 3 | 1 | 0.1 | – | – | 40 | 7 | – | 500 |
| Buds growth and proliferation medium (BGPM) | – | 0.1 | – | – | – | 0.1 | – | 0.1 | 80 | 7 | – | 500 |
| Elongation medium (EM) Liquid medium | – | 1 | 1 | – | – | 1 | – | 1 | 80 | – | 500 | 500 |
| Rooting medium (RM1) | – | – | 1 | – | – | – | – | – | 80 | 7 | – | 500 |
| Rooting medium (RM2) Liquid medium | – | – | – | – | – | – | 0.01 | – | 80 | – | – | 500 |
| Pre-weaning medium (PWM) Sugar-free medium Liquid medium | – | – | – | – | – | – | 0.01 | 1 | 80 | – | – | 500 |

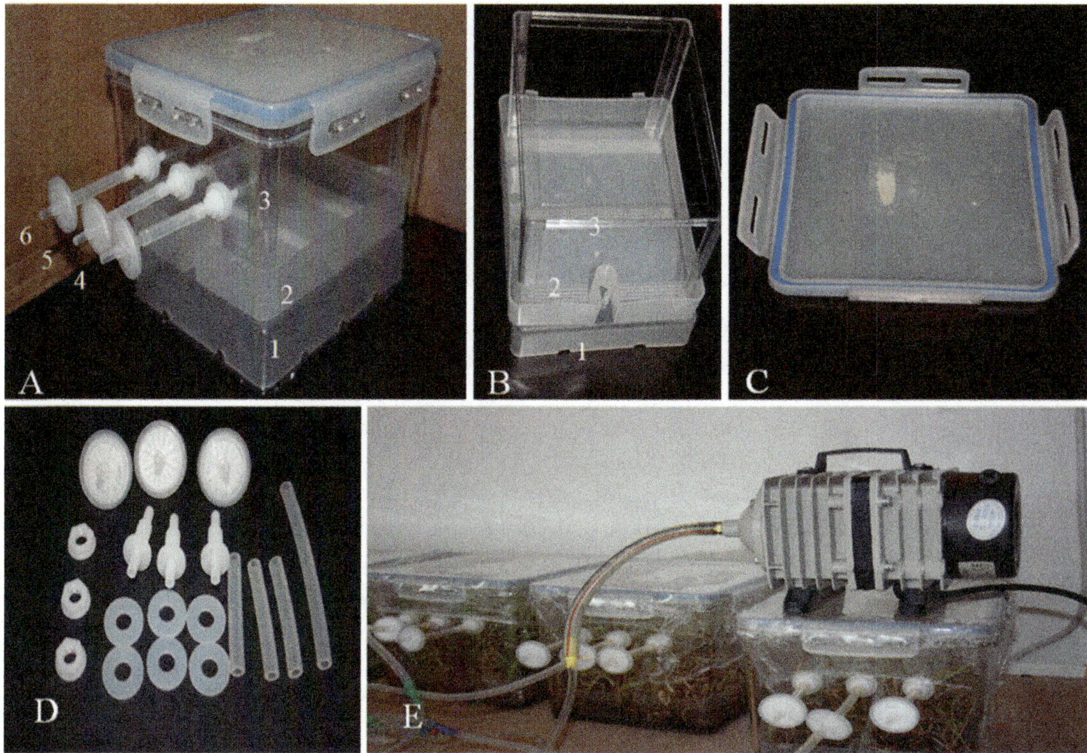

**Fig. 1** Bioreactor and its different parts numbered from *1* to *6*. (**a**) Outer container with three inlets/outlets for gas exchange (*4–6*); (*5*) shows the middle filter connected to a plastic tube on the inner chamber. (**b**) (*1*) Inner chamber with connection to the middle filter, (*2*) basket with three rows of small holes, (*3*) frame with four legs. (**c**) Lid with four flaps and an inner silicon ring. (**d**) Filters, plastic tubes, nuts, clamps, and silicon rings to be connected to the three inlets/outlets on the outer container. (**e**) Fish tank pump

*2.4 Equipment*

1. Tissue culture jars: Standard G7, 250 mL plastic plant tissue culture container.
2. Plantform bioreactor: *see* Fig. 1.

## 3  Methods

*3.1 Explant Preparation and Sterilization*

1. Separate healthy offshoots weighing 5–6 kg of selected cultivar Barhee from field-grown trees (Fig. 2a).
2. Remove offshoot leaves one by one from the outer ring toward the center (Fig. 2b).
3. When the soft white tissue area is reached, this part should be handled carefully to avoid breaking the shoot tip, because it is located deep in the soft meristematic area. Remove the outer leaves and expose the shoot tip region, about 3–4 cm in width and 6–8 cm in length (Fig. 2c).

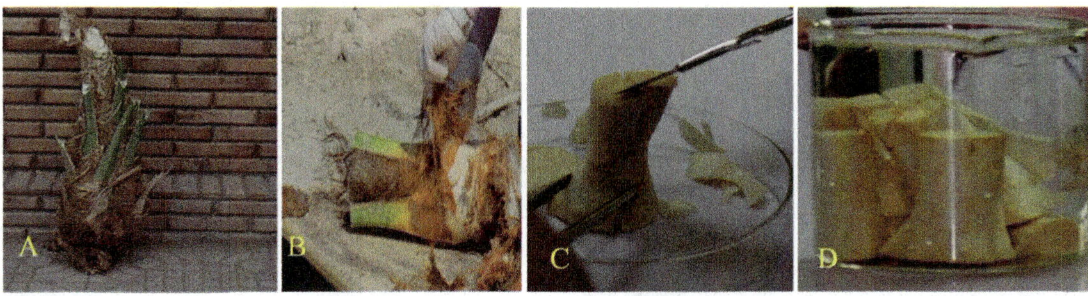

**Fig. 2** Explant preparation. (**a**) Excised offshoot weighing 5–6 kg ready for dissection. (**b**) Removing of offshoot leaves. (**c**) Offshoot core or heart. (**d**) Shoot tip dipped in chilled antioxidant solution

4. Transfer the shoot tips with some of the soft white tissues attached to sterilized beakers containing chilled, sterilized antioxidant solution prepared with deionized water supplemented with 150 mg/L citric acid and 100 mg/L ascorbic acid to reduce browning of explant tissues (Fig. 2d).

5. Immerse explants in 0.01% mercuric chloride solution for 80 min with some swirled or inverted by hand every 5 min.

6. In a laminar flow hood, rinse the explants five times with sterilized distilled water.

*3.2 Indirect Somatic Embryogenesis*

1. For initial callus induction, culture shoot tip explants on CIM solid medium and incubate the cultures in darkness at 27 °C ± 2 for 3 months (*see* **Notes 1–3**).

2. Subculture responsive explants in CPM at 4- to 6-month interval to get sufficient embryonic callus (*see* **Note 4**).

3. For partial desiccation treatment, transfer 500 mg embryogenic callus to sterile bioreactor container containing 550 mL medium, stop immersion cycle, and increase aeration cycle to 3 min/h for 15 days (*see* **Notes 5**).

4. Transfer desiccated callus to SEM1 medium. Keep cultures in light, 100 μmol/m$^2$/s and 16 h photoperiod for 8 weeks (*see* **Notes 5** and **6**).

5. Transfer embryogenic callus directly without desiccation to the plant basket of the Plantform bioreactor for the scaling-up of somatic embryogenesis. Use liquid SUM medium (550 mL) at an immersion frequency of 3 min every 8 h under the fluorescent lights (100 μmol/m$^2$/s, 16 h photoperiod) (*see* **Note 7**).

6. To increase the numbers of somatic embryos, shift the germinated embryos obtained in **step 4** to Plantform bioreactors containing LEM liquid medium (*see* **Note 10**).

*3.3 Direct Somatic Embryogenesis*

1. To induce direct somatic embryogenesis without intervening the callus, culture shoot tip explants on DSIM media for 3–6 months. Cultures should be incubated in the dark at

$27 \pm 2$ °C. Re-culture the explants without cutting them up to the same medium at 8-week intervals (*see* **Note 8**).

2. Transfer the somatic embryos obtained in **step 1** to SEM2 medium (*see* **Note 9**).

3. To increase the numbers of somatic embryos, shift the germinated embryos to Plantform bioreactors containing LEM liquid medium (*see* **Note 10**).

*3.4 Direct Organogenesis*

1. Culture shoot tip segments on DOM medium, and regularly subculture to fresh medium at 8-week intervals up to 9–10 months. Discard non-responsive cultures and maintain cultures in the dark at $27$ °C $\pm 2$ (*see* **Note 11**).

2. Subculture small clumps with up to four initial buds directly onto BGPM (Table 2); divide larger clumps with minimal wounding before subculturing. Incubate cultures under light intensity (7 $\mu$mol/m$^2$/s) for 16 h photoperiod at $27$ °C $\pm 2$ (*see* **Note 12**).

3. Transfer actively growing buds to fresh BGPM at 6–8 week intervals until sufficient numbers (30–40) exist for Plantform bioreactor use (*see* **Note 12**).

4. For obtaining elongated shoots in a short period, transfer cultures from **steps 2** and **3** to Plantform bioreactor plant basket, and using EM medium, incubate in the Plantform bioreactor at $27 \pm 2$ °C, with a light intensity of 7 $\mu$mol/m$^2$/s and with a photoperiod of 16 h. Immersion cycle subculturing of clumps is for 3 min at 8-h intervals (*see* **Note 13**).

5. Repeat **step 4** using a modified Plantform bioreactor by removing the baskets, inner chamber, and frame with four legs and aerate the medium through the polyethylene tube.

6. Alternatively use the stainless steel meshes over the basket to discharge the shoots very easily (*see* **Notes 14–19**).

7. Transfer the separated shoots obtained from **steps 4** and **5** to RM1 solid medium (*see* **Note 20**).

8. Transfer the unrooted shoots from **step 6** to RM2 liquid medium (*see* **Note 20**).

9. Rooting and shoot development: Subculture individual rooted shoots into liquid PWM either using tall glass vessels or the Plantform bioreactor (*see* **Note 21**).

## 4  Notes

1. Induction and continuous proliferation of embryogenic callus requires 2,4-D or NAA and 2iP, BAP, and kinetin. The issues of low embryo numbers per culture, synchrony of embryos, and

limited maturation and germination frequencies of somatic embryos are major hindrances that increase production costs and add to planning difficulties [10, 11].

2. Callus is induced on shoot tip segments after 1–2 months of culture on CIM medium.

3. Embryogenic callus is induced after 4–6 months by transferring the initial callus to CPM medium.

4. Embryogenic callus is proliferated and maintained on the same medium CPM.

5. Embryogenic callus subjected to desiccation in a sterile bioreactor for 15 days increased somatic embryo formation and germination by transferring to SEM1 medium (Fig. 3a, b) as compared with control. However, the number of somatic embryos decreases with only 5 days of desiccation.

6. The somatic embryogenesis process and plant regeneration are influenced by a number of factors such as genotype, medium composition (auxins, cytokinins, sugar, amino acids, abscisic acid, and retardant), and physical status changes caused by solidifying agents in the medium and desiccation. Partial desiccation has been reported to promote somatic embryo differentiation and development in many plants including date palm [12, 13].

7. Transferring undesiccated embryogenic callus to the Plantform bioreactor containing 550 mL sterilized (SUM) liquid medium (Table 1) at an immersion frequency of 3 min every 8 h increases the formation of somatic embryos in 2 months as compared to culture of this callus to solid medium (Fig. 4).

8. Some shoot tip explants produced direct somatic embryos without intervening callus phase after 16 weeks of culturing

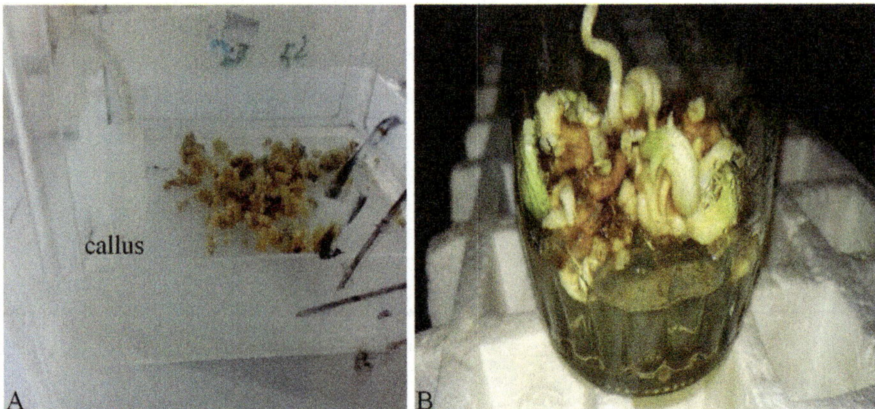


**Fig. 3** Using Plantform bioreactor for (**a**) callus desiccation and (**b**) maturation and germination of somatic embryos after subjecting embryogenic callus to desiccation for 15 days in Plantform bioreactor

**Fig. 4** Using a Plantform bioreactor with temporary immersion system for the induction of somatic embryos from embryogenic callus

**Fig. 5** Direct somatic embryogenesis

(Fig. 5) at rates of 15, 25, and 15 and 25% only on DSIM media (Table 1).

9. The efficiency of growth regulator combinations on somatic embryo development is determined based on the somatic embryo germination rate, shoots, and roots (numbers and length). High somatic embryo germination, increasing root and shoot numbers and shoot length are obtained with the treatment of SEM2 (88.88%) (2.57 and 3.25, 3.87 cm and 3.53 cm, respectively).

10. Germinated somatic embryos obtained by direct or indirect somatic embryogenesis mode (Subheadings 3.1 and 3.2) mentioned above are transferred separately to Plantform bioreactors containing LEM (Table 1) liquid medium with a high level of sucrose (90 g/L) for further production of somatic embryos. Secondary embryogenesis is induced within 8 weeks (Fig. 6).

11. Initiation stage: Buds initiated on DOM medium (Table 2) after 8–10 months (Fig. 7).

12. Bud growth stage: Active and slower growing buds are transferred to BGPM (Table 2) for further proliferation. When slower growing buds do not respond, re-culture on the same medium (1–2 times).

**Fig. 6** The number of somatic embryos increased by using MS medium supplemented with 0.1 mg/L benzylaminopurine (BAP), 0.1 mg/L 2iP, and 0.1 mg/L kinetin (KN) with 60 g/L of sucrose (LEM)

**Fig. 7** Direct organogenesis occurred after 8–10 months on DOM medium

13. Shoot elongation and individualization stage: Use the Plantform bioreactor with a fine mesh (200 holes/in.) in the basket. Fine, delicate shoots should be transferred to (EM) liquid medium. To avoid vitrification or hyperhydricity problems, add high concentration of $Ca(NO_3)_2.4H_2O$ (1–2.5 g/L) in the culture medium (Table 2). Additionally, activated charcoal and PVP must be included in the medium, 0.5 g/L each to avoid browning. Grow shoots under light 7 μmol/m$^2$/s with a photoperiod of 16 h. Low light aids in shoot individualization and elongation; whereas, higher light intensity does not give good results. Shoots are elongated within 8 weeks.

14. Plantform bioreactor is efficient for both bud and shoot elongation and root development. The main problem is that roots penetrate the small holes of the basket and create difficulties in removing shoots from the basket container (Fig. 8a).

15. Modify Plantform bioreactor by removing the baskets, inner chamber, or frame with four legs and aerate the medium through a polyethylene tube.

16. Alternatively, use the stainless steel meshes over the basket to prevent roots from penetrating the basket holes thus making shoot discharge very easy (Fig. 8b, c).

17. By using both methods, high-quality and easy-to-handle plants are produced, and keep oxygenation of the medium and gas exchange under control through stationary medium. Additionally, LEM with higher levels of sucrose (over 50 g/L) increases shoot length and overcomes slow growth problems. The harvesting and detaching of the shoots from the bioreactor become very simple.

18. Shoot elongation and individualization are important in the in vitro stage of organogenesis of date palm to permit

**Fig. 8** In vitro date palm plantlets ready for soil transfer and acclimatization. Roots developed from shoots at elongation phase penetrate the small holes of basket in the TIS bioreactor (**a**). Vessel with successfully developed shoots in stationary liquid medium (**b, c**)

conversion of bud shoots to plantlets. This stage depends mainly on the genotype and media composition [14].

19. Detachment of individual shoots from clusters in the multiplication phase is difficult, and attached plantlets need more than 4–6 months in culture to be ready for separation. Liquid medium improves shoot elongation and increases survival rate of date palm plantlets in the greenhouse [15, 16].

20. Rooting stage: Roots are induced in elongated shoots, 15 cm in length, after 8–12 weeks of culturing on (RM1) and (RM2). Smaller shoots less than 10 cm fail to root.

21. The pre-weaning stage is important to increase the percentage of plantlet survival. Main root and branch root numbers increase when cultured on PWM for 4 weeks. Transfer plantlets from this medium to soil.

## References


1. El-Juhany L (2010) Degradation of date palm trees and date production in Arab countries: causes and potential rehabilitation. Aust J Basic Appl Sci 4(8):3998–4010

2. Mazri MA, Meziani R (2015) Micropropagation of date palm – a review. Cell Dev Biol 4 (3):1–5

3. Abahmane L (2013) Recent achievements in date palm (*Phoenix dactylifera* L.) micropropagation from inflorescence tissues. Emir J Food Agric 25(11):863–874

4. Jatoi MA, Abul-Soad A, Markhand GS, Solangi N (2015) Establishment of an efficient protocol for micropropagation of some Pakistani cultivars of date palm (*Phoenix dactylifera* L.) using novel inflorescence explants. Pak J Bot 47(5):1921–1927

5. Welandar M, Sayegh A (2012.) A temporary immersion bioreactor method and relative product patent. A WO patent WO 2012044239 A1 5 April 2012. https:// patentscope.wipo.int/search/en/detail.jsf? docId=WO2012044239&redirectedID=true

6. Kärkönen A (2001) Plant tissue culture as models for tree physiology: somatic embryogenesis of *Tilia cordata* and lignin biosynthesis in *Picea abies* suspension cultures as case studies. University of Helsinki, Helsinki

7. Berthouly M, Etienne H (2005) Temporary immersion system: a new concept for liquid medium in mass propagation. In: Hvoslef-Eide AK, Walet P (eds) Liquid culture systems for in vitro plant propagation. Springer, Dordrecht, pp 165–185

8. Ibraheem Y, Pinker I, Böhme M (2013) A comparative study between solid and liquid cultures relative to callus growth and somatic embryo formation in date palm (*Phoenix dactylifera* L.) cv. Zaghlool. Emir J Food Agric 25 (11):883–898

9. Murashige T, Skoog F (1962) A revised medium for rapid growth and bioassay with Tobacco tissue cultures. Phys Plant 15:473–497

10. Djibril S, Frédérique AB, Léopold IDD, Badara G, Abdourahman D, Maurice S et al (2012) Influence of growth regulators on callogenesis and somatic embryo development in date palm (*Phoenix dactylifera* L.) Sci World J 2012:1–8

11. Othmani A, Mzid R, Bayoudh C, Trifi M, Drira N (2011) Bioreactors and automation in date palm micropropagation. In: Jain SM, Al-Khayri JM, Johnson DV (eds) Date palm biotechnology. Springer, Dordrecht, pp 119–137

12. Othmani A, Bayoudh C, Drira N, Trifi M (2009) *In vitro* cloning of date palm *Phoenix dactylifera* L., cv. Deglet Bey by using embryogenic suspension and temporary immersion bioreactor (TIB). Biotech Biotchcal Equip 23 (2):1181–1188

13. Ibrahim AI, Mona M, Rania AT (2012) Partial desiccation improves plant regeneration of *in vitro* date palm cultures. J Agric Res 1 (6):208–214

14. Mazri MA, Meziani R (2012) An improved method for micropropagation and regeneration

of date palm (*Phoenix dactylifera* L.) J Plant Biochem Biotechnol 22(2):178–184

15. Mazri MA (2014) Effect of liquid medium and *in vitro* pre-acclimatization stage on shoot elongation and acclimatization of date palm (*Phoenix dactylifera* L.) cv. Najda. J Ornam Hort Plant 2(4):225–231

16. Okere AU, Odofin WT, Aladele SE, Fajimi O, Adetunji OA, Gidado RM, Solomon, BO (2010) Cultivation of *Phoenix dactylifera* L. (date palm) for combating desertification and enhanced livelihood: NACGRAB R and D Focus. Acta Hortic 882:81–87

# Part V

# Genetic Transformation

# Chapter 22

# Genetic Transformation of Date Palm Via Microprojectile Bombardment

Mousa Mousavi, Amir Mousavi, Ali A. Habashi, Kazem Arzani, Bahareh Dehsara, and Mohsen Brajeh

## Abstract

Efficient protocols for date palm embryogenic callus and somatic embryo transformation with *uid*A gene are described in this chapter. The embryogenic callus transformation procedure is 1.6 μm gold particle size coated with 2.5 μg DNA (p*Act1*-D plasmid), 1100 psi helium pressure, 9 cm target distance, 26 inHg vacuum pressure, 3 mm distance between the rupture disk and macrocarrier, and osmotic pretreatment with 0.4 M mannitol followed by 60 min air desiccation. The somatic embryo transformation procedure is 0.6 μm gold particle size coated with 2.5 μg DNA (p*Act1*-D plasmid), 1350 psi helium pressure, 6 cm target distance, 28 inHg vacuum pressure, 3 mm distance between the rupture disk and macrocarrier, and osmotic pretreatment with 0.4 M mannitol followed by 60 min air desiccation. Protocols for analysis of the transgenic plantlets have also been described.

**Key words** Particle bombardment, Particle delivery, Transformation, Transgenic

## 1 Introduction

Date palm is well adapted for cultivation in the subtropical and tropical dry hot regions where it provides nutritional fruits and holds important socioeconomic, ecological, and traditional values [1]. There are limited available date palm genetic resources for improvement through traditional breeding [2, 3]. In addition, the demand for date palm fruit increases day by day [4]. Some of the limitations of tree improvement such as long life cycle, long juvenility phase [5], and high levels of heterozygosity [6] can be resolved by genetic transformation. Biotechnology can contribute positively to enhance improvement of date palm like gaining resistance to biotic and abiotic stress agents, modifying growth habits, and increasing quality of the fruits [7]. Genetic manipulation has paved the way for a gene transfer in a single generation [8], which otherwise cannot be done with classical breeding. Therefore, the development of genetic transformation methods is very important

Jameel M. Al-Khayri et al. (eds.), *Date Palm Biotechnology Protocols Volume 1: Tissue Culture Applications*, Methods in Molecular Biology, vol. 1637, DOI 10.1007/978-1-4939-7156-5_22, © Springer Science+Business Media LLC 2017

to improve date palm with desirable traits. Among the several methods reported, biolistic has proven most effective for direct gene transfer in various plants [9, 10]. The *uid*A (GUS) gene is a common reporter gene that can be used to set conditions before the final genetic transformation.

The physical and biological parameters for β-glucuronidase (GUS) and green fluorescent protein (GFP) gene transfer were successfully accomplished in oil palm [11, 12] and in banana [13–15].

In this chapter, we describe optimal conditions for gene delivery to date palm tissue with high expression rates. All the procedures described are based on our work [16–18].

## 2 Materials

### 2.1 Plant Material

Healthy 1- to 2-year-old date palm offshoots of Estamaran cultivar.

### 2.2 Medium Requirements for Tissue Culture and Particle Bombardment

1. Callus induction medium: MS medium containing salts and vitamins [19] (Table 1) supplemented with 100 mg/L 2,4-D, 3 mg/L 2iP, 40 mg/L adenine, 30 g/L sucrose, and 3 g/L activated charcoal (Table 2).

2. Callus proliferation medium: MS medium (Table 1) containing salts and vitamins amended with 10 mg/L 2,4-D, 3 mg/L 2iP, 40 mg/L adenine, and 30 g/L sucrose and 3 g/L activated charcoal (Table 2).

3. Somatic embryo induction medium: MS medium (Table 1) supplemented with 10 mg/L 2,4-D, 3 mg/L 2iP, 40 mg/L adenine, and 20 mg/L glutamine (Table 2).

4. Root production medium: MS medium (Table 1) containing 1 mg/L NAA, 3 g/L activated charcoal, and 30 g/L sucrose (Table 2).

5. Osmoticum medium before tissue bombardment: MS medium (Table 1) containing 10 mg/L 2,4-D, 3 mg/L 2iP, 40 mg/L adenine, and 0.4 M mannitol.

### 2.3 Plasmid Type and DNA Coating Mixture

1. Plasmid p*Act1*-D [20] harboring the *uid*A gene with the 5′ region of the rice actin1 promoter (*see* **Note 1**).

2. Mixture for coating gold particles with plasmid DNA: 50 μL particle solution (prepared in 50% glycerol), 10 μL DNA, 25 μL 0.1 M spermidine, and 50 μL 2.5 M $CaCl_2$.

### 2.4 DNA Extraction from Plant Tissue

1. Extraction buffer: Contains 50 mM Tris-HCl (pH 8), 300 mM mannitol, 3 mM EDTA (pH 8), 1% (w/v) polyethylene glycol (PEG, MW 6000), and 0.1% (v/v) bovine serum albumin (BSA).

**Table 1**
**Components of MS basal culture medium [19]**

| Medium composition | Stock concentration (mg/L) | Final concentration in culture medium (mg/L) |
|---|---|---|
| *Stock I: Major inorganic nutrients (20× stock) use 50 mL to prepare 1 L of medium* | | |
| $NH_4NO_3$ | 33,000 | 1650 |
| $KNO_3$ | 38,000 | 1900 |
| $CaCl_2 \cdot 2H_2O$ | 8800 | 440 |
| $MgSO_4 \cdot 2H_2O$ | 7400 | 370 |
| $KH_2PO_4$ | 3400 | 170 |
| $NaH_2PO_4 \cdot H_2O$ | 3400 | 170 |
| *Stock II: Minor inorganic nutrients (200× stock) use 5 mL to prepare 1 L of medium* | | |
| KI | 166 | 0.83 |
| $H_3BO_3$ | 1240 | 6.2 |
| $MnSO_4 \cdot 2H_2O$ | 4460 | 22.3 |
| $ZnSO_4 \cdot 7H_2O$ | 1720 | 8.6 |
| $Na_2.MoO_4 \cdot 2H_2O$ | 50 | 0.25 |
| $CuSO_4 \cdot 5H_2O$ | 5 | 0.025 |
| $CoCl_2 \cdot 6H_2O$ | 5 | 0.025 |
| *Stock III: Iron source (200× stock) use 5 mL to prepare 1 L of medium* | | |
| $FeSO_4 \cdot 7H_2O$ | 5560 | 27.8 |
| $Na_2EDTA \cdot 2H_2O$ | 7460 | 37.3 |
| *Stock IV: Vitamins (200× stock) use 5 mL to prepare 1 L of medium* | | |
| *myo*-Inositol | 25,000 | 125 |
| Nicotinic acid | 200 | 1 |
| Pyridoxine·HCl | 200 | 1 |
| Thiamine·HCl | 200 | 1 |
| Glycine | 400 | 2 |
| Biotin | 200 | 1 |
| Adenine sulfate | 8000 | 40 |
| Pantothenic acid | 200 | 1 |
| *Antioxidants* | | |
| Citric acid | – | 100 |
| Ascorbic acid | – | 100 |
| *Gelling agent* | | |
| Agar | – | 7500 |

**Table 2**
**Summary of various media additives required for date palm in vitro regeneration stages**

| Medium | Components | | | | | | |
|---|---|---|---|---|---|---|---|
| | 2, 4-D (mg/L) | 2iP (mg/L) | Adenine (mg/L) | Sucrose (g/L) | Charcoal (g/L) | Glutamine (mg/L) | NAA (mg/L) |
| Callus induction | 100 | 3 | 40 | 30 | 3 | – | – |
| Callus proliferation | 10 | 3 | 40 | 30 | 3 | – | – |
| Somatic embryo | 10 | 3 | 40 | – | – | 20 | – |
| Root production | – | – | – | 30 | 3 | – | 1 |

2. Phenol-chloroform (1:1 v/v).

3. 20% sodium dodecyl sulfate (SDS).

4. 3 M sodium acetate (pH 8).

5. Isopropanol.

6. 70% ethanol.

7. RNase A (10 mg/L).

***2.5 Histochemical GUS Assay***

1. GUS assay staining solution (100 mL): 500 μL 1 M $NaH_2PO_4$, 10 mg chloramphenicol, 100 μL triton X-100, 2 mL methanol, and 88 mg X-gluc dissolved in DMSO (pH 8).

***2.6 PCR Analysis and Southern Blot Hybridization***

1. PCR reaction (25 μL): 1 μL template DNA, 0.5 μL of each primer, 0.5 μL dNTPs, 0.7 μL $MgCl_2$, 2.5 μL 10× buffer, 0.5 μL *Taq* polymerase, and 18.8 μL deionized water.

2. GUS primers:

   GUS+2: 5′-GGT GGT CAG TCC CTT ATG TTA CG-3′ and GUS−4: 5′-CCG GCA TAG TTA AAG AAA TCA TG-3′.

3. 6× loading buffer: 0.25% bromophenol blue, 0.25% xylene cyanol FF, 30% glycerol, 50 mM EDTA, and 0.4% sucrose.

4. 1× TBE buffer: 0.45 mM Tris-HCl, 0.45 mM boric acid, and 1 mM EDTA (pH 8).

5. Restriction enzymes: *Eco*RI and *Xba*I.

6. Agarose.

7. DNA ladder (1 kb).

8. Ethidium bromide (2 μg/mL).

| | |
|---|---|
| ***2.7  Equipment*** | 1. Glassware: Petri dish, Erlenmeyer flask, graduated cylinder, tissue culture jars, test tubes, Petri dishes (2.5 cm), beaker, mortar and pestle, plastic vials, falcons, syringe, and wash bottles. |

1. Glassware: Petri dish, Erlenmeyer flask, graduated cylinder, tissue culture jars, test tubes, Petri dishes (2.5 cm), beaker, mortar and pestle, plastic vials, falcons, syringe, and wash bottles.

2. Tools: Forceps, scalpel, scissor, spatula, and Bunsen burner.

3. Instruments: Balance, vortex, autoclave, laminar airflow cabinet, water bath, spin, centrifuge and micro centrifuge, thermal cycler, gel documentation system, electrophoresis power pack and tank, spectrophotometer, shaker incubator, biolistic apparatus (PDS-1000/He), biolistic optimization kit, binocular microscope, plant growth incubator, pH meter, magnetic stirrer, $-20\ ^{\circ}\mathrm{C}$ freezer and $-80\ ^{\circ}\mathrm{C}$ freezers, and UV illuminator.

# 3  Methods

***3.1  Establishment of Embryogenic Callus and Somatic Embryos***

1. Initiate embryogenic callus from offshoot shoot tip meristem explants (Estamaran cv.), 1- to 2-year-old, on callus induction medium. Incubate cultures in the dark at $26 \pm 1\ ^{\circ}\mathrm{C}$.

2. Transfer the explants to a fresh medium at every 4-week interval for 16–20 weeks.

3. For the proliferation of embryogenic callus, subculture it on callus proliferation medium at every 4-week interval.

4. Somatic embryos are produced after maintaining the embryogenic calli on somatic embryo induction medium for 30–45 days.

5. After transfer of the somatic embryos to glutamine-free medium (callus proliferation medium in Table 2), somatic embryos germinate and form shoots with tiny rootlets.

6. For root development, transfer shoots to root induction medium (Table 2), and incubate for 30–45 days at $26 \pm 1\ ^{\circ}\mathrm{C}$ and 16 h photoperiod ($40\ \mu\mathrm{mol/m^2/s}$).

***3.2  Washing of Microparticles***

1. Washing the particles: Weigh 9 mg gold particle and transfer to 1.5 mL tube, add 500 µL 70% ethanol, and vortex for 5 min.

2. Allow tube to stand for 15 min at room temperature and spin for 6 s.

3. Remove 70% ethanol, and add 500 µL sterile deionized water.

4. Vortex for 1 min, stand it for 1 min, and then spin for 6 s.

5. Remove water and again add 500 µL sterile deionized water.

6. Vortex for1 min, let it stand for 1 min, and then spin for 6 s.

7. Wash again in deionized water by repeating **steps 5** and **6**.

8. Remove water and add 120 μL glycerol 50%, vortex for 6 s and store particle solution at 4 °C until coating with DNA (*see* **Note 2**).

***3.3 DNA Coating of Microparticles***

1. Precipitate DNA on microcarriers according to the procedure given for the PDS-1000/He system by the supplier. While continuously vortexing, add particle solution, DNA, spermidine, and $CaCl_2$ in a tube (*see* Subheading 2.2), and vortex for 2 min.

2. Allow gold particles to settle for 4 min and then pelletize by spinning for 3 s.

3. Remove the supernatant, and wash the pellet twice with 140 μL of 70% and 100% ethanol, respectively.

4. Add 82 μL 100% ethanol and resuspend the pellet by vortexing.

5. Add 12.5 μL DNA-coated gold particles suspension load onto the center of a macrocarrier, and desiccate until ethanol completely evaporates and ready to use for bombardment.

***3.4 Preparation of Embryogenic Callus and Somatic Embryos for Bombardment***

1. Embryogenic callus: Arrange the embryogenic callus in the center of Petri dishes, 2 cm diameter (Fig. 1i) containing osmoticum medium (*see* Subheading 2.1). Incubate in the darkness for 24 h at 26 ± 1 °C.

2. Somatic embryos: Arrange the 15–20 somatic embryos in the center of Petri dishes (2.5 cm diameter) (Fig. 1j), containing osmoticum medium (*see* Subheading 2.1). Incubate in the dark for 24 h.

3. Partially desiccate the plant tissue (the embryogenic callus or the somatic embryos) by removal of the Petri dish cover in a laminar airflow hood for 60 min prior to bombardment.

***3.5 Disinfect Biolistic Apparatus before Bombardment***

1. Disinfect the main biolistic apparatus (PDS-1000/He) by spray ethanol 70% then wipe with cotton wool.

2. Disinfect all components of PDS-1000/He apparatus before bombardment via immersion in ethanol 70% for 15 min.

3. Remove excess ethanol with sterile filter papers inside the airflow hood.

***3.6 Bombardment Conditions and Procedures***

1. Embryogenic callus bombardment conditions: 1100 psi, 9 cm target distance (distance between target tissue and stopping screen), 1.6 μm gold particles coated with 2.5 μg plasmid DNA, 26 inHg vacuum pressure, 3 mm distance between the rupture disk and macrocarrier for one bombardment (Table 3) (*see* **Note 3**).

2. Somatic embryos bombardment conditions: 1350 psi, 6 cm target distance, 0.6 μm gold particles coated with 2.5 μg of

**Fig. 1** Stages in date palm gene transformation by particle bombardment. (**a**) 1- to 2-year-old offshoot. (**b**) Remove outer leaves. (**c, d**) Remove additional leaves. (**e**) Surface sterilization. (**f**) Establishment of callus induction culture. (**g**) Incubation cultures in darkness at 26 ± 1 °C. (**h**) Embryogenic callus formation. (**i, j**)

**Table 3**
**Summary of the bombardment conditions used for each target tissues**

| Target tissue | Bombardment parameters | | | | |
| --- | --- | --- | --- | --- | --- |
| | Rupture disk (psi) | Target distance (cm) | Chamber vacuum (inHg) | Rupture disk to macrocarrier distance (mm) | Gold particle size (μm) |
| Embryogenic calli | 1100 | 9 | 26 | 6 | 1.6 |
| Somatic embryos | 1350 | 6 | 28 | 3 | 0.6 |

plasmid DNA, 28 inHg vacuum pressure, 3 mm distance between the rupture disk and macrocarrier for one bombardment (Table 3) (*see* **Note 4**).

3. Turn on the PDS-1000/He system, open valve of the helium gas cylinder, and set the regulator on 200 psi more than required pressure.

4. Put the rupture disk in the disk rating cap, and place the macrocarrier holder with the macrocarrier into a micro-carrier launch assembly.

5. Insert the target shelf at the desired distance position (6 cm for somatic embryos or 9 cm for embryogenic callus), and transfer the Petri dish containing the target tissue on the shelf (*see* **Note 5**).

6. Close the chamber door and turn on vacuum pressure by keeping the key in VAC position (until vacuum inside attain 26 or 28 inHg).

7. Push HOLD key, push and keep FIRE key until bombardment takes place, then put the VAC key in VENT position.

***3.7 Transient GUS Expression Assay***

1. Incubate bombarded embryogenic calli and somatic embryos on the same plates at room temperature for 72–94 h before being assayed for GUS expression.

**Fig. 1** (continued) Arrangement of date palm embryogenic callus and somatic embryos before particle bombardment. (**k**) Partial desiccation. (**l**) PDS-1000/He system. (**m**) Transient GUS expression in bombarded embryogenic callus. (**n**) Transient GUS expression in bombarded somatic embryos. (**o**) Shoot production from bombarded somatic embryo. (**p**) Rooting. (**q**) GUS staining of transformed and no transformed plantlets. (**r**) PCR analysis of regenerated plantlets of date palm bombarded with p*Act1*-D: (*1*) 1 kb DNA molecular marker, (*2*) negative control, (*3*) DNA amplified GUS gene for callus genomic DNA, (*4*) DNA amplified GUS gene for somatic embryos genomic DNA

2. Prepare GUS assay staining solution (100 mL): Mix 500 μL 1 M $NaH_2PO_4$, 10 mg chloramphenicol, 100 μL triton X-100, 2 mL methanol, and 88 mg X-gluc dissolve in DMSO. Adjust solution to pH 8 [21].

3. Transfer the bombarded and control tissues to GUS assay staining solution for 24 h at 37 °C.

4. To assess transient GUS activity, record the number of blue spots (irrespective of size) in each tissue sample under a binocular microscope, and then take a photograph (Fig. 1m, n).

5. Use the same method to examine the plantlet leaves (Fig. 1q).

***3.8 DNA Extraction***

Extract total genomic DNA for PCR analysis from green leaves of each regenerated plantlet (Fig. 1o, p) according to the procedure described by Quenzar et al. [22] with some modifications. Prepare extraction buffer by mixing 50 mM Tris-HCl (pH 8), 300 mM mannitol, 3 mM EDTA (pH 8), 1% PEG 6000, and 0.1% BSA (*see* **Note 6**).

1. Collect 0.2 g plant tissue (embryogenic callus or leaf tissue from regenerated plantlets) and place into a prechilled cold mortar, add 800 μL extraction buffer, and crush with a pestle to a fine homogenate.

2. Transfer the homogenate into 2 mL micro tube and add 800 μL phenol-chloroform, 100 μL 20% SDS, and 80 μL 3 M sodium acetate, and then mix by inversion for 5 min at room temperature

3. Incubate at 50 °C for 5 min in water bath and shake several times.

4. Centrifuge in 8000 × $g$ for 10 min at room temperature.

5. Transfer 800 μL from supernatant to 2 mL tube, add 800 μL isopropanol, mix by inversion, and keep for 2 h on ice.

6. Centrifuge at 8000 × $g$ for 10 min at room temperature. Remove supernatant and wash the pellet in 70% ethanol

7. Resuspend the pellet in 50 μL sterile deionized water, add 1 μL RNase A (10 mg/L), and incubate at 37 °C for 30 min.

8. Add one equal volume 1:1 phenol-chloroform solution, and centrifuge at 8000 × $g$ for 10 min at room temperature.

9. Transfer the supernatant into 2 mL tube and add 1/10 equal volume 3 M sodium acetate and 2.5 equal volume 100% ethanol, and maintain at −20 °C for 20 min.

10. Centrifuge at 8000 × $g$ for 10 min at room temperature. Remove supernatant, and dry the pellet by inverting the tubes on filter papers.

11. Resuspend the DNA pellet in 50 μL sterile deionized water, and keep at −20 °C until use.

**3.9  PCR Analysis**

1. PCR amplification takes place in a 25 μL total reaction volume containing 1 μL template DNA, 0.5 μL of each primer, 0.5 μL dNTPs, 0.7 μL MgCl$_2$, 2.5 μL 10× buffer, 0.5 μL *Taq* polymerase, and 18.8 μL deionized water.

2. Amplification protocol: one cycle of denaturation at 95 °C for 4 min, followed by 30 cycles of (94 °C for 1 min, annealing at 55 °C for 1 min, extension at 72 °C for 1 min), and then final extension at 72 °C for 5 min.

3. Electrophoresis of the PCR products: aliquots (5 μL) of amplified products with 6× loading buffer are electrophoresed on 0.8% agarose gels in 1× TBE buffer. Electrophoresis is performed in 12 cm long gels and run at 80 V for 90 min. In each run, 4 μL 1 kb DNA ladder is used. The gels are stained in 2 μg/mL ethidium bromide for 30 min, destained in 1× TBE buffer for 30 min, and photographed under UV illumination.

4. Screen the transformed plantlets carrying the GUS gene by detection of the bands on agarose gel as expected molecular size equals 520 bp (Fig. 1r).

**3.10  Southern Blot Hybridization**

1. Digest 20 μg DNA with *Eco*RI or *Xba*I restriction enzyme at 37 °C.

2. Separate the DNA restriction fragments on 0.8% (w/v) agarose gel by electrophoresis. Briefly, aliquots (10 μL) of digested products with 6× loading buffer are electrophoresed on 1.2% agarose gels in 1× TBE buffer. Electrophoresis is performed in 12 cm long gels and run at 60 V for 120 min. Use 4 μL of 1 kb DNA ladder for fragment size comparison and 1 μg pBI221 DAN as the positive DNA control.

3. Blot to a positively charged nylon (Hybond N$^+$), and hybridize with a transgene-specific probe. The probe for *uidA* is a 560 fragment prepared by PCR from pBI221 using GUS−4 and GUS+2 primers.

4. Probe labeling is performed according to the instructions for the DIG Probe DNA Labeling and Detection kit I.

5. Prehybridization and hybridization are performed using a standard method [23]. Hybridizing bands corresponding to the *uidA* gene are monitored as follows:
   (a) Look for the hybridized band on the positive control lane.
   (b) Look for any hybridized bands at positions either the same level (means same length) or lower level (means shorter length) than the positive control.

## 4    Notes

1. Promoter CaMV35s is a constitutive promotor, which expresses well in many plants; however, our results show that this promotor does not have the capacity to transform a gene in different date palm tissue. Therefore, we suggest using *Act*1 promoter.

2. It is recommended to use freshly washed microparticles for DNA coating.

3. Despite the high gene transfer rate in embryogenic callus, plant regeneration ability is reduced due to bombardment shock stress.

4. Use somatic embryos smaller than 0.5 cm in length, which are less affected by bombardment pressure and regenerate well.

5. The gene delivery efficiency into different date palm tissues vary, and the least amount of transformation is observed in root and leaf, respectively. Somatic embryos and embryogenic callus are well suited for date palm genetic transformation.

6. The buffer components used for DNA extraction is similar to the method described by Quenzar et al. [22] except for β-mercaptoethanol.


## References

1. Jain SM (2012) Date palm biotechnology: current status and prospective – an overview. Emir J Food Agric 24(5):386–399

2. Jaradat AA (2015) Biodiversity, genetic diversity, and genetic resources of date palm. In: Al-Khayri JM, Jain SM, Johnson DV (eds) Date palm genetic resources and utilization, vol 1, Africa and the Americas. Springer, Dordrecht, pp 19–71

3. Mathew LS, Spannagl M, Al-Malki A, George B, Torres MF, Al-Dous EK, Mohamoud YA (2014) A first genetic map of date palm (*Phoenix dactylifera*) reveals long-range genome structure conservation in the palms. BMC Genomics 15:1–10

4. Abass MH (2013) Microbial contaminants of date palm (*Phoenix dactylifera* L) in Iraqi tissue culture laboratories. Emir J Food Agric 25:875–882

5. Chao CCT, Krueger RR (2007) The date palm (*Phoenix dactylifera* L.): overview of biology, uses, and cultivation. Hortscience 42(5):1077–1082

6. Gomez-Lim MA, Litz RE (2004) Genetic transformation of perennial tropical fruits. In Vitro Cell Dev Biol Plant 40(5):442–449

7. Laimer M, Mendoca D, Maghuly F, Marzban G, Leopold S, Khan M et al (2005) Biotechnology of temperate fruit trees and grapevines. Acta Biochim Pol 52:673–678

8. Sartoretto LM, Cid LPB, Brasileiro ACM (2002) Biolistic transformation of *Eucalyptus grandis* × *E. urophylla* callus. Funct Plant Biol 29:917–924

9. Altpeter F, Baisakh N, Beachy R, Bock R, Capell T, Christou P, Fauquet C (2005) Particle bombardment and the genetic enhancement of crops: myths and realities. Mol Breed 15:305–327

10. Matthews BF, Saunders JA, Gebhardt JS, Lin JJ, Koehler SM (1995) Plant cell electroporation and electrofusion. In: Nickoloff JA (ed) Reporter genes and transient assays for plants, vol 55. Humana, Totowa, NJ, pp 147–162

11. Parveez GKA, Chowdhury MKU, Saleh NM (1997) Physical parameters affecting transient gus gene expression in oil palm (*Elaeis guineensis* Jacq.) using the biolistic device. Ind Crop Prod 6:41–50

12. Majid NA, Parveez GKA (2007) Evaluation of green fluorescence protein (*gfp*) as a selectable marker for oil palm transformation via transient

expression. Asia-Pac J Mol Biol Biotechnol 15 (1):1–8

13. Sreeramanan S, Maziah M, Abdullah MP, Rosli NM, Xavier R (2006) Potential selectable marker for genetic transformation in banana. Biotechnology 5(2):189–197

14. Sreeramanan S, Maziah M, Abdullah MP, Sariah M, Xavier R, Noraini MF (2005) Pysical and biological parameters affecting transient *gus* and *gfp* expression in banana via particle bombardment. Asia-Pac J Mol Biol Biotechnol 13(1):35–57

15. Becker DK, Dugdale B, Smith MK, Harding RM, Dale JL (2000) Genetic transformation of Cavandish banana (*Musa* spp. AAA group) cv 'Grand Nain' via microprojectile bombardment. Plant Cell Rep 19:229–234

16. Mousavi M, Mousavi A, Habashi AA, Dehsara B (2014) Genetic transformation of date palm (*Phoenix dactylifera* L. cv. 'Estamaran') via particle bombardment. Mol Biol Rep. doi:10.1007/s11033-014-3720-6

17. Mousavi M, Mousavi A, Habashi AA, Arzani K, Dehsara B (2014) Transient transformation of date palm via agrobacterium-mediated and particle bombardment. Emir J Food Agric 26:497–510

18. Mousavi M, Mousavi A, Habashi A, Arzani K (2009) Optimization of physical and biological parameters for transient expression of *uid*A gene in embryogenic callus of date palm (*Phoenix dactylifera* L.) via particle bombardment. Afr J Biotechnol 8(16):3722–3730

19. Murashige T, Skoog F (1962) A revised medium for rapid growth and bioassays with tobacco cultures. Physiol Plant 15:473–497

20. McElroy D, Zhang W, Cao J, Wu R (1990) Isolation of an efficient actin promoter for use in rice transformation. Plant Cell 2 (2):163–171

21. Jefferson RA, Kavanagh TA, Bevan MW (1987) GUS fusions: beta-glucuronidase as a sensitive and versatile gene fusion marker in higher plants. EMBO J 6(13):3901–3907

22. Quenzar B, Hartmann C, Rode A, Benslimane AA (1998) Date palm DNA mini-preparation without liquid nitrogen. Plant Mol Biol Rep 16:263–269

23. Sambrook J, Russel DW (2001) Molecular cloning, a laboratory manual. Cold Spring Harbor Laboratory Press, Cold Spring Harbor, NY, pp 12.1–12.114

# Microprojectile Bombardment Transformation of Date Palm Using the Insecticidal Cholesterol Oxidase (*ChoA*) Gene

## Mai A. Allam and Mahmoud M. Saker

### Abstract

The overall objective of this work is to optimize the transformation system for date palm as a first step toward production of date palm clones resistant to noxious pests. A construct harboring the cholesterol oxidase (*ChoA*) gene, which renders plant resistance against insect attack, is introduced into embryogenic date palm callus using the PDS-1000/He particle bombardment system. The process involves the establishment of embryogenic callus cultures as well as immature embryo-derived microcalli that are used as target tissues for shooting and optimization of transformation conditions. This chapter in addition explains molecular and histochemical assays conducted to confirm gene integration and expression.

**Key words** Biolistic transformation, DNA, Gene gun, In vitro regeneration, Microprojectile bombardment, Somatic embryogenesis

## 1  Introduction

Date palm (*Phoenix dactylifera* L.) is well suited for cultivation in arid and semiarid regions due to its tolerance of high temperatures, drought, and salinity [1]. However, date palm is the target for many noxious pests and diseases [2]. Therefore, there is a great demand for a powerful tool to produce new date palm cultivars resistant to pests. Gene transformation using biolistic (particle bombardment) has been used to address this problem, a well-proven and effective tool used in other tree species like oil palm [3, 4].

Particle bombardment technology is a direct physical method aiming to introduce nucleic acids, e.g., DNA, RNA, or some other biological molecules directly into the cells. It was introduced by John Sanford at Cornell University in 1987 as an alternative procedure for *Agrobacterium*-mediated transformation, which faced difficulties of moving bacteria across grain cell walls [5]. The biolistic device PDS-1000/He™ system is commonly used to deliver DNA to monocot plant cells. The underlying theoretical idea depends upon complementary work between two factors; first is

Jameel M. Al-Khayri et al. (eds.), *Date Palm Biotechnology Protocols Volume 1: Tissue Culture Applications*,
Methods in Molecular Biology, vol. 1637, DOI 10.1007/978-1-4939-7156-5_23, © Springer Science+Business Media LLC 2017

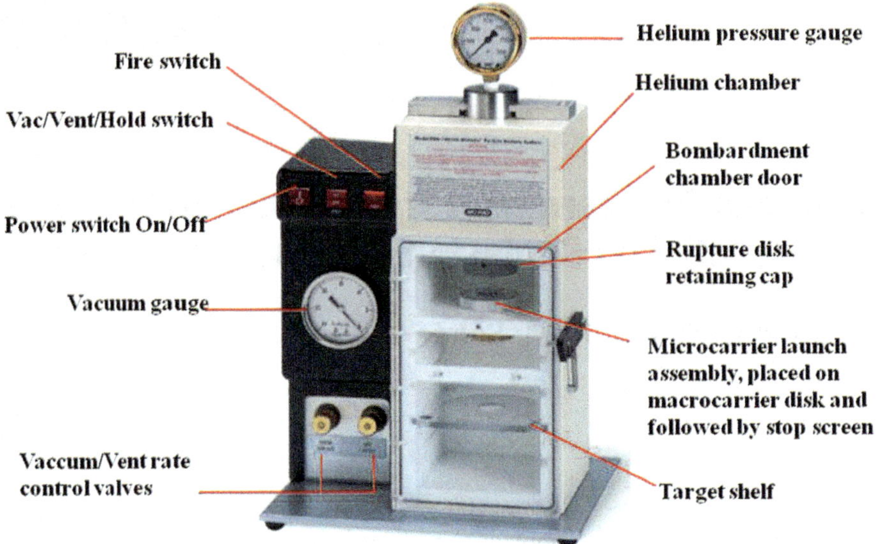


**Fig. 1** Construction of the biolistic device showing various mechanical parts involved in operation

the high-pressure helium pulse, which accelerates the particle movement, and second is a partial vacuum, which reduces the fractional drag of the particles as they are accelerated. As the particles start to move in the direction of the targeted cells, they will collide first with the rupture disks that burst at a defined helium pressure (Fig. 1). The sample to be transformed is placed on the shelf below the rupture disk and the launch assembly. When the instrument is fired, the helium gas moves in the gas acceleration tube until the specific pressure of the rupture disk is reached. The strong pressure breaks the rupture disk, and the helium wave moves to reach the macrocarrier disk (plastic disks carrying the coated microcarriers, the coated gold, or tungsten nanoparticles). The microcarriers are dropped at high speed on a metal net called a stopping screen that prevents the macrocarrier disk from going farther and allows the coated microparticles to fly through the net holes and continue their way to the target tissue [6, 7].

The efficiency of biolistic transformation depends on several factors such as explant type, genotype, in vitro culture capacity, explant regeneration, plant adaptation to ex vitro conditions, seed production capacity, magnitude of helium pressure, and particle size [8]. It was not until 2007, the work on date palm transformation was published by using *Agrobacterium* transformation and microparticle bombardment [9–11]. Recently, publications have addressed a complete transformation system in date palm from Iran, Bangladesh, and India [12–15].

In this chapter, we describe detailed steps of in vitro propagation of date palm, optimal transformation conditions, and molecular confirmation analysis. Such a system can be used for future gene transfer for biopesticide proteins, which confer resistance against date palm pests to recipient plants.

## 2   Materials

### 2.1   Plant Material

Healthy date palm offshoots of Siwy cv., 2- to 3-year-olds, weighing 5–8 kg, as a source of shoot tip explants (Fig. 2).

### 2.2   Antioxidant and Disinfection Solutions

1. Antioxidant solutions: 150 mg/L of each ascorbic acid and citric acid; 0.1 mg/L polyvinylpyrrolidone (PVP) in $ddH_2O$.

2. Ethanol (EtOH 70%): 70 mL absolute ETOH in 30 mL double distilled water $(ddH_2O)$.

3. Sodium hypochlorite (NaOCl) (30% Clorox): 30 mL Clorox in 70 mL $ddH_2O$ with three drops of Tween 20.

4. Mercuric chloride $(HgCl_2)$ (0.1%): 100 mg $HgCl_2$ in 100 mL $ddH_2O$.

5. Sodium hypochlorite (NaOCl) (50% Clorox): 50 mL Clorox in 50 mL $ddH_2O$ containing 0.1 mL Tween 20.

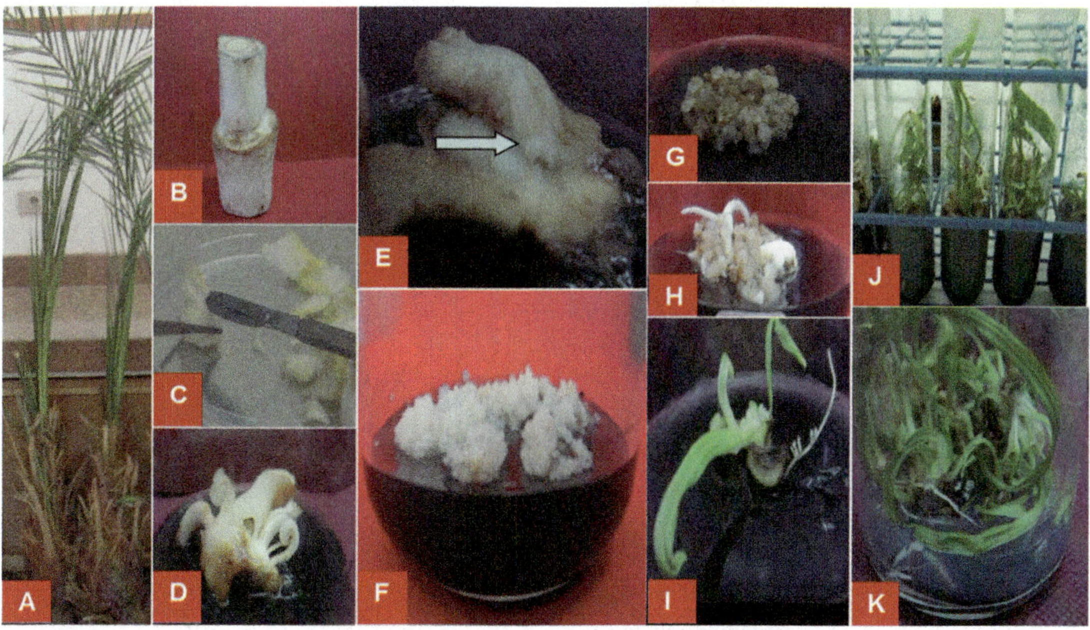

**Fig. 2** Propagation system of date palm Siwy cv. (**a**) Selected offshoots for in vitro culture, (**b**) Offshoots after removing the outer leaves, (**c**) Preparation of shoot tip and leaf primordial explants, (**d**) Shoot tip after cultivation on callus induction medium, (**e**) Appearance of friable callus on shoot tip explants, (**f**) Mass propagation of callus cells after several subcultures, (**g**) Induction of embryogenic callus, (**h**) Development of somatic embryos, (**i**) First plantlet formation, (**j–k**) Shoot and root proliferation

**2.3 Plant Culture Medium**

1. Basal culture medium: Murashige and Skoog (MS) medium [16] with modifications shown as shown in Table 1.

2. Plant growth regulator stock solutions (100 mg/mL each): 2-Isopentyladenine (2iP) and 2,4-dichlorophenoxyacetic acid (2,4-D), naphthaleneacetic acid (NAA).

**Table 1**
**Composition of the basal culture medium consisting of MS medium [16] with modifications**

| Component | Final concentration |
|---|---|
| *Macroelements* | *mg/L* |
| $CaCl_2$ | 332.02 |
| $KH_2PO_4$ | 170 |
| $KNO_3$ | 1900 |
| $MgSO_4$ | 180.54 |
| $NH_4NO_3$ | 1650 |
| *Microelements* | *mg/L* |
| $CoCl_2 \cdot 6H_2O$ | 0.025 |
| $CuSO_4 \cdot 5H_2O$ | 0.025 |
| FeNaEDTA | 36.7 |
| $H_3BO_3$ | 6.2 |
| KI | 0.83 |
| $MnSO_4 \cdot H_2O$ | 16.9 |
| $Na_2MoO_4 \cdot 2H_2O$ | 0.25 |
| $ZnSO_4 \cdot 7H_2O$ | 8.6 |
| *Vitamins and amino acids* | *mg/L* |
| Glycine | 2 |
| myo-Inositol | 100 |
| Nicotinic acid | 0.50 |
| Pyridoxine HCl | 0.50 |
| Thiamine HCl | 0.8 |
| Glutamine | 200 |
| Adenine sulfate | 80 |
| *Other additives* | *g/L* |
| Sucrose | 30 |
| Activated charcoal | 2 |
| Phytagel | 2 |

3. Antibiotics stock solutions (100 mg/mL each): Working concentrations are 50 mg/L hygromycin and 100 mg/L kanamycin.

4. Medium additives to various culture stages are shown in Table 3 including Medium 1 (callus induction and somatic embryogenesis), Medium 2 (shoot and root development), Medium 3 (osmotic stress treatment), and Medium 4 (selection of transgenic tissue).

### 2.4 Plasmid DNA

A molecular construct (pBC4), harboring *cholesterol oxidase* gene (insect resistance gene), GUS reporter marker gene, and kanamycin resistance gene under the control of 35 S promoter and a NOS terminator [17].

### 2.5 Reagents for Microparticle Preparation

1. Glycerol solution (50% v/v): 50 mL glycerol and 50 mL ddH$_2$O. Filter sterilized (0.2 μm microfilter).

2. Calcium chloride (CaCl$_2$) solution (2.5 M): 277.45 g CaCl$_2$ in 1 L ddH$_2$O (or 2.775 g in 10 mL ddH$_2$O).

3. 0.1 M spermidine solution: 14.552 g in 1 L ddH$_2$O (or 0.146 g in 10 mL ddH$_2$O). Prepare freshly and filter-sterilize using 0.2 μm microfilter.

4. Ethanol (75%): 75 mL absolute ethanol and 25 mL ddH$_2$O.

### 2.6 Reagents for GUS Assay Buffer

(a) 87.10 mg dipotassium phosphate (50 mM K$_2$HPO$_4$).

(b) 68.043 mg monopotassium phosphate (50 mM KH$_2$PO$_4$).

(c) 29.224 mg ethylenediaminetetraacetic acid (10 mM EDTA).

(d) 1.646 mg potassium ferricyanide (0.5 mM C$_6$N$_6$FeK$_3$).

(e) 1.842 mg potassium ferrocyanide (0.5 mM C$_6$N$_6$FeK$_4$).

(f) 10 μL Tween 20.

(g) 10 mg X-glucuronide (X-gluc, pH 6.8).

(h) ddH$_2$O up to 10 mL.

### 2.7 Reagents for ELISA Buffers

1. GEP solution (100 mL) containing the following:

   (a) 50 mM dipotassium phosphate (Na$_2$HPO$_4$): 871 mg/ 100 mL.

   (b) 10 mM DTT or ß-mercaptoethanol: 78.13 mg/100 mL.

   (c) 1 mM EDTA: 29.224 mg/100 mL.

   (d) 0.1% N-lauroylsarcosine (100 μL/100 mL).

   (e) 0.1% Triton X-100 (100 μL/100 mL).

2. MUG solution (100 mL): 20 mg MUG (4-methylumbelliferyl-β-galactopyranoside) in 50 mL GEP solution.

**2.8 DNA Analysis Reagents**

1. DNA isolation: Commercial DNA isolation kit.

2. PCR components: *Taq* polymerase, 10× amplification buffer, dNTPs mix, forward and reverse GUS gene primers (the expected size of the amplified band is 930 bp):

   Forward primer: 5′ CCT GTA GAA ACC CCA ACC CG 3′.
   Reverse primer: 5′ TGG CTG TGA CGC ACA GTT CA 3′.

3. Electrophoresis agarose gel (2%): 2 g agarose in 1× TAE and containing 0.2 μg/L ethidium bromide.

4. TAE buffer 50× (Tris–acetate–EDTA buffer, pH 8.3): 242 g Tris-base and 57.1 mL 100% acetic acid.

5. TAE buffer (1×): 20 mL of 50× TAE buffer in 980 mL ddH$_2$O.

6. EDTA: 100 mL 0.5 M (14.612 g) Na-EDTA in 1 L ddH$_2$O.

**2.9 Equipment**

1. Glassware: Culture jars (150 mL), graduated cylinders, beakers, flasks, and 9 cm Petri dishes.

2. Instruments: Micropipettes (10–100 and 100–1000 μL), thermal cycler, magnetic stirrer, pH meter, autoclave, vortex, centrifuge, microprojectile bombardment device (biolistic), and DNA electrophoresis system.

3. Supplies: Microcentrifuge tube (0.2 and 1.5 mL), sterile microfilters (2 μm), sterile mortars and pestles, Falcon centrifuge tubes, micropipette tips, DNeasy Plant Mini Kit, and microprojectile bombardment optimization kit.

# 3    Methods

**3.1 Offshoot Preparation and Sterilization**

1. Dissect the offshoots using a serrated knife, and gradually peel away the mature leaves, the outer brown cover, and outer layers of the shoot tip explants.

2. Keep the shoot tip explants, 5 × 3 cm, in antioxidant solutions for 30 min.

3. Transfer the shoot tip to the air laminar flow hood for surface sterilization.

4. Add 70% sodium hypochlorite on the shoot tip explants in glass jars, and shake them at 50 rpm for 25 min.

5. Wash the shoot tips with sterilized ddH$_2$O for 5 min, and then remove the soft outer primordial leaf layer of the shoot tip explants.

6. Immerse in 0.1% mercuric chloride (HgCl$_2$), shake for 25 min, and then repeat **step 5**.

7. Soak in 30% sodium hypochlorite with few drops of Tween 20, keep on a shaker for 20 min, and then repeat **step 5**.

8. Wash the explants three times with 0.1 mg/L polyvinylpyrrolidone (PVP) solution.

9. Soak the explants in 50% Clorox, shake for 15 min, and then repeat **step 5**.

10. Gradually remove the primordial leaves one after another, cut the shoot tip base after sterilization, until obtaining the main shoot tip, and then cut into small sections of 2 mm$^3$ in size to use as explants as shown in Fig. 1.

***3.2 Medium Preparation***

1. Prepare 100 mg/mL stock solutions of plant growth regulators and antibiotics: Weigh 1 g from each reagent and dissolve separately with 0.1 mL solvent (ddH$_2$O, HCl, NaOH or ethanol) as described in Table 2. Raise the volume of dissolved stock to 10 mL with ddH$_2$O (*see* **Note 1**). Filter-sterilize antibiotic solutions using 0.2 μm microfilter and store at −20 °C.

2. Weigh 4.4 g MS medium commercial powder (Table 1) and place in 1 L beaker containing 500 mL ddH$_2$O, and stir on a hot plate with magnetic stirrer (40 rpm, 30 °C).

3. Add other components including plant growth regulators, according to the culture stages, as described in Table 3. Bring the final volume to 1 L.

**Table 2**
**Solvents used for preparing the medium additive stock solutions (100 mg/mL) and the required amounts to obtain the indicted final concentrations in the culture medium**

| Reagent | Solvent | Final concentration in medium (mg/L) | Required amount from the stock solution (μL/L) |
|---|---|---|---|
| Potassium dihydrogen phosphate | H$_2$O | 170 | 1700 |
| Adenine sulfate | H$_2$O | 80 | 800 |
| Thiamine HCl | H$_2$O | 70 | 700 |
| Glutamine | H$_2$O | 200 | 2000 |
| 2,4-D | EtOH 1% | 10 | 100 |
| 2iP | NaOH 1% | 3 | 30 |
| NAA | NaOH 1% | 0.4 | 4 |
| Hygromycin | H$_2$O | 50 | 500 |
| Kanamycin | H$_2$O | 100 | 1000 |

**Table 3**
**Additive incorporated in the basal culture medium (Table 1) for different tissue culture and transformation steps**

| Medium code | Culture stage | Composition |
| --- | --- | --- |
| Medium 1 | Callus induction and somatic embryogenesis | Basal medium +10 mg/L 2,4-D + 3 mg/L 2iP |
| Medium 2 | Shoot and root development | Basal medium +0.4 mg/L NAA |
| Medium 3 | Osmotic stress | Basal medium +0.2 M (36.434 g/L) D-sorbitol +0.2 M (36.4344 g/L) D-mannitol |
| Medium 4 | Selection of transgenic tissue | Basal medium containing 10 mg/L 2,4-D + 3 mg/L 2iP + 50 mg/L hygromycin and 100 mg/L kanamycin |

4. Adjust the pH on $5.8 \pm 0.1$ using 1 N NaOH and 1 N HCl. Add 2 g/L Phytagel and heat to dissolve.

5. Dispense the medium in 150 mL culture jars (20 mL medium per jar). Seal the jars tightly with polyvinylpropylene caps and autoclave for 20 min at 121 °C and 1.1 kg/cm$^2$ (*see* **Note 2**).

6. For the selection medium (Medium 4, Table 3), add the filter-sterilized antibiotic solutions (50 mg/L hygromycin and 100 mg/L kanamycin) to the culture medium after autoclaving. Dispense in 9 cm Petri dishes (20 mL per dish).

*3.3 Induction of Embryogenic Callus*

1. Place the shoot tip and primordial leaf explants on callus induction medium (Medium 1, Table 2) at $27 °C \pm 1$ in complete darkness.

2. Transfer the explants into fresh medium at every 3-week interval for at least 12 subcultures until the first friable callus formation.

3. Separate the formed callus from the mother explants and transfer to the fresh medium with same components for other six subcultures at every 3-week interval.

4. The friable callus then starts to increase; embryogenic callus is obtained as a first stage of somatic embryo formation. Use embryogenic callus for transformation.

5. Transfer the embryogenic callus on a sterile filter paper placed on the surface of the osmotic stress medium (Medium 3, Table 3) in Petri dish; the cultures are now ready for the transformation process.

**3.4  Bombardment Transformation**

*3.4.1  Preparing Microparticles*

1. Weigh 60 mg tungsten particles and place in a 1.5 mL microcentrifuge tube. Vortex the tube vigorously for 10 min after adding 1 mL 100% ethanol above the particles.

2. Leave the particles in ethanol at room temperature for 15 min, and then centrifuge for 15 min at 4472 × $g$.

3. Remove the supernatant and wash the particles three times in sterile ddH$_2$O.

4. Suspend and mix the particles in 2 mL 50% (v/v) glycerol and divide the total amount of the mixture into 1.5 mL microcentrifuge tubes each containing 125 μL aliquots (3.75 mg particles).

*3.4.2  Plasmid DNA Coating of Microparticles*

1. Add 9 μL plasmid DNA (pBC4) in 1.5 mL microcentrifuge tube containing 125 μL aliquots, and vortex gently for 3 min.

2. Add to the mixture 125 μL 2.5 M CaCl$_2$ and vortex gently for 3 min.

3. Add 50 μL 0.1 M spermidine, vortex gently for 3 min, and keep on ice for 15 min.

4. Centrifuge for 20 s at 440 × $g$ and then remove the supernatant.

5. Add 0.5 mL 75% ethanol to the pellet, vortex for 3 min, and centrifuge for 20 s at 440 × $g$.

6. Remove supernatant and add 0.5 mL 100% ethanol.

*3.4.3  Bombardment*

1. Fit the appropriate rupture disk (1300 psi) in the retaining cap.

2. Spread 6 μL solution containing particles in the center of the macrocarrier disk (*see* **Note 3**).

3. Fix the shelf carrying the target tissues at an appropriate flight distance (9 cm), and adjust the chamber vacuum to 25 inHg. Then activate the shot button to carry the coated plasmid directly on to the plant tissue (Fig. 3).

**3.5  Selection and Confirmation of Putative Transformants**

*3.5.1  Antibiotics Selection*

1. Transfer the transformed embryogenic callus to callus induction medium (Medium 1, Table 3), 2 weeks after bombardment.

2. Subculture the transformed explants on a selection medium (Medium 4, Table 3) containing 50 mg/L hygromycin and 100 mg/L kanamycin (*see* **Note 4**). Subculture twice at a 4-week interval.

3. Transfer the surviving (tentatively transformed) somatic embryos to Medium 2 (Table 3) for shoot and root development.

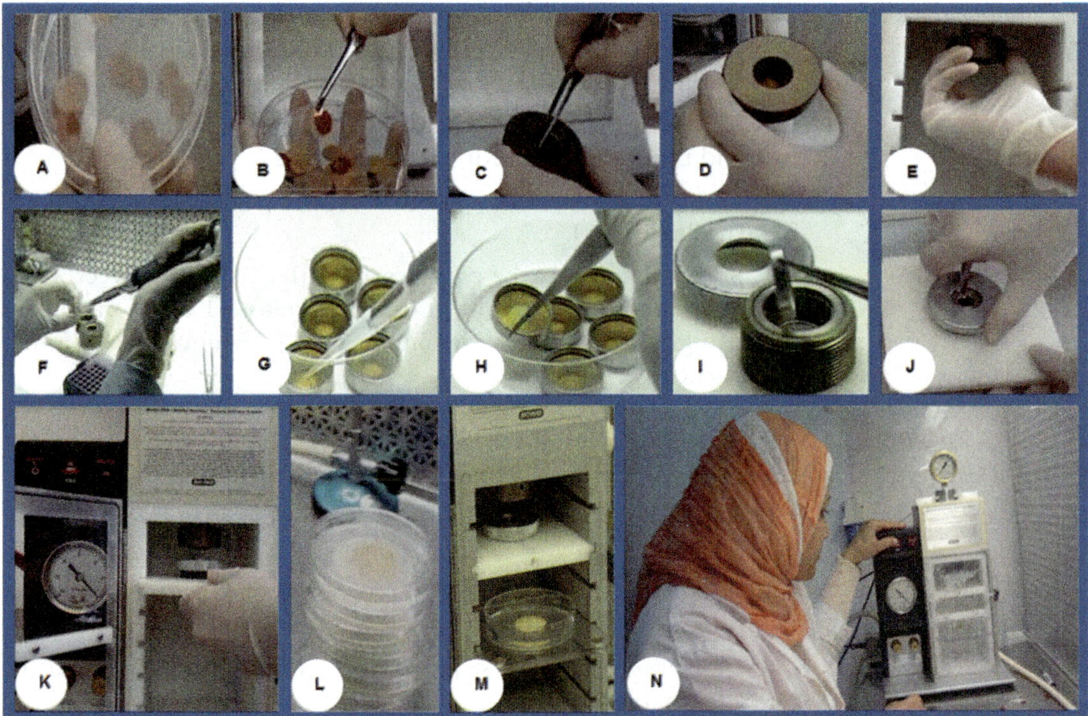

**Fig. 3** Practical ordered steps for biolistic transformation. (**a–e**) Prepare the suitable rupture disk, (**f–j**) Place the coated particles on the macrocarrier disks and settled upside down above the stopping screen and close the metal cover, (**k–n**) Place the rupture disk shelf, then the target cells in the following shelf, vacuum and then fire the microcarriers

*3.5.2  Transient GUS Assay*

Transient expression level is assayed for expression of the GUS-intron gene in the callus, 3 days after transformation. β-Glucuronidase activity is detected histochemically as described previously [18] with some modifications according to the following procedures:

1. Transfer the 0.5 g callus samples to Petri dish containing 4 mL GUS buffer and incubate overnight at 37 °C in the darkness.

2. Evaluate transformation frequency by counting the total number of blue spots observed under binocular microscope. Count the number of spots that are approximately ≤1 mm in diameter which represent one or a few GUS-expressed cells (Fig. 4).

*3.5.3  Enzyme-Linked Immunosorbent Assay (ELISA)*

For rapid screening of putative transgenic calli, GUS expression is assayed using a fluorescence detection method as follows:

1. Add the embryogenic microcallus samples into wells of an ELISA plate containing 100 μL GEP solution.

2. Add 100 μL MUG solution and incubate the soaked callus samples overnight at 37 °C.

3. Visualize under UV light. The positive samples show a strong fluorescence (Fig. 4).

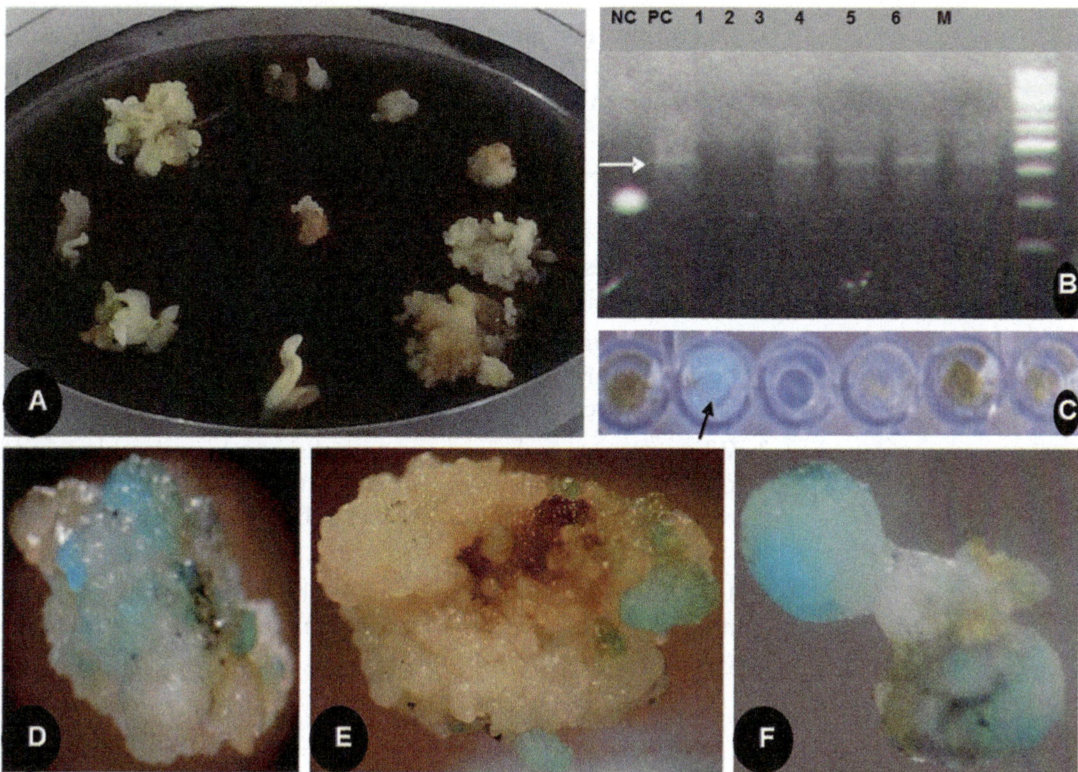

**Fig. 4** (**a**) Putative transformed embryogenic date palm calli onto selective medium, (**b**) PCR amplification Arrows point to positive GUS expression (*bright blue*) and amplicon of interest. *Lane 1* (positive control, i.e. plasmid used in bombardment), *lane 2* (Negative control, i.e. non-transformed), *lanes 3–5* transformed clones (GUS positive) and *lane 6* (positive control), (**c**) ELISA fluorescence, (**d–f**) Histochemical GUS assay in putative transgenic date palm in friable callus, embryogenic callus and somatic embryos colonies, respectively

*3.5.4  PCR Analysis*

1. Isolate DNA from transformed plant tissue using DNeasy Plant Mini Kit according to manufacturer instructions (https://www.qiagen.com/us/shop/sample-technologies/dna/dneasy-plant-mini-kit/#orderinginformation).

2. Add 40 ng from the isolated DNA template in addition to 2 μL 10× amplification buffer, 200 μM dNTPs mixture, 10 pmol each of forward and reverse primers (targeting GUS gene), 1 U *Taq* polymerase, and complete to 20 μL using sterilized ddH$_2$O.

3. Place the PCR mixture in PCR machine for 35 cycles using a preheated thermal cycler.

4. Follow temperature profile for denaturation at 94 °C for 1.5 min, annealing at 65 °C for 1.5 min and extension at 72 °C for 2 min, and finally incubation at 72 °C for 5 min.

5. Electrophorese amplification products in 2% agarose using TAE buffer 70 V, and visualize under UV light after staining in 0.2 μg/mL ethidium bromide.

## 4   Notes

1. Prepare all solutions using sterilized ddH$_2$O and use high-purity chemicals.

2. It is recommended to prepare the culture media about a week before culturing, in order to check the contamination possibility of the prepared media.

3. When preparing the macrocarrier disks, it is preferable to add the coated particles onto the disk after they are placed on the round metal part and not before. This ensures less movement of the coated particles.

4. In the first selection step using antibiotics, hygromycin use is not recommended before shoot formation because it is known to inhibit shoot development.

## References


1. Al-Khayri JM, Jain SM, Johnson DV (eds) (2015) Date palm genetic resources and utilization, vol 1, Africa and the Americas. Springer, Dordrecht, p 546

2. Al-Khayri JM, Niblett CL (2012) Envision of an international consortium for palm research. Emir J Food Agric 24(5):470–479

3. Parveez GKA, Bahariah B, Ayub NH, Abdul Masani MY, Abdul Rasid O, Tarmizi AH, Ishak Z (2015) Production of polyhydroxybutyrate in oil palm (*Elaeis guineensis* Jacq.) mediated by microprojectile bombardment of PHB biosynthesis genes into embryogenic calli. Front Plant Sci 6:1–12

4. Izawati AMD, Abdul Masani MY, Ismanizan I, Parveez GKA (2015) Evaluation on the effectivness of 2-deoxyglucose-6-phosphate phosphatase (DOG$^R$1) gene as a selectable marker for oil palm (*Elaeis guineensis* Jacq.) embryogenic calli transformation mediated by *Agrobacterium tumefaciens*. Front Plant Sci 6:1–10. doi:10.3389/fpls.2015.00727

5. Sanford JC (1990) Biolistic plant transformation. Physiol Plant 79:206–209

6. Bio-Rad. Biolistic particle delivery systems brochure. Bulletin #5443. http://www.bio-rad.com

7. Décima CO, González G, Lewi D (2010) Biolistic maize transformation: improving and simplifying the protocol efficiency. Afr J Agric Res 5(25):3561–3570

8. Zhang W, Subbarao S, Addae P, Shen A, Armstrong C, Peschke V, Gilbertson L (2003) Cre/lox-mediated marker excision in transgenic maize (*Zea mays* L.) plants. Theor Appl Genet 107:1157–1168

9. Saker M, Ghareeb H, Kumlehn J (2009) Factors influencing transient expression of *Agrobacterium*-mediated transformation of *Gus* gene in embryogenic callus of date palm. Adv Hort Sci 23(3):150–157

10. Saker MM, Allam MA, Gomaa AH, Abd El-Zaher MH (2007) Optimization of some factors affecting genetic transformation of semi-dry Egyptian date palm cultivar (Sewi) using particle bombardment. J Genet Eng Biotechnol 5(1):57–62

11. Saker MM (2011) Transgenic date palm. In: Jain SM, Al-Khayri JM, Johnson DV (eds) Date palm biotechnology. Springer, Dordrecht, pp 631–650

12. Mousavi M, Mousavi A, Habashi AA, Arzani K (2009) Optimization of physical and biological parameters for transient expression of *uidA* gene in embryogenic callus of date palm (*Phoenix dactylifera* L.) via particle bombardment. Afr J Biotechnol 8(16):3721–3730

13. Hassan L (2013) Successful genetic transformation in date palm (*Phoenix dactylifera*). J Bangladesh Agric Univ 11(2):171–176

14. Mousavi M, Mousavi A, Habashi AA, Dehsara B (2014) Genetic transformation of date palm

(*Phoenix dactylifera* L. cv. 'Estamaran') via particle bombardment. Mol Biol Rep 41:8185–8194

15. Aslam J, Khan SA, Azad MA (2015) *Agrobacterium*-mediated genetic transformation of date palm (*Phoenix dactylifera* L.) cultivar "Khalasah" via somatic embryogenesis. Plant Sci Today 2(3):93–101

16. Murashige T, Skoog F (1962) A revised medium for rapid growth and bioassays with tobacco tissue cultures. Physiol Plant 15:473–497

17. Cho H, Choi K, Yamashita M, Morikawa H, Murooka Y (1995) Introduction and expression of the *Streptomyces* cholestrol oxidase gene (*ChoA*), a potent insecticidal protein active against boll weevil larvae, into tobacco cells. Appl Microb Biotechnol 44:133–138

18. Jefferson RA, Kavanagh TA, Bevan MW (1987) GUS fusions: *ß-glucuronidase* as a sensitive and versatile gene fusion marker in higher plants. EMBO J 6:3901–3909

# Chapter 24

# Transient GUS Gene Expression in Date Palm Fruit Using Agroinjection Transformation Technique

**Mohei El-Din M. Solliman, Hebaallah A. Mohasseb, Abdullatif A. Al-Khateeb, Jameel M. Al-Khayri, and Suliman A. Al-Khateeb**

## Abstract

Transient expression of foreign genes in plant tissue is a valuable tool for testing the efficacy of transformation methods. In this work, we present, for the first time, the utilization of agroinjection as an efficient transformation system for gene delivery in date palm fruit. The research utilized *Agrobacterium tumefaciens* strain LBA4404 harboring the binary vector pRI201-AN-GUS carrying the beta-glucuronidase (GUS) gene, under the control of a CaMV 35S and kanamycin (NPTII) as an antibiotic gene under the control of a NOS promoter. Based on histochemical assay of agroinjected fruit for the GUS gene expressions, this protocol has proved to be an efficient and reliable tool for transgene expression in date palm. PCR for plasmid DNA, extracted from the transformed *Agrobacterium*, demonstrated the generation of the expected amplicon, corresponding to the GUS gene using GUS primers.

**Key words** Agroinjection, *Agrobacterium*, Beta-glucuronidase GUS gene, Genetic transformation

## 1 Introduction

Biotechnology offers a reliable approach to develop improved date palms (*Phoenix dactylifera* L.) [1]. This economically important fruit crop is a target of several insects and diseases. Gene silencing or RNA interference has great potential to control disease and insect pests of date palm [2]. However, the successful application of this technology requires the availability of a reliable transformation system providing efficient gene expression.

In planta gene expression can be accomplished using *Agrobacterium tumefaciens* expression vector (agroinfiltration) or with viral expression vector (agroinfection) [3]. Agroinjection is the process of forcing *Agrobacterium tumefaciens* suspension into plant tissues using a medical syringe, as demonstrated in tomato fruit [4]. These techniques are rapid and well suited for gene expression, as well as gene function analysis in transgenes.

Jameel M. Al-Khayri et al. (eds.), *Date Palm Biotechnology Protocols Volume 1: Tissue Culture Applications,*
Methods in Molecular Biology, vol. 1637, DOI 10.1007/978-1-4939-7156-5_24, © Springer Science+Business Media LLC 2017

The efficiency of transgene expression differs among monocotyledon and dicotyledon plants and is affected by tissue types, plasmid constructs, and bacterial-host compatibility [5–7]. Although numerous plasmids have been constructed for that purpose, there is no recommended plasmid for trees including date palm.

Agroinjection has proved useful for measuring gene expression of various gene constructs before embarking on the lengthy process of stable gene transformation. It is convenient to assert transient gene expression to confirm functionality and stability before using the larger-scale regeneration of transgenic plants [8, 9]. The transient gene expression systems are quick and unaffected by positional effects of the genes, thus overcoming biased gene expression [10, 11].

Agroinjection by directly injecting the fruit was recently demonstrated in date palm [12]. Previous work on date palm genetic transformation involved *Agrobacterium tumefaciens*-mediated genetic transformation [13, 14] and microprojectile bombardment of embryogenic callus and somatic embryos [15–17] based on the beta-glucuronidase gene (GUS) insertion.

This chapter describes a protocol for the genetic transformation of date palm based on agroinjection using *A. tumefaciens* carrying a plasmid (pRI201-AN-GUS) expressing a reporter *udiA* gene in addition to describing transient expression evidence based on histochemical and molecular assays.

## 2  Materials

### 2.1  Plant Material

1. Date palm fruits from different plants of Khalas cv. at the khalal fruit stage (*see* **Note 1**).

### 2.2  Agrobacterium Strain and Plasmid

1. *A. tumefaciens* strain LB4404 (*see* **Note 2**).
2. Binary plasmid pRI201-AN-GUS (Fig. 1, *see* **Note 3**).

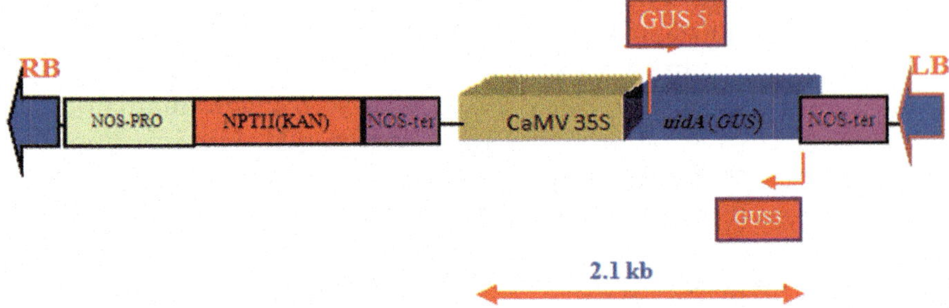


**Fig. 1** Plasmid map of the transformation vector 35S β-glucuronidase (GUS), pRI201-AN-GUS DNA (TaKaRa Comp)

**2.3 Bacterial Culture Medium**

1. Luria-Bertani (LB) medium: 10 g/L tryptone, 5 g/L yeast extract, 2.5 g/L NaCl, and 10 g/L mannitol, pH 7.5.

2. Antibiotic solutions: 50 mg kanamycin in 1 mL double-distilled water (ddH$_2$O), 25 mg streptomycin in 1 mL ddH$_2$O.

3. Acetosyringone stock solution (0.2 M): 0.7848 g acetosyringone in 20 mL dimethyl sulfoxide (DMSO).

4. Liquid nitrogen.

5. 1 M mannitol stock solution: 182 g/L and ddH$_2$O up to 1 L.

**2.4 Histochemical Analysis**

1. Reagent A: 2 M sodium phosphate (pH 7) and 4 mM EDTA in 500 mL ddH$_2$O.

2. Reagent B: 1 M potassium ferrocyanide (C$_6$N$_6$FeK$_4$), 4.224 g potassium ferrocyanide in 100 mL ddH$_2$O.

3. Reagent C: 1 M potassium ferricyanide (C$_6$N$_6$FeK$_3$), 3.292 g potassium ferricyanide in 100 mL ddH$_2$O.

4. 2-(N-Morpholino)ethanesulfonic acid (1 M, MES): 42.64 g MES in 200 mL ddH$_2$O, pH 5.6.

5. X-GlcA substrate (1 mM, X-GlcA): 25 mg 5-bromo-4-chloro-3-indolyl β-D-glucuronide in 2 mL methanol.

6. Staining solution:

| | |
|---|---|
| Reagent A | 25 mL |
| Reagent B | 100 μL |
| Reagent C | 100 μL |
| Triton X-100 | 2 mL |

ddH$_2$O to a final volume of 100 mL.
Store the staining solution at 4 °C in the dark until use within 2–3 days.

7. Substrate solution (X-GlcA):

| | |
|---|---|
| Reagent A | 25 mL |
| Reagent B | 100 μL |
| Reagent C | 100 μL |
| Triton X-100 | 2 mL |
| Methanol | 20 mL |
| X-GlcA solution | 20 μL |

ddH$_2$O to a final volume of 100 mL.
Store at 4 °C in the dark until use within 2–3 days.

8. Fixative solution: 1 M stock solution MES, 20 mM MES (pH 5.6), 1 M stock solution mannitol, and 0.6% formaldehyde.

| 1 M MES | 200 µL (20 mM final concentration) |
| 1 M mannitol | 6 mL (600 mM final concentration) |
| Formaldehyde | 600 µL |

   ddH$_2$O to a final volume of 100 mL.
9. Ethanol, 70%.
10. 50% glycerol.
11. pH adjustment solutions: 0.1 N HCl and 0.1 N NaOH.

**2.5 Polymerase Chain Reaction (PCR) Reagents**

1. Tris–Borate–EDTA (TBE) buffer (5×): 54 g Tris base, 27.5 g boric acid, 20 mL 0.5 M EDTA, and up to 1 L ddH$_2$O, pH 8.
2. PCR buffer (10×): 50 mM Tris-HCl (pH 8.5), 1.5 mM MgCl$_2$, 50 mM KCl, and 0.1% Triton X-100.
3. GUS primers:

   GUS forward 5′ primer (CGCGGATCCATGTTAGCTCCT GTAGAAACCCC) and GUS reverse 3′ primer (CGCGGAT CCTCATTGTTTGCCTCCCTGCTG).
4. PCR reaction mixture: 10× PCR buffer, 200 mM of each dNTP (deoxynucleotide), 0.4 mM of each primer, and 1 U Taq DNA polymerase.
5. Agarose gel: 1.0 g of agarose into a 100 mL 1× TBE buffer.
6. 1× TBE buffer (Tris-acetate electrophoresis buffer): 40 mM Tris-OH, 20 mM acetic acid, pH 7.8.
7. Ethidium bromide solution: 5 mg ethidium bromide in 1 mL ddH$_2$O. Store at 4 °C.

**2.6 Equipment**

1. Glassware and plastic ware: Petri dishes (size 90 mm), 10-mL sterile medical syringe with needle, and 250-mL Erlenmeyer flasks.
2. Instruments: Vacuum desiccator, incubator, microcentrifuge, vortex mixer, biological safety cabinet, PCR thermal cycler, and rotary shaker.
3. Tools: Forceps, scalpel, and 0.2-µm sterile nylon microfilter.

## 3   Methods

### 3.1   Preparation of Fruit Explant

1. Collect fresh date palm fruits from Khalas cv. at the khalal fruit stage.

2. Clean the fruits with a liquid detergent and thoroughly wash in tap water.

### 3.2   Preparation of LB Medium and Antibiotic Solutions

1. Dissolve 10 g tryptone, 5 g yeast extract, 2.5 g NaCl, and 10 g mannitol in 1 L ddH$_2$O.

2. Adjust to pH 7.5 using 0.1 N HCl and 0.1 N NaOH. Dispense 100 mL aliquots in screw cap bottles, and autoclave for 20 min at 121 °C and 1.04 kg/cm$^2$. Store at 28 °C (*see* **Note 1**).

3. Prepare antibiotic solutions: Dissolve separately 50 mg kanamycin in 1 mL ddH$_2$O and 25 mg streptomycin in 1 mL ddH$_2$O. Sterilize with 0.2-µm sterile microfilter and store at −20 °C.

4. In each 100 mL medium aliquots, add 100 µL kanamycin solution (final concentration 50 µg/mL), and add 25 µL streptomycin solution (final concentration 25 µg/mL).

### 3.3   Preparation of Agrobacterium tumefaciens Culture

1. Transform the *Agrobacterium tumefaciens* (strain LB4404) by adding 1 µg pRI201-AN-GUS plasmid to 100 µL *Agrobacterium* competent cells in an Eppendorf tube, mix gently, and immediately freeze in liquid nitrogen for 2 min (*see* **Note 2**).

2. Incubate the Eppendorf tube at 37 °C in water bath for 5 min to thaw the cells.

3. Add 1 mL LB medium (antibiotic free) and incubate at 28 °C for 6 h with slow shaking at 40 rpm.

4. Transfer the bacterial suspension to a 250-mL Erlenmeyer flask containing 50 mL liquid LB medium supplemented with antibiotic solutions.

5. Store the *A. tumefaciens* harboring the pRI201-AN-GUS plasmid in 50% glycerol at −20 °C.

6. Inoculate *A. tumefaciens* from the glycerol stock into LB medium liquid cultures supplemented with antibiotic solutions. Incubate overnight at 28 °C on a rotary shaker (120 rpm).

7. Pellet the bacteria by centrifugation at 4000 × *g* for 6 min at room temperature.

8. Resuspend bacteria in LB liquid medium containing antibiotic solutions, and incubate at 28 °C on a rotary shaker (120 rpm) for 24 h to allow growth (*see* **Note 2**).

9. Measure the optical density (OD) of the overnight bacterial culture using a spectrophotometer at 600 nm. Adjust bacterial suspension density to $OD_{600}$ equals 0.7 (*see* **Note 3**).

10. Grow *A. tumefaciens* on a rotary shaker shaking (120 rpm) at 28 °C for 4 h, in LB medium containing antibiotic solutions, 150 μg/mL acetosyringone (150 μL 0.2 M stock solution).

11. Prepare 0.2 M acetosyringone solution: Dissolve 0.2 g acetosyringone in 1 mL 70% ethanol. Raise volume to 20 mL by ddH$_2$O. Sterilize it with 0.2-μm sterile microfilter and store at −20 °C.

12. Pellet the bacterial cells by centrifugation for 5 min at $4000 \times g$, and gently resuspend the pellet in 1 mL LB medium containing antibiotic solutions, and adjust concentration to the required final $OD_{600}$ value, and incubate for 1 h at 28 °C (*see* **Note 4**).

13. Resuspend the bacterial pellet in 10 mL LB medium containing antibiotic solutions, 150 μg/mL acetosyringone, and incubate the bacterial suspension culture on a rotary shaker shaking (120 rpm) for 3 h at 28 °C (*see* **Note 5**).

**3.4 Agroinjection**

1. Transfer plant tissue (date palm fruits) to a Petri dish.

2. Fill a sterile 10-mL plastic medical syringe, fitted with a detachable stainless steel needle, with the bacterial suspension. Inoculate the date palm fruit by inserting the needle through the epicarp and injecting the bacterial suspension into the mesocarp (fruit flesh) at 0.5 mL per fruit (*see* **Note 6**).

3. Cover the plate and incubate the fruits at 28 °C for 24 h to allow time for gene expression.

4. Degas the Petri dish plates for 2–3 min in a vacuum desiccator to remove the air for better infiltration of bacterial suspension into the tissue.

**3.5 GUS Histochemical Assay**

1. Slice the fruits longitudinally into halves in a Petri dish using sterile forceps and scalpel.

2. Add 10 mL staining solution and assure that the date palm fruit tissue is in contact with the solution (*see* **Note 7**).

3. Add 2 mL substrate solution (X-GlcA) for developing GUS expression colors, followed by brief vacuum infiltration in a vacuum desiccator (*see* **Note 8**).

4. Cover Petri dish with lid and incubate at 37 °C for 4–24 h; blue-stained spots develop within 4 h (*see* **Note 9**).

5. Add fixative solution to the fruit slices assuring that all tissues are covered by the solution and incubate at 28 °C for 4 h (*see* **Note 10**).

6. Remove the fixative solution and wash in 70% ethanol four times, at 1 min each, before visual examination (*see* **Note 11**).

**3.6  PCR Procedures**

1. Add the following reaction mixture components in a Eppendorf tube (0.25 mL size) for a final volume of 50 µL for each reaction:

| | |
|---|---|
| Taq DNA polymerase (5 U/µL) | 1 µL (final concentration 0.1 U/µL) |
| 10× PCR buffer | 5 µL (final concentration 1×) |
| dNTPs nucleotides mix (5 mM) | 5 µL (final concentration 0.1 mM) |
| Forward GUS 5′ primer (5 µM) | 1 µL (final concentration 0.1 µM) |
| Reverse GUS 3′ primer (5 µM) | 1 µL (final concentration 0.1 µM) |
| Sterile water | 36 µL |
| Plasmid DNA sample | 1 µL |

2. Run the amplification cycles according to the following program: first denature DNA once for 5 min at 94 °C, followed by 35 cycles of PCR amplification. For each cycle, denature the DNA for 30 s at 94 °C, anneal for 30 s at 50 °C, and elongate for 2 min at 72 °C (*see* **Note 12**).

3. Prepare 1× TBE (1 L): Dissolve 10.8 g Tris and 5.5 g boric acid in 900 mL ddH$_2$O. Add 4 mL 0.5 M Na$_2$EDTA (pH 8). Adjust volume to 1 L. Store at room temperature.

4. Prepare 1% agarose gel: Weigh 1 g agarose and place into a 250-mL conical flask. Add 100 mL of 1× TBE, and boil for 2 min in the microwave to dissolve the agarose. Add 10 µL ethidium bromide solution to the dissolved agarose and mix (*see* **Note 13**).

5. Separate the DNA amplification products on the agarose gel. Visualize the gel under 300 nm UV light (*see* **Note 13**).

# 4  Notes

1. Glassware used for preparation of culture medium should be cleaned with detergent, rinsed with ddH$_2$O, and dried at 28 °C for 1–4 h before use.

2. The efficient introduction of constructed plasmids into *A. tumefaciens* is of great practical importance. *Agrobacterium tumefaciens* (LBA 4404) cells transformed with recombinant pRI201-AN-GUS clones by freeze-thaw method [18]. Chill *Agrobacterium* competent cells in an Eppendorf tube, by immediately freezing in liquid nitrogen before heat shock.

3. Transformation with the calcium chloride treatment or the freeze-thaw treatment method [4], which has been successfully used in dicots. The results are successful in depicting transformation of date palm fruits (Fig. 1).

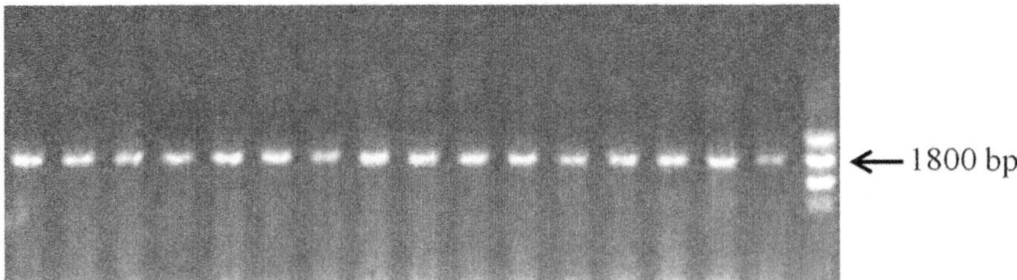



**Fig. 2** Detection of GUS gene in constructed plasmids (pRI 201-AN-GUS DNA) by PCR in transformed *A. tumefaciens* using GUS primers. *M* is DNA ladder marker; *lanes 1–16* are transformed *A. tumefaciens*

4. Adjust *A. tumefaciens* cultures carrying the pRI201-AN-GUS construct to an optical density ($OD_{600}$) equals 1. The exposure of fruit explants to *A. tumefaciens* inoculum of 0.8 $OD_{600}$ for 24 h results in 90% transformation efficiency.

5. The acetosyringone is involved in plant-pathogen recognition when added together with *A. tumefaciens*.

6. The thinness of the injection needle, besides minimizing the wound damages, allows very fine control of the injections in fruits through the epicarp. Injection is made such to avoid the locules where the bacteria could concentrate, thus making infection of the fruit less effective.

7. GUS expression is observed in transformed *A. tumefaciens* strain LB4404 but never in the *A. tumefaciens* control. PCR amplification of plasmid DNA, extracted from the transformed *A. tumefaciens*, demonstrates the generation of the expected amplicon, corresponding to the GUS gene (Fig. 2). These results validate the successful transformation of the *A. tumefaciens*.

8. GUS histochemical assay for screening the transgenic date palm fruits is according to Spolaore et al. [11] and with some modifications by Kapila et al. [19]. It verifies the expression of *uidA* gene in date palm fruits. The date palm fruits developing blue spots are recorded (Fig. 3).

9. The efficiency of DNA (GUS gene) delivery is assessed by agroinjection transient GUS expression in date palm fruits tissue. The blue spots produced by the GUS histochemical assay showed that all agroinjection tissues are affected by the transient GUS expression (Fig. 3). Results of a mean comparison show that GUS transient expression increases by increasing the time of agroinjection and with all fruits used for GUS staining of date palm fruits, which is reported for the first time in Saudi Arabia [12].

**Fig. 3** Agroinjection of date palm fruit of Khalas cv. (**a**) Date palm fruit at Khalal stage and *A. tumefaciens* suspension strain LBA4404 harboring the binary vector pRI201-AN-GUS carrying the beta-glucuronidase (GUS) plasmid. (**b**) The process of injecting date fruit flesh using a medical syringe filled with *A. tumefaciens* suspension. (**c**) GUS assay showing no blue spots on the non-agroinjected control fruit slices (extreme *right* and *left* of the photo), whereas *blue color* is visible in the transformed tissue

10. GUS histochemical assay for screening the transformed date palm fruits and keeping the date palm fruits in 70% ethanol for clear expression of *uidA* gene in agroinjected date palm fruits [8, 11].

11. The GUS gene under control of the 35S-promotor, with an intron, is commonly used as the reporter gene [19–22]. We also used the genetically engineered *A. tumefaciens* strain LBA4404 as a vector for infection in the transformation experiment, which contains plasmid pRI201-AN-GUS.

12. The reaction conditions of PCR amplification are total number of cycles to be run and the temperature and duration of each step in those cycles.

13. Detection of GUS gene in the PCR amplification band in constructed plasmids (pRI 201-AN-GUS DNA) in transformed *A. tumefaciens* using GUS forward and reverse primers Fig. 3.

## References


1. Jain SM, Al-Khayri JM, Johnson DV (eds) (2011) Date palm biotechnology. Springer, Dordrecht, p 743

2. Niblett CL, Bailey AM (2012) Potential applications of gene silencing or RNA interference (RNAi) to control disease and insect pests of date palm. Emir J Food Agric 24(5):462–469

3. Du J, Rietman H, Vleeshouwers VGAA (2014) Agroinfiltration and PVX Agroinfection in potato and *Nicotiana benthamiana*. J Vis Exp 83:e50971. doi:10.3791/50971

4. Orzaez D, Sophie M, Willemien WH, Granell A (2006) Agroinjection of tomato fruits: a tool for rapid functional analysis of transgenes directly in fruit. Plant Phys 140(1):3–11

5. Shah KH, Almaghrabi B, Bohlmann H (2013) Comparison of expression vectors for transient expression of recombinant proteins in plants. Plant Mol Biol Rep 31(6):1529–1538

6. Hellens RP, Allan AC, Friel EN, Bolitho K, Grafton K, Templeton MD et al (2005) Transient expression vectors for functional genomics, quantification of promoter activity and RNA silencing in plants. Plant Methods 1:13. doi:10.1186/1746-4811-1-13

7. Lee MW, Yang Y (2006) Transient expression assay by agroinfiltration of leaves. In: Salinas J, Sanchez-Serrano JJ (eds) *Arabidopsis* protocols. Humana Press, New York, NY, pp 225–229

8. Prabhavathi VR, Rajam MV (2007) Polyamine accumulation in transgenic eggplant enhances tolerance to multiple abiotic stresses and fungal resistance. Plant Biotechnol 24:273–282

9. Hansen G, Wright MS (1999) Recent advances in transformation of agricultural plants. Trends Plant Sci 4:226–231

10. Lorito M, Scala F (1999) Microbial genes expressed in transgenic plants to improve disease resistance. J Plant Pathol 81(2):73–88

11. Spolaore S, Trainotti L, Casadoro G (2001) A simple protocol for transient gene expression in ripe fleshy fruit mediated by *Agrobacterium*. J Exp Bot 52:845–850

12. Solliman MM (2017.) Towards production of transgenic date palm using agroinjection for transient expression of plant and *Agrobacterium-mediated* transformation. Plant Cell Biotech Mol Biol http://www.ikpress.org/articles-press/34 (in press)

13. Aslam J, Saeed AK, Abul Kalam MA (2015) *Agrobacterium-mediated* genetic transformation of date palm (*Phoenix dactylifera* L.) cultivar "Khalasah" via somatic embryogenesis. Plant Sci Today 2(3):93–101

14. Saker M, Ghareeb H, Kumlehn J (2009) Factors influencing transient expression of *Agrobacterium* mediated transformation of GUS gene in embryogenic callus of date palm. Adv Hort Sci 23:150–157

15. Saker M, Allam MA, Goma AH (2007) Optimization of some factors affecting genetic transformation of semi-dry Egyptian date palm cultivar (Sewi) using particle bombardment. J Genet Eng Biotechnol 5:1–6

16. Habashi AA, Kaviani M, Mousavi A, Khoshkam S (2008) Transient expression of β-glucuronidase reporter gene in date palm (*Phoenix dactylifera* L.) embryogenic calli and somatic embryos via microprojectile bombardment. J Food Agric Environ 6(2):160–163

17. Mousavi M, Mousavi A, Habashi AA, Dehsara B (2014) Genetic transformation of date palm (*Phoenix dactylifera* L. cv. 'Estamaran') via particle bombardment. Mol Biol Rep 41:8185–8194

18. Holsters M, de Waele D, Depicker A, Messens E, van Montagu M, Schell J (1978) Transfection and transformation of *Agrobacterium tumefaciens*. Mol Gen Genet 163:181–187

19. Kapila J, DeRycke R, VanMontagu M, Angenon G (1997) An *Agrobacterium*-mediated transient gene expression system for intact leaves. Plant Sci 122:101–108

20. Scholthof HB, Scholthof KB, Scholthof A, Jackson O (1996) Plant virus gene vectors for transient expression of foreign proteins in plants. Annu Rev Phytopathol 34:299–323

21. Jefferson RA (1987) Assaying chimeric genes in plants: the GUS gene fusion system. Plant Mol Biol Rep 5:387–405

22. Solliman MM, Mohasseb HA, Al-Khateeb AA, Al-Khateeb SA (2016) *In vitro* reliable regeneration protocol of tomato plants and efficient transformation system for transient gene expression in transgenic tomato varieties resistance to fungi. J Food Agric Environ 14 (3/4):29–36

# Part VI

# Secondary Metabolites and Abiotic Stress

# Chapter 25

# Bioreactor Steroid Production and Analysis of Date Palm Embryogenic Callus

## Sherif El-Sharabasy and Maiada El-Dawayati

## Abstract


Several compounds and families of compounds of date palm secondary metabolites have been investigated. The analysis of date palm tissue has shown the abundance of secondary metabolites including phytosterols, e.g., steroids, an important group of pharmaceutical compounds. Biotechnology offers the opportunity to utilize cells, tissues, and organs grown in vitro and manipulated to obtain desired compounds. This chapter presents a protocol for the production, determination, and identification of steroids in date palm callus tissue. The addition of 0.01 mg/L pyruvic acid as a precursor to MS liquid culture medium enhances steroid production. In addition, the chapter describes the sterol analytical techniques based on gas-liquid chromatography and gas chromatography-mass spectrometry.

**Key words** Bioreactor, Callus, Cell suspension, Gas-liquid chromatography (GLC), Gas chromatography-mass spectrometry (GC-MS), Phytosterols, Precursor, Secondary metabolites, Steroid


## 1   Introduction

Secondary metabolites of date palm (*Phoenix dactylifera* L.) have received special attention given their health-benefit claims and potential uses in the booming functional food and nutraceutical industries [1]. Investigations of date palm secondary metabolites led to the identification of several compounds such as phenolics, polyamines, phenolamides, phytosterols, and tocopherols [2–4]. Phytosterols, referred to plant steroids, which are diverse in their structures and biological functions and provide numerous medicinal applications including synthetic drugs used as antihormone contraceptives, anticancer, cardiovascular, osteoporosis, antibiotics, anesthetics, anti-inflammatories, and anti-asthmatics [5–7].

Phytosterols are abundant in shoot tips and pollen grains of date palm [8–10]. El-Sharabasy [10] separated and identified the steroids, cholesterol, and β-sitosterol from the tissue of two cultivars of date palm, Zagloul and Siwy, by thin layer


Jameel M. Al-Khayri et al. (eds.), *Date Palm Biotechnology Protocols Volume 1: Tissue Culture Applications*,
Methods in Molecular Biology, vol. 1637, DOI 10.1007/978-1-4939-7156-5_25, © Springer Science+Business Media LLC 2017

chromatography (TLC). The biosynthesis of secondary metabolites in plant cell culture is affected by precursors, physical environmental factors such as light and temperature, as well as nutrient medium composition [11]. El-Sharabasy [12] showed that 0.01 mg/L pyruvic acid was the best precursor, as compared to squalene and cholesterol, to stimulate steroid formation in callus.

Abdel-Ail [13] illustrated that using full MS salt strength nutrient culture medium gave the most effective results to promote steroid formation in date palm callus cv. Sakkoty. Furthermore steroid formation in date palm cv. Sakkoty callus culture could be induced by the addition of tryptophan amino acid at 50 mg/L [13].

Several different types of bioreactors provide ideal conditions for cell mass accumulation and expression of bioactive substances, including submerged stirred bioreactors, bubble columns, airlift bioreactors and their modifications, and gas-phase bioreactors [14, 15]. El-Sharabasy [12] found that bioreactor systems produce steroids from embryogenic callus of date palm 14 times faster than on the solid medium. Commercial production of steroids from date palm callus as important medicinal compounds can be achieved successfully by using bioreactor systems [12].

This chapter presents a protocol for the determination and identification of total steroids in date palm tissues, induced by the addition of pyruvic acid as a precursor in Murashige and Skoog (MS) medium supplemented with plant growth regulators for callus proliferation. Cell suspension cultures are established in a submerged stirred bioreactor, and the sterols are analyzed by gas-liquid chromatography (GLC) and gas chromatography-mass spectrometry (GC-MS) analytical techniques.

## 2  Materials

### 2.1  Plant Materials and Sterilization

1. Healthy offshoots, 3–5 years old and weigh 5–7 kg, from mature date palm Siwy cv.

2. Antioxidant solution: Ascorbic acid (150 mg/L) and citric acid (100 mg/L).

3. Disinfectant solution 1: 0.5% (w/v) sodium hypochlorite solution (10% v/v Clorox, commercial bleach) containing three drops of Tween 20 per 100 mL.

4. Disinfectant solution 2: 2.6% (w/v) sodium hypochlorite solution (50% v/v Clorox, commercial bleach) containing three drops of Tween 20 per 100 mL.

### 2.2  Culture Medium

1. Basal culture medium: Stock solutions of Murashige and Skoog (MS) culture medium [16] (Table 1).

**Table 1**
**Chemical composition of the MS nutrient medium [16]**

| Medium composition | Stock concentration (mg/L) | Final concentration in culture medium (mg/L) |
|---|---|---|
| *Stock I: Major inorganic nutrients (20× stock) use 50 mL to prepare 1 L of medium* | | |
| $NH_4NO_3$ | 33,000 | 1650 |
| $KNO_3$ | 38,000 | 1900 |
| $CaCl_2 \cdot 2H_2O$ | 8800 | 440 |
| $MgSO_4 \cdot 2H_2O$ | 7400 | 370 |
| $KH_2PO_4$ | 3400 | 170 |
| $NaH_2PO_4 \cdot H_2O$ | 3400 | 170 |
| *Stock II: Minor inorganic nutrients (200× stock) use 5 mL to prepare 1 L of medium* | | |
| KI | 166 | 0.83 |
| $H_3BO_3$ | 1240 | 6.2 |
| $MnSO_4 \cdot 2H_2O$ | 4460 | 22.3 |
| $ZnSO_4 \cdot 7H_2O$ | 1720 | 8.6 |
| $Na_2.MoO_4 \cdot 2H_2O$ | 50 | 0.25 |
| $CuSO_4 \cdot 5H_2O$ | 5 | 0.025 |
| $CoCl_2 \cdot 6H_2O$ | 5 | 0.025 |
| *Stock III: Iron source (200× stock) use 5 mL to prepare 1 L of medium* | | |
| $FeSO_4 \cdot 7H_2O$ | 5560 | 27.8 |
| $Na_2EDTA \cdot 2H_2O$ | 7460 | 37.3 |
| *Stock IV: Vitamins (200× stock) use 5 mL to prepare 1 L of medium* | | |
| *myo*-Inositol | 25,000 | 125 |
| Nicotinic acid | 200 | 1 |
| Pyridoxine·HCl | 200 | 1 |
| Thiamine·HCl | 200 | 1 |
| Glycine | 400 | 2 |
| Biotin | 200 | 1 |
| *Carbon source* | | |
| Sucrose | – | 30,000 |
| *Antioxidants* | | |
| Glutamine | – | 200 |
| Citric acid | – | 100 |
| Ascorbic acid | – | 100 |

**Table 2**
**Different culture stages and their corresponding additives supplemented to the MS medium**

| Culture stage | 2,4-D, mg/L | 2iP, mg/L | NAA, mg/L | Agar, g/L | Activated charcoal, g/L | Sucrose, mg/L | Pyruvic acid, mg/L |
|---|---|---|---|---|---|---|---|
| (I) Culture induction | 100 | 3 | | 6 | 3 | 30 | |
| (II) Embryogenic callus formation | 10 | 3 | | 6 | 3 | 30 | 0.01 |
| (III) Cell suspension | | | 0.1 | | | 30 | 0.01 |

2. Hormone stock solutions (1 mg/mL each): 2,4-dichlorophenoxyacetic acid (2,4-D), 2-isopentenyladenine (2iP), and naphthaleneacetic acid (NAA).

3. Culture medium additives according to culture stages (Table 2): Callus induction medium (I), embryogenic callus formation medium (II), and cell suspension (III).

*2.3  Cell Culture and Bioreactor*

1. Bioreactor: 2 L flask containing 1.5 L liquid medium.
2. Orbital shaker.

*2.4  Analysis of Total Steroids and Sterol Content*

1. Diethyl ether solution.
2. Hydrochloric acid solution 5% w/v.
3. Potassium hydroxide solution in alcohol 5% (90% v/v).
4. Potassium hydroxide aqueous solution 3% w/v.
5. Phenolphthalein solution.
6. Anhydrous sodium sulfate.
7. Glacial acetic acid solution.
8. β-sitosterol.
9. Sulfuric acid solution.
10. Mercury oxide ($HgO_2$).

*2.5  Equipment*

1. Instruments: Gas-liquid chromatography (GLC) (PYE UNICAM PRO-GC), gas chromatography-mass spectrometer (GC-MS), bioreactor type Virtis Omni (submerged stirred bioreactor), spectrophotometer, hot air oven, orbital shaker, and water bath.

2. Glassware: Erlenmeyer flasks (500 mL), volumetric flasks (1000 mL), separator funnel, sintered glass Buckner filter (porosity grade G4, 5–15 μm), and small glass bowls (50 mL Gooch crucibles).

3. Tools and supplies: Scalpel, forceps, spatula, and Whatman No. 1 filter paper.

## 3   Methods

### 3.1   Preparation of Explant

1. Remove all layers of hard, fibrous leaf bases surroundeding the apical regions with machete or hacksaw till the softer core tissues are exposed with three or four young leaf bases surrounding the entire meristem and leaf primordial.

2. Wash the tissue in running tap water for 1–2 h and then immerse in sterile antioxidant solution for 1 h.

3. Surface sterilize by dipping the shoot tip tissue in 70% ethanol for 3 min and then rinse once with sterile distilled water.

4. Remove the outer young leaf from the shoot tip tissue with a sterile scalpel.

5. Place the shoot tip in the disinfection solution 1 for 15 min and then thoroughly wash once with sterile distilled water.

6. Place the shoot tip explants in the disinfection solution 2 for 25 min, and finally wash with sterile distilled water three times.

7. Remove outer soft leaves and excise the meristem tip surrounded with 4–6 leaf primordia, 2–2.5 cm in length.

### 3.2   Preparation Culture Medium

1. Prepare stock solutions of plant growth regulators by dissolving the 2,4-D, NAA in 95% ethanol, or 1 N NaOH, whereas 2iP using 1 N HCl makes the required volume by adding double distilled water. Store in the refrigerator at 4 °C up to 1 month.

2. Mix (MS) salt (Table 1) solutions with other components for each nutrient culture medium.

3. Adjust pH to 5.7 with 0.1 and 1 N NaOH and 0.1 and 1 N HCl before adding agar.

4. Add the agar, and heat the mixture until fully dissolved. Distribute the medium in small culture jars (200 mL), 30 mL per jar.

5. Cap culture jars with polypropylene closure and autoclave for 20 min at 121 °C and 1.1 kg/cm$^2$.

### 3.3   Callus Initiation and Embryonic Callus Formation

1. Cut the sterilized shoot tip longitudinally into four sections and culture initially on nutrient medium for callus induction (I) (Table 2) and subcultures to fresh culture medium at each 6-week interval. Maintain under total darkness at 27 ± 1 °C incubation conditions.

2. Transfer the callus tissues to nutrient medium for embryogenic callus formation containing 0.01 mg/L pyruvic acid (II) (Table 2) (*see* **Note 1**).

3. Subculture at each 6-week interval until obtaining the embryogenic callus mass, and maintain under total darkness at 27 ± 1 °C.

**Fig. 1** Bioreactors for steroid production from date palm cell suspension. (**a**) Bioreactor apparatus used to produce steroid compounds and cell suspension culture from embryonic callus of date palm, (**b**) date palm cell suspension culture from bioreactor culturing embryonic callus

### 3.4 Bioreactor Operation

1. For aggregating embryogenic callus volume to increase steroid production, replace previously obtained embryogenic callus batch to the bioreactor (submerged stirred bioreactor type; *see* **Note 2**) with a 2 L flask containing 1.5 L of cell suspension nutrient medium (III) (Table 2) which is supplemented with 0.01 mg/L pyruvic acid to induce sterol production.

2. Maintain on orbital shaking at 150 rpm.

3. Sterilize aeration, through air filter 0.2 mL and sterilized fiber cotton, is 0.6 volume of air/volume of fermenter/min (VVm).

4. Adjust temperature through warm water pump at $27 \pm 1\,^{\circ}\mathrm{C}$, light intensity by cool-white florescent light ($40\ \mu\mathrm{mol/m^2/s}$) (Fig. 1a).

### 3.5 Determination of Total Steroids

1. Dry 0.5 g weight of embryogenic callus sample, in an oven at 75 °C for 48 h.

2. Place dried embryogenic callus sample in a clean flask, and add 100 mL of 5% potassium hydroxide solution in alcohol (90% v/v) to it. Heat in a water bath at 50 °C for 2 h and then cool it for 5 min.

3. Transfer the callus sample to a separator funnel. Wash the flask with 100 mL water followed by 100 mL diethyl ether. Transfer the washings into the same separator funnel and shake slowly by hand for 3 min (Fig. 2a).

4. Separate the formed layer by removing the aqueous phase from separator funnel (Fig. 2b).

**Fig. 2** Determination of total steroids. (**a**) The separator funnel, (**b**) separated layer which appeared

5. Wash this layer in a separator funnel four times with 100 mL diethyl ether and place in a clean flask.

6. Wash the received ethereal extracts with successive portions of 40 mL water (shake gently to avoid emulsions), 40 mL 5% w/v hydrochloric acid, and 40 mL 3% w/v potassium hydroxide aqueous solution.

7. Wash with successive portions of 40 mL water (each wash) until the washings become neutral to phenolphthalein solution (use 2 drops 1% phenolphthalein in 70% ethanol and 2 N NaOH until rose color is stable).

8. Add one drop of 0.1 N HCl to sample and rapidly mix until the rose color disappears.

9. Add 100 mg anhydrous sodium sulfate powder, shake well, and filter the mixture through folded Whatman filter paper.

10. Evaporate the resulted solution in water bath at 50 °C until fully dry.

11. Add 100 mL glacial acetic acid to the residue and stir for 30 min in small glass bowl.

12. Take 2 mL of previous resulted solution, then transfer to a 20 mL volumetric flask, and dilute to 20 mL with glacial acetic acid (test solution).

13. Dissolve 40 mg β-sitosterol in 100 mL glacial acetic acid then take 5 mL of this solution and dilute it to 50 mL with glacial acetic acid (reference solution).

14. Prepare Deniges reagent as two solutions: (solution A) 100 mL sulfuric acid and 50 mL glacial acetic acid and (solution B) 5 g mercury oxide ($HgO_2$), 100 mL water, and 20 mL sulfuric acid. Take 100 mL solution (A) and 1 mL solution (B), mix them, and filter through a sintered glass filter (grade G4) before use (*see* **Note 3**).

15. Take 5 mL from Deniges reagent in a tube filled with 1 mL (test solution) and 1 mL (reference solution) previously obtained in **steps 12** and **13**.

16. Carry out the blank in 1 mL glacial acetic acid instead of the sample in a test tube. Both tubes are left to stand in the dark for 15 min. The absorbance is read using a spectrophotometer at 510 nm against the blank. The amount of steroids is calculated as β-sitosterol from a standard curve prepared by dissolving 40 mg of β-sitosterol in 10 mL glacial acetic acid. Series of standards are prepared as 5, 10, 20, and 40 mg/100 mL, respectively; 1 mL of each is mixed with 5 mL Deniges reagent, and read at 510 nm against the blank. The absorbance of each concentration is plotted against the absorbance obtained from the standard curve.

### 3.6 Identification of Steroid Composition Using Gas–Liquid Chromatography (GLC)

1. Use capillary column OV 17 (methyl phenyl silicone) 1.5 m × 4 mm.

2. Adjust conditions of GLC to separate unsaponifiable materials according to the following parameters (*see* **Note 4**):

   (a) Temperature programming:
   - Initial 70 °C, upper 270 °C, rate 10 °C/min.
   - Injector 250 °C, $N_2$ carrier.
   - Detector 300 °C, $H_2$ flame ionization detector (FID).

   (b) Flow rate of gases:
   - $N_2$: 30 mL/min.
   - $H_2$: 33 mL/min.
   - Air: 330 mL/min.

   (c) Chart speed: 0.50 cm/min.

### 3.7 Identification of Steroids by Gas Chromatography-Mass Spectrometry (GC-MS)

To identify the steroidal components of the unsaponifiable fraction obtained from date palm embryogenic callus, use the GC-MS technique (*see* **Note 5**) under the following conditions:

- Column: Capillary column Hp.1 (cross-linked) methyl silicone gum phase. 12 m × 0.2 mm × 0.33 mm film thickness, 100% dimethylpolysiloxane gum.
- Carrier gas: Helium.
- Oven temperature: 50–325 °C.
- Injector temperature: 250 °C.
- Initiate temperature: 50 °C.
- Final temperature: 270 °C.
- Program rate: 10 °C/min.
- Initial time: 1 min.
- Total time: 23 min.
- Ionizing voltage: 70 eV.

## 4  Notes

1. Adding 0.01 mg/L pyruvic acid precursor to the culture medium is necessary to stimulate steroid formation.

2. Cell culture systems could be used for the large-scale culturing of plant cells from which secondary metabolites can be extracted. For plant cell culture techniques to become economically viable, it is important to develop methods for consistent high yields of products from cultured cells. Typical bioreactors for plant cell and tissue cultures are made of glass or stainless steel. Bioreactor types include stirred reactors, rotating drum reactors, airlift reactors, bubble columns, fluidized bed reactors, packed bed reactors, and trickle bed reactors. Culture volumes vary and may reach up to 75 m$^3$ in bioreactor types for commercial production processes.

3. Deniges reagent should be prepared immediately before use.

4. The steroid composition in both in vitro culture tissue and their diffusers in MS basal medium treated with precursor (pyruvic acid) as separated and identified by GLC revealed the presence of compounds cholesterol, estrone, ethinylestradiol, ethisterone, ostrial, stigmasterol, and sitosterol (Fig. 3a, b).

5. The GC-MS technique is used to identify the steroidal components of the unsaponifiable fraction obtained from date palm embryogenic callus. Steroid compounds such as estrogen, androstene, hydroxyprogesterone, and progesterone are identified by GC-MS instrument in the unsaponifiable fraction obtained from date palm embryogenic callus.

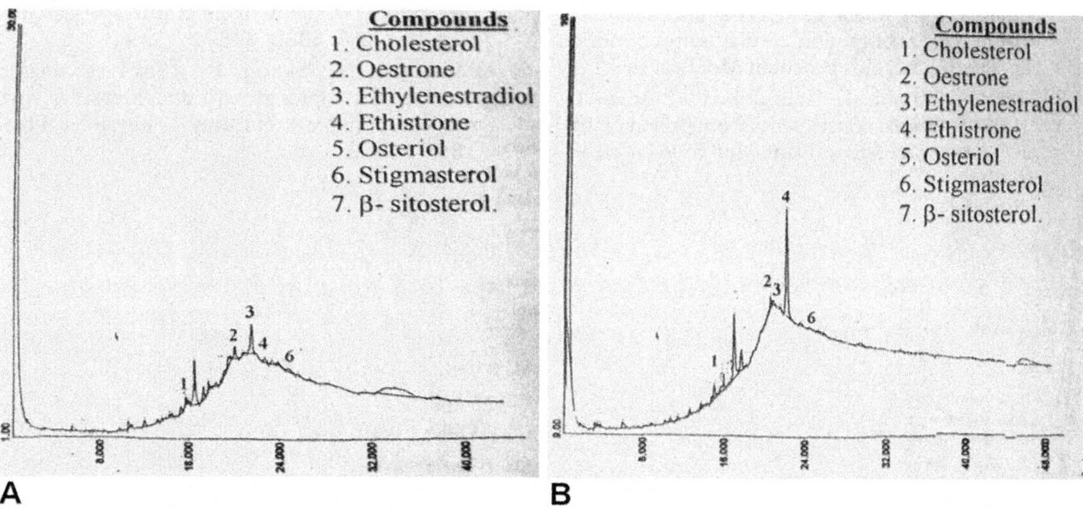


**Fig. 3** Identification of steroid composition using GLC. (**a**) Gas-liquid chromatogram of steroids separated from callus without precursor, (**b** (gas-liquid chromatogram of steroids separated from callus treated with 0.01 mg/L pyruvic acid precursor

## Acknowledgment

The author wishes to acknowledge researchers Mr. Walid Badwy and Mr. Mahmoud Hamza for their vital help to complete this work.

## References

1. Biglari F, AlKarkhi A-F, Easa AM (2008) Antioxidant activity and phenolic content of various date palm (*Phoenix dactylifera* L.) fruits from Iran. Food Chem 107:1636–1641

2. Mossa J, Hifnawy M, Mekkawi G (1986) Phytochemical and biological investigations on date seeds (*Phoenix dactylifera* L.) produced in Saudi Arabia Arab Gulf J Sci Res 4(2):495–507

3. Daayf F, El Bellaj M, El Hassni M (2003) Elicitation of soluble phenolics in date palm (*Phoenix dactylifera*) callus by *Fusarium oxysporum* f. sp. *albedinis* culture medium. Environ Exp Bot 49:41–47

4. El Hadrami A, Daayf F, El Hadrami I (2011) Secondary metabolites of date palm. In: Jain SM, Al-Khayri JM, Johnson DV (eds) Date palm biotechnology. Springer, Dordrecht, pp 653–674

5. Sultan A, Raza AR (2015) Steroids: a diverse class of secondary metabolites Med Chem 5 (7):310–317

6. Jovanovic Santa SS, Petri ET, Klisuric OR, Szecsi M, Kovacevic R (2015) Antihormonal potential of selected D-homo and D-secoestratriene derivatives. Steroids 97:45–53

7. Rattanasopa C, Phungphong S, Wattanapermpool J, Bupha-Intr T (2015) Significant role of estrogen in maintaining cardiac mitochondrial functions. J Steroids Biochem Mol Biol 147:1–9

8. Zaki A, Schmidt J, Hammouda F, Adam G (1993) Steroidal constituents from pollen grains of *Phoenix dactylifera*. Plant Med 59:613–614

9. Bennett R, KOS HE (1966) Isolation of estrone and cholesterol from the date palm. Phytochemistry 5:231–235

10. El-Sharabasy SF (2000.) Studies on the production of secondary metabolites from date palm by using tissue culture technique. Ph.D. Thesis, Fac Agric, Al-Azhar University, Cairo, p 200

11. Christian M (2015) Steroids – chemical constituents of *Phyllanthus fraternus* Webster through TLC and HPTLC. Int Res J Chem 2321–2399

12. El-Sharabasy SF (2004) Effects of some precursors on development of secondary products in tissues and media of embryogenic callus of date palm cv. Sewi Arab. J Biotechnol 7 (1):83–90

13. Abdel-Ail W ( 2011.) Some factors affecting secondary metabolites production in date palm by using plant tissue and cell culture technique. M.Sc. Thesis, Aromatic Plants Dept, Fac Agric, Al-Azhar University, Cairo, p 147

14. Eibl R, Eibl D (2008) Design of bioreactors suitable for plant cell and tissue cultures. Phytochem Rev 7:593–598

15. Filová A (2014) Production of secondary metabolites in plant tissue cultures research. J Agric Sci 46(1):336–345

16. Murashige T, Skoog F (1962) A revised medium for rapid growth and bioassays with tobacco tissue cultures. Physiol Plant 15:473–497

# Extraction and Estimation of Secondary Metabolites from Date Palm Cell Suspension Cultures

**Poornananda M. Naik and Jameel M. Al-Khayri**

## Abstract

The health benefits of dates arise from their content of phytochemicals, known for having pharmacological properties, including flavonoids, carotenoids, phenolic acids, sterols, procyanidins, and anthocyanins. In vitro cell culture technology has become an attractive means for the production of biomass and bioactive compounds. This chapter describes step-by-step procedures for the induction and proliferation of callus from date palm offshoots on Murashige and Skoog (MS) medium supplemented with plant growth regulators. Subsequently cell suspension cultures are established for optimum biomass accumulation, based on the growth curve developed by packed cell volume as well as fresh and dry weights. The highest production of biomass occurs at the 11th week after culturing. Moreover, this chapter describes methodologies for the extraction and analysis of secondary metabolites of date palm cell suspension cultures using high-performance liquid chromatography (HPLC). The optimum level of catechin, caffeic acid, apigenin, and kaempferol from the cell suspension cultures establishes after the 11th and 12th weeks of culture. This protocol is useful for scale-up production of secondary metabolites from date palm cell suspension cultures.

**Key words** Biomass, Cell suspension, Date palm, Growth curve, High-performance liquid chromatography (HPLC), Secondary metabolites

## 1 Introduction

Date palm (*Phoenix dactylifera* L.) is a monocotyledonous angiosperm diploid species ($2n = 36$) which belongs to the Arecaceae family [1]. Date palm has high socioeconomic importance due to its food value and provides many other products such as shelter, fiber, clothing, aesthetics, and furniture [2]. Dates represent a high-energy food source besides being rich in dietary fibers [1]. Date fruits are rich in phenolics [2]. The phenolic compounds play an important role in the defense mechanism of plants against biotic and abiotic stresses [3, 4]. Date palm cultivars exhibit distinct levels and profiles of phenolic acids such as gallic, protocatechuic, *p*-hydroxybenzoic, vanillic, caffeic, syringic, *p*-coumaric, ferulic, and *o*-coumaric acids. Other studies have reported an abundance

Jameel M. Al-Khayri et al. (eds.), *Date Palm Biotechnology Protocols Volume 1: Tissue Culture Applications,*
Methods in Molecular Biology, vol. 1637, DOI 10.1007/978-1-4939-7156-5_26, © Springer Science+Business Media LLC 2017

of flavonoids [5–7]. Flavonoids such as luteolin, quercetin, and kaempferol, as well as polymerized proanthocyanidins, are also found in abundant quantities in dates [2].

The phenolics of dates are known to exhibit various pharmacological activities including antiaging, anticancer, antioxidant, antiviral, and antimicrobial properties, making them a remedy for certain diseases and prevention of chronic inflammations. In addition, they prevent oxidative damages caused by lymphocytes phagocytosis activity of invasive pathogens and pests [8]. In addition, phenolics inhibit $\alpha$-amylase and $\alpha$-glucosidase activities and are widely used for the control of blood glucose level in type II diabetes [8–10].

Alpha-tocopherol is claimed to reduce the risks of various cancers, Alzheimer's and Parkinson's diseases, and other dementias. Vitamin E is also used against allergies, asthma, and other respiratory problems, as well as digestive or circulatory diseases. Additionally, vitamin E is used topically against dermatitis, aging skin, and granuloma annulare as well as in preventing skin ulceration often caused as a consequence of using chemotherapeutic drugs [7].

The production of dates by field cultivation is a slow process and needs several years from planting to the fruiting stage for harvest. Furthermore, environmental conditions, pests, and diseases limit the production of dates. The phenolic content of dates varies depending on the age of the plant, season of the harvest, and cultivar as well as the extraction method [2]. Therefore, in vitro cell and organ cultures have become attractive alternatives for the production of date palm biomass and secondary metabolites. Recently, bioreactor technology is playing a vital role for the large-scale production of plant cells providing a source for harvesting secondary metabolites. Bioreactors allow monitoring and controlling important process variables such as pH, temperature, dissolved oxygen, oxygen uptake rate, carbon dioxide evolution rate, temperature, and specific gravity [11]. Additionally, bioreactor technology provides uninterrupted production of biomass and phenolic compounds with reduced cost and time [12]. Recently, date palm cell suspension cultures have been established from callus [13–15]. These cell suspension cultures demonstrated sustainable production of phenolic compounds [12, 16] and showed that secondary metabolite accumulation was influenced by ammonium nitrate, and total phenolic contents ranged from $0.653 \pm 0.002$ to $2.115 \pm 0.010$ mg/g gallic acid equivalents (GAE) induced per explant [17]. The present chapter describes procedures for the induction of callus and establishment of cell suspension cultures for optimum biomass accumulation and the production of secondary metabolites from date palm cultivar Shishi. It also describes the extraction methodology as well as the protocol for the quantification of secondary metabolites (catechin, caffeic acid, apigenin, and kaempferol) using high-performance liquid chromatography (HPLC).

## 2  Materials

### 2.1  Plant Materials and Sterilization

1. Date palm cv. Shishi offshoots: Shoot tip is isolated, disinfected, and cut into small pieces to serve as explants following the procedures described in subheading 3.

2. Antioxidant solution: ascorbic acid and citric acid, 150 mg/L each.

3. Disinfectant solution: 1.6% (w/v) sodium hypochlorite solution (30% v/v Clorox, commercial bleach) containing 3 drops of Tween 20 per 100-mL solution.

### 2.2  Culture Medium

1. The stock solutions of Murashige and Skoog (MS) [18] medium (MS stocks I, II, III, and IV) (Table 1).

2. Hormone stock solutions: 2,4-dichlorophenoxyacetic acid (2,4-D, 5 mg/mL), 2-isopentenyladenine (2iP, 1 mg/mL), and naphthaleneacetic acid (NAA, 1 mg/mL) (*see* **Note 1**).

3. Solutions to adjust pH: 0.1 and 1 N NaOH and 0.1 and 1 N HCl.

### 2.3  Medium Additives Used for Various Culture Stages (Table 2)

1. Culture initiation (CI): MS medium containing 100 mg/L 2,4-D, 3 mg/L 2iP, and 1.5 g/L activated charcoal with 8 g/L agar.

2. Callus proliferation (CP): MS medium containing 6 mg/L 2iP, 10 mg/L NAA, and 1.5 g/L activated charcoal with 8 g/L agar.

3. Callus maintenance (CM): MS medium containing 1.5 mg/L 2iP and 10 mg/L NAA with 8 g/L agar.

4. Cell suspension (CS): MS liquid medium containing 1.5 mg/L 2iP and 10 mg/L NAA.

### 2.4  Extraction and HPLC Analysis of Secondary Metabolites

1. Mobile phase solution: Methanol, acetonitrile, and water in a 40:15:45 (v/v/v) compositions, respectively, and are supplemented with 1% glacial acetic acid (*see* **Note 2**).

2. HPLC standard compounds: catechin, caffeic acid, apigenin, and kaempferol.

### 2.5  Equipment

1. Instruments: autoclave, weighing balance, refrigerator, magnetic stirrer, microwave oven, hot air oven, water bath, rotary evaporator, centrifuge machine, orbital shaker, HPLC instrument, and ultrasonic bath.

2. Glassware: 1000-mL conical flask, 1000-mL measuring cylinder, 250-mL conical flask, 200-mL tissue culture jar, and 15-mL sterile graduated centrifuge tube.

3. Tools: pestle and mortar, Whatman Grade 1 filter paper, 0.45-μm membrane filter, and syringe.

**Table 1**
**Components of MS basal medium [18] and additives used for date palm in vitro culture stages. Hormones, agar, and activated charcoal are added according to the culture stage as shown in Table 2**

| Medium composition | Stock concentration (mg/L) | Final concentration in culture medium (mg/L) |
| --- | --- | --- |
| *Stock I: Major inorganic nutrients (20× stock) use 50 mL to prepare 1 L of medium* | | |
| $NH_4NO_3$ | 33,000 | 1650 |
| $KNO_3$ | 38,000 | 1900 |
| $CaCl_2 \cdot 2H_2O$ | 8800 | 440 |
| $MgSO_4 \cdot 2H_2O$ | 7400 | 370 |
| $KH_2PO_4$ | 3400 | 170 |
| $NaH_2PO_4 \cdot H_2O$ | 3400 | 170 |
| *Stock II: Minor inorganic nutrients (200× stock) use 5 mL to prepare 1 L of medium* | | |
| KI | 166 | 0.83 |
| $H_3BO_3$ | 1240 | 6.2 |
| $MnSO_4 \cdot 2H_2O$ | 4460 | 22.3 |
| $ZnSO_4 \cdot 7H_2O$ | 1720 | 8.6 |
| $Na_2.MoO_4 \cdot 2H_2O$ | 50 | 0.25 |
| $CuSO_4 \cdot 5H_2O$ | 5 | 0.025 |
| $CoCl_2 \cdot 6H_2O$ | 5 | 0.025 |
| *Stock III: Iron source (200X stock) use 5 mL to prepare 1 L of medium* | | |
| $FeSO_4 \cdot 7H_2O$ | 5560 | 27.8 |
| $Na_2EDTA \cdot 2H_2O$ | 7460 | 37.3 |
| *Stock IV: Vitamins (200× stock) use 5 mL to prepare 1 L of medium* | | |
| *myo*-Inositol | 25,000 | 125 |
| Nicotinic acid | 200 | 1 |
| Pyridoxine·HCl | 200 | 1 |
| Thiamine·HCl | 200 | 1 |
| Glycine | 400 | 2 |
| Biotin | 200 | 1 |
| *Carbon source* | | |
| Sucrose | – | 30,000 |
| *Antioxidants* | | |
| Glutamine | – | 200 |
| Citric acid | – | 100 |
| Ascorbic acid | – | 100 |

**Table 2**
**Different culture stages and their corresponding hormonal and activated charcoal additives supplemented to the MS medium (Table 1)**

| Culture stage | Hormones, agar, and activated charcoal additives | | | | |
| --- | --- | --- | --- | --- | --- |
| | 2,4-D, mg/L | 2ip, mg/L | NAA, mg/L | Agar, g/L | Activated charcoal, g/L |
| Culture initiation (CI) | 100 | 3 | – | 8 | 1.5 |
| Callus proliferation (CP) | – | 6 | 10 | 8 | 1.5 |
| Callus maintenance (CM) | – | 1.5 | 10 | 8 | – |
| Cell suspension (CS) | – | 1.5 | 10 | – | – |

## 3  Methods

### 3.1  Culture Medium Preparation

1. Soak the glassware in a liquid detergent for 1 h and thoroughly wash with hot water. Rinse the glassware with double distilled water and air-dry.

2. MS stocks: Weigh the MS stocks I, II, III, and IV components individually by using weighing balance, and dispense each stock in a 1000-mL conical flask separately; dissolve each component in 800-mL distilled water with the help of magnetic stirrer and make up the volume of 1000 mL by using 1000-mL measuring cylinder. Transfer the stocks into reagent bottles and store the solutions in a refrigerator at 4 °C until use (*see* **Note 3**).

3. Add MS stocks I, II, III, and IV components (*see* Table 1) and mix the ingredients for each medium as mentioned in Subheading 2.3 (*see* Table 2) in a 1000-mL conical flask separately. Make up the medium volume of 800-mL using distilled water, and adjust the medium to pH 5.8 using the solution 0.1 and 1 N NaOH and 0.1 and 1 N HCl. Finally make up the medium volume of 1000-mL using distilled water and 1000-mL measuring cylinder. Heat the medium in a microwave oven until the agar completely dissolves.

4. Dispense 40-mL medium in a 200-mL tissue culture jar with screw cap and 50-mL medium in a 250-mL conical flasks for cell suspension and cap using aluminum foil and rubber bands. Sterilize all the media using an autoclave for 15 min at 121 °C and 1.1 kg/cm$^2$. Store the medium flasks in a room temperature, maximum of 2 weeks.

### 3.2  Explant Preparation

1. Separate 3- to 4-year-old offshoots from mother trees. Remove the outer leaves exposing the shoot-tip region (about 8 cm long) and immediately dip in a beaker containing chilled antioxidant solution to prevent oxidation-induced browning (Fig. 1a, b).

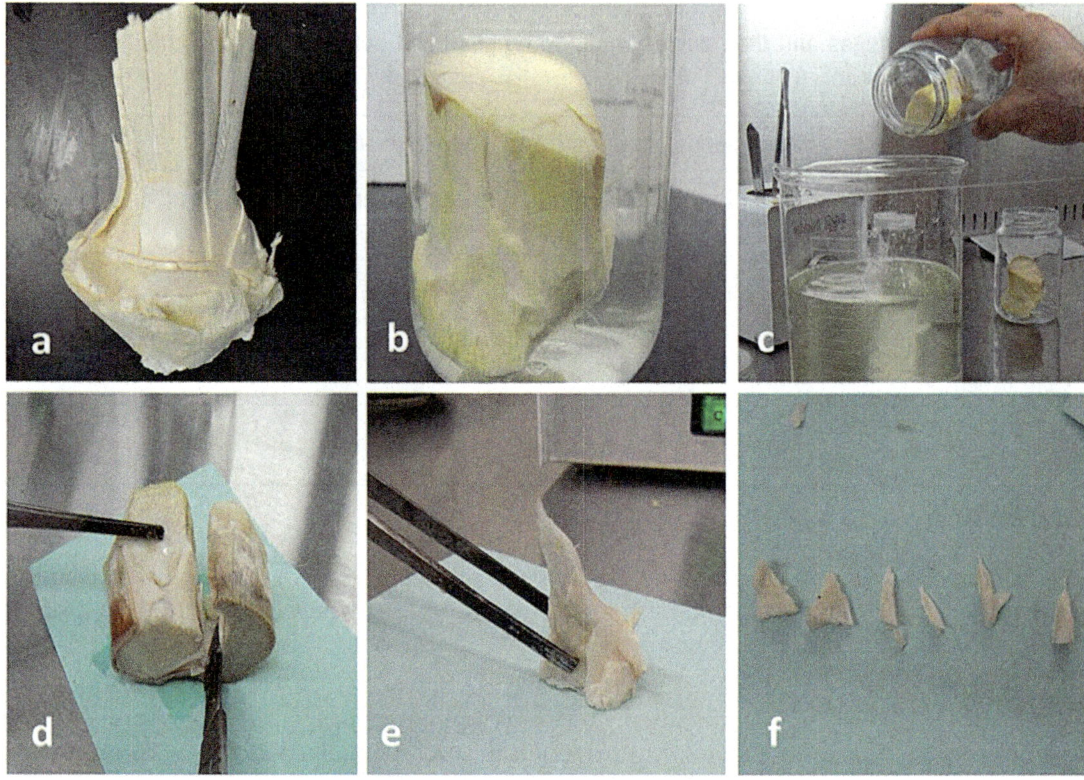

**Fig. 1** Date palm explant preparation: (**a**) shoot-tip region, (**b**) shoot-tip region in the antioxidant solution, (**c**) disinfection of shoot-tip region, (**d**) dissection of shoot-tip region, (**e**) prominent shoot-tip portion, (**f**) explants isolated from the shoot-tip region

2. Surface-sterilize the shoot-tip tissue in 70% ethanol for 1 min. And then in disinfectant solution for 15 min. (Fig. 1c).

3. Rinse the tissue in sterilized distilled water 3–5 times for 5 min each, and then remove the tissue surrounding the shoot-tip terminal until the shoot-tip terminal is exposed, about 1 cm long. Excise the tip and section longitudinally into 6–10 small sections (0.8 cm) inside the laminar flow hood (Fig. 1d–f).

### 3.3 Callus Induction from Shoot-Tip Explants

1. For callus culture initiation, place shoot-tip explants on the surface of CI medium (Fig. 2a; Table 2).

2. Incubate cultures at 25 ± 2 °C for 12 weeks in the darkness. Subculture the callus cultures to a fresh medium at a 4-week interval. Explants start to expand in the initial 4 weeks of culture (Fig. 2b). After 8–10 weeks of culture, callusing initiates from the shoot-tip explants.

3. Transfer callus to the CP medium and maintain for an additional 12 weeks to obtain sufficient growth for subsequent experiments (Fig. 2c; Table 2).

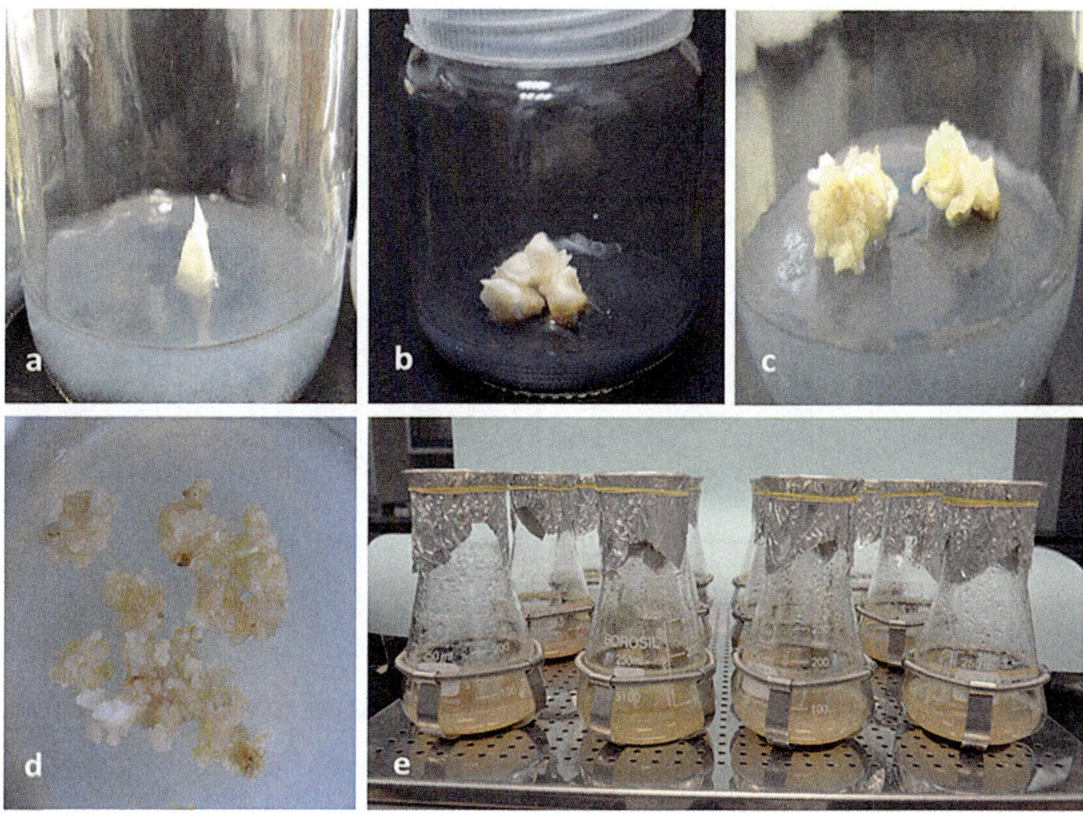

**Fig. 2** Stages of date palm cell suspension culture: (**a**) explants on culture initiation medium, (**b**) expansion of explant after 4 weeks, (**c**) callus formation in the proliferation medium, (**d**) callus in the maintenance medium, (**e**) cell suspension culture on an orbital shaker

***3.4 Maintenance of Callus and Establishment of Cell Suspension***

1. The friable callus collected from the proliferation medium is subcultured on the CM medium (Fig. 2d; Table 2) at 4-week interval for prolific callus growth as desired. The calli were subcultured 3 times before transferring it to the cell suspension medium.

2. Collect actively growing friable callus from the solid medium cultures and inoculate 500-mg fresh biomass per 250-mL conical flask containing 50-mL cell suspension medium (Fig. 2e; Table 2).

3. Place the culture flasks containing cell suspension medium on a rotary shaker set at 150 rpm under a 16-h photoperiod of cool-white florescent light (40 $\mu$mol//m$^2$/s) and 25 $\pm$ 2 °C. Every 2 weeks, the cultures are filtered by using a funnel fitted with Whatman Grade 1 filter paper, and filtrate is inoculated into two conical flasks containing 50-mL fresh CM medium. This also serves as cell suspension maintenance medium.

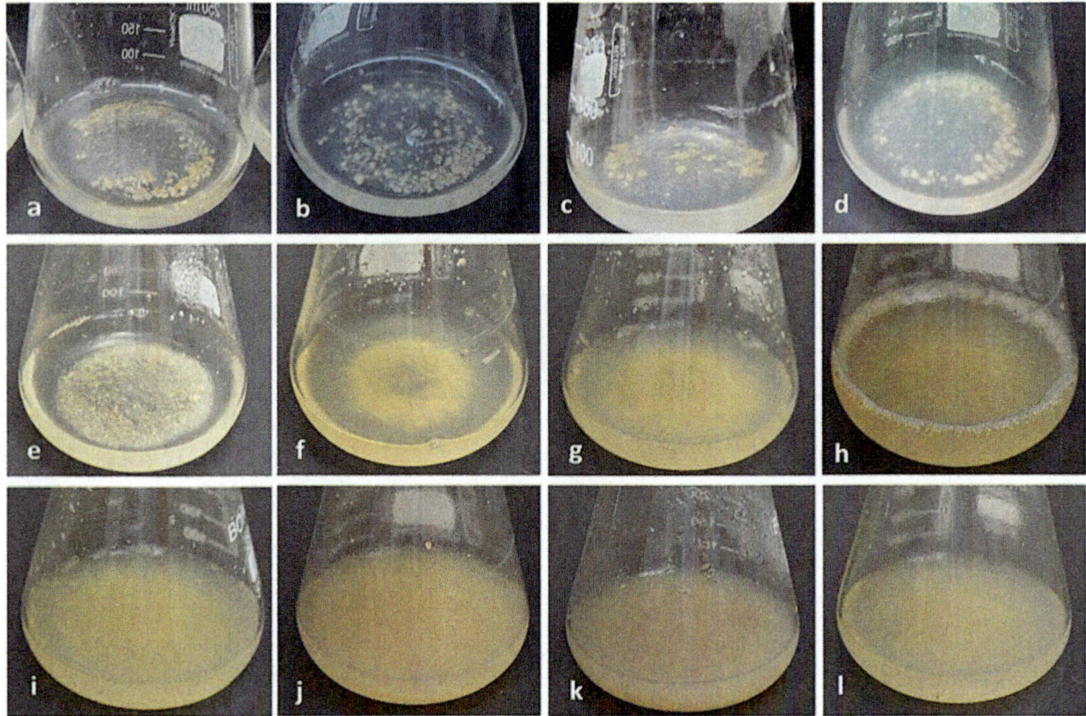

**Fig. 3** Cell suspension cultures showing the accumulation of biomass over time: (**a** and **b**) 1- and 2-week-old suspension culture showing lag phase, (**c–f**) 3- to 6-week-old suspension culture showing exponential phase, (**g–i**) 7- to 9-week-old suspension culture showing linear phase, (**j–l**) 10- to 12-week-old suspension culture showing stationary phase

### 3.5 Cell Suspension Cultures for Determination of Growth Curve

1. For the determination of cell growth curve, collect 500-mg cells from the established cell suspension cultures and scalpel-macerated to initiate the fine cell suspension cultures in 250-mL conical flask (*see* **Note 4**).

2. Place the culture flask on the orbital shaker and set the speed at 150 rpm. Incubate the cultures under 16-h photoperiods provided by cool-white florescent light (40 $\mu mol//m^2/s$) and 25 $\pm$ 2 °C. Under these conditions, the cells will proliferate.

3. Assess the growth of cell suspensions after 1–12 weeks of culturing (Fig. 3) in terms of packed cell volume, fresh weight, and dry weight.

*3.5.1 Determination of Packed Cell Volume (PCV)*

1. To determine the PCV, place 10-mL cell suspension in a 15-mL sterile graduated centrifuge tube and centrifuge at 3600 $\times$ $g$ for 5 min.

2. Record the PCV as percentage cell mass of the total centrifuged volume (*see* **Note 5**).

3. The PCV values plotted in relation to time to obtain growth curve reflecting various phases of cell growth (Fig. 4) (*see* **Note 6**).

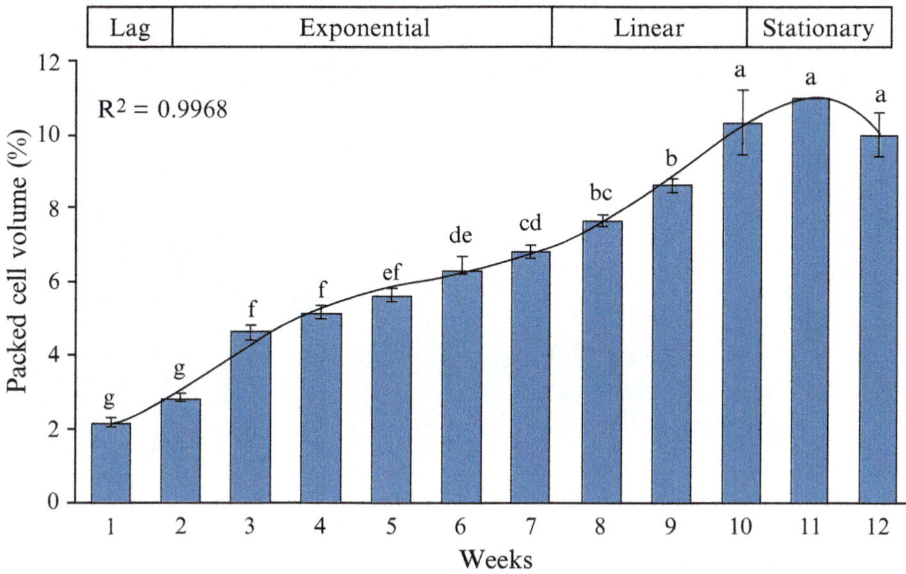


**Fig. 4** Growth curve of the date palm cell suspension culture showing PCV (%) in relation to time. Standard error is represented as bar in the graph. Mean values following the same letter in the graph are not significantly different, according to Duncan's multiple range ($p \leq 0.05$) test

*3.5.2 Determination of Cell Biomass*

1. Filter the cells from the suspension culture medium by using a funnel fitted with Whatman Grade 1 filter paper.

2. Record the fresh weight accurately using a weighing balance.

3. Plot the fresh weight values in relation to time expressed in weeks (Fig. 5).

4. Record the dry weight after drying the cells for 24 h in a hot air oven set at 60 °C.

5. Plot the graph of dry weight values against time expressed in weeks (Fig. 6).

***3.6 Sample and Standard Preparation for HPLC Analysis***

1. Grind the dried date palm cells and make it into powder using a pestle and mortar.

2. Take 100-mg powder sample in a 15-mL sterile graduated centrifuge tube and add 10-mL 80% methanol.

3. Incubate the sample in a water bath at 60 °C for 2 h, and shake the tube every 15 min.

4. Centrifuge the tube at $3600 \times g$ for 20 min. The clear supernatant solution with dried sample powder settles at the bottom of the centrifuge tube (*see* **Note 7**).

5. Evaporate the supernatant solution using vacuum rotary evaporator (*see* **Note 8**).

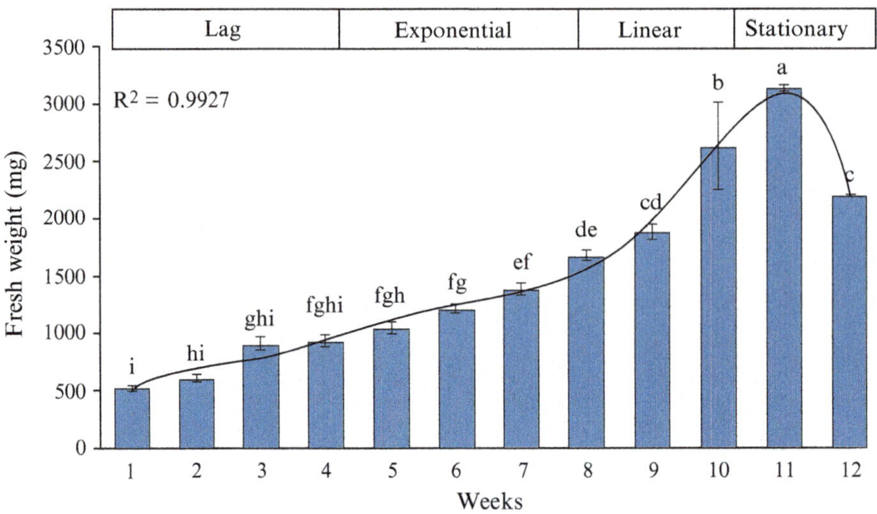

**Fig. 5** The graph representing the production of fresh weight from the date palm cell suspension culture in relation to time. Standard error is represented as bar in the graph. Mean values following the same letter in the graph is not significantly different, according to Duncan's multiple range ($p \leq 0.05$) test. The mean fresh weights are obtained from 50-mL of cell suspension culture

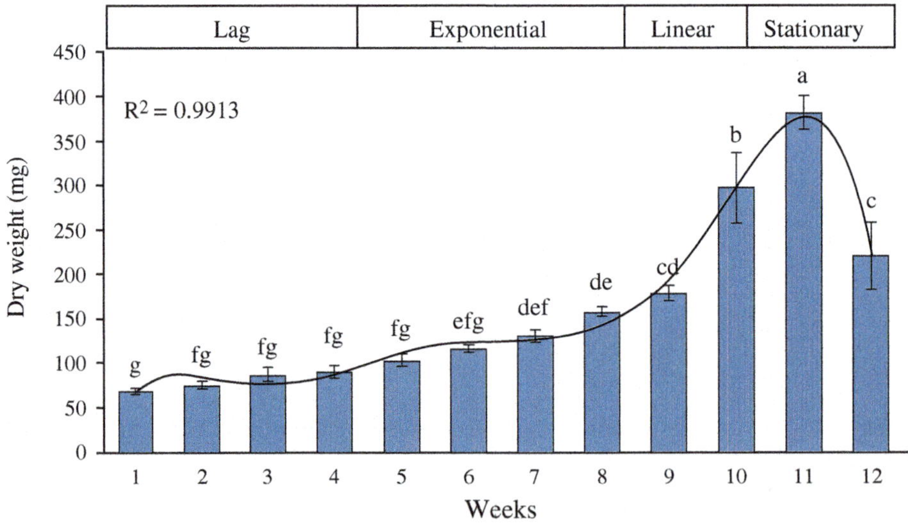

**Fig. 6** The graph representing the production of dry weight from the date palm cell suspension culture in relation to time. Standard error is represented as bar in the graph. Mean values following the same letter in the graph is not significantly different, according to Duncan's multiple range ($p \leq 0.05$) test. The mean dry weights are obtained from 50-mL of cell suspension culture

6. Add 1-mL mobile phase solution to the residue present in the round bottom flask, and rinse the round bottom flask until the residue of the extract dissolves.

7. Filter the extract into HPLC sample vial using 0.45-μm membrane filter and syringe.

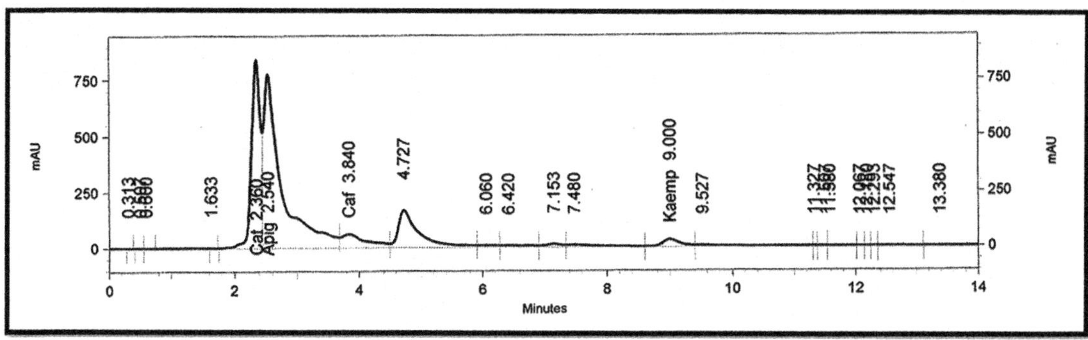

**Fig. 7** HPLC chromatogram of extract from the date palm cell suspension culture showing the retention time of the secondary metabolites: catechin, apigenin, caffeic acid, and kaempferol

8. Prepare the HPLC standards catechin, caffeic acid, apigenin, and kaempferol stock solutions in absolute ethanol of HPLC grade at 1 mg/mL (*see* **Note 9**).

**3.7  HPLC Analysis of Secondary Metabolites**

1. Use the filtrate for the HPLC analysis. By using mobile phase solution, wash the HPLC instruments containing different parts such as degasser, quaternary pump, thermostated column compartment, and diode array detector (DAD).

2. Carry out the chromatographic analysis of secondary metabolites by reversed-phase Nucleosil C18 column (4.6 mm × 250 mm, 5 μm); the oven temperature is set at 25 °C.

3. Quantify the compounds by RP-HPLC using DAD, following separation at 200–400 nm. The run time is 15 min; flow rate and injection volume are 1 mL/min and 35 μL, respectively.

4. Confirm the chromatographic peaks of the analytes (Fig. 7) by comparing their retention time with those of the reference standards. Carry the quantification by the integration of the peak using external standard method (Fig. 8; *see* **Note 10**). Perform all the chromatographic operations at ambient temperature.

# 4  Notes

1. Prepare the sock solutions of plant growth regulators by dissolving the 2,4-D and NAA in 95% ethanol or 1 N NaOH and 2iP using 1 N HCl, and make the required volume by adding double distilled water. Store in the refrigerator at 4 °C for a month.

2. The mobile phase deaerated ultrasonically for 30 min by using ultrasonic bath prior to use.

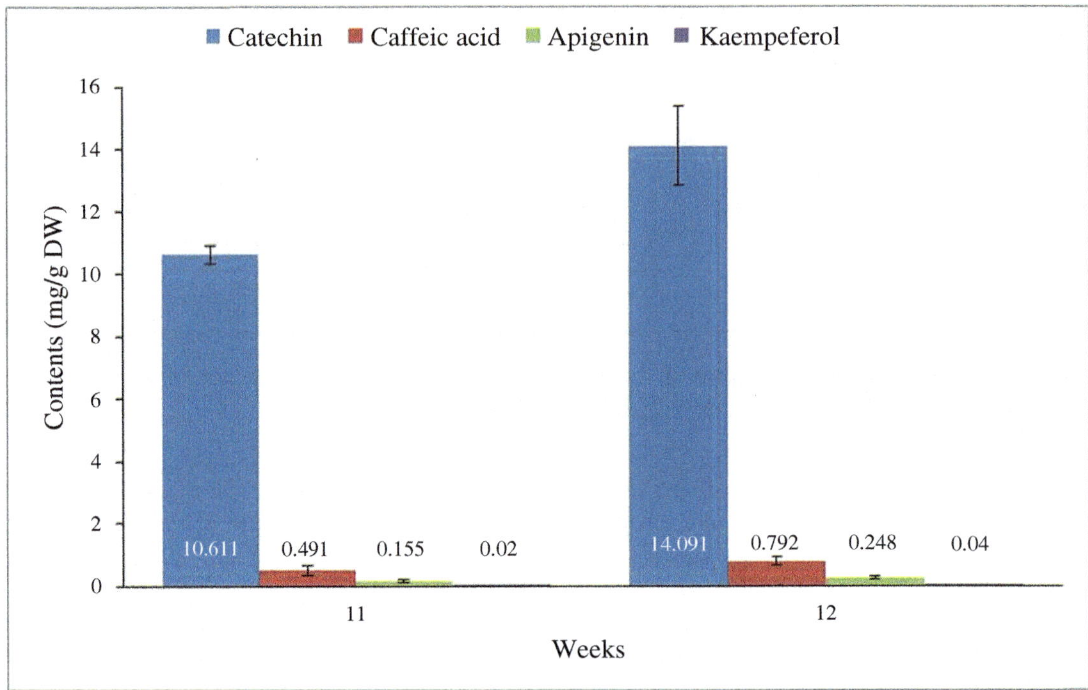



**Fig. 8** Secondary metabolite contents in the date palm cell suspension at 11 and 12 weeks of culture extract. Standard error is represented as error bar in the graph

3. Prepare the MS medium (Table 1) stock solutions such as major, minor inorganic nutrients, iron source, and vitamins (organic supplements) separately, add stock solution separately while preparing media to avoid precipitation. The stock solutions are stored at 4 °C. Store the stock solutions for 2–3 months. There are two main reasons to replace stock solutions after 2–3 months: (a) To avoid the contamination with microorganisms or precipitation of the stock solutions. (b) The stock solution chemicals are stable at 4 °C, but thawing and freezing affect the stability of the stock solutions.

4. To determine the growth curve, 500-mg of scalpel-macerated fresh cells from the cell suspension cultures were used. Depending on the results obtained from the PCV, fresh weight, and dry weight, the growth curve has been defined.

5. After recording the PCV, filter the cells from the 15-mL sterile graduated centrifuge tube through Whatman No.1 filter and dispense the cells to their respective flasks.

6. The growth curve reflected lag, exponential, linear, and stationary phases of cell growth. The PCV value was found optimum at 11 weeks cell culture after this stationary phase started.

7. The supernatant solution filtered through 0.45-μm membrane filter and collected in the 125-mL round bottom flask.

8. The filtrate dried at 50 °C with the rotation speed of 280 rpm under reduced pressure in a rotary evaporator.

9. All the HPLC standard stock solutions are filtered through 0.45-μm membrane filter and collected in a HPLC sample vial, and stored in the refrigerator at 4 °C.

10. The catechin, caffeic acid, apigenin, and kaempferol contents present in the 11 and 12 weeks of culture extract were shown in Fig. 8. Time course accumulation of secondary metabolites was highest in the initial phase of growth, but biomass was very less at these time interval, whereas in the 11 and 12 weeks of culture, the optimum quantity of metabolites yield obtained with respect to the culture biomass.

## Acknowledgment

The authors wish to express gratitude to King Abdulaziz City for Science and Technology (KACST), Saudi Arabia, for the financial support of this research project (Grant No. AT-34-63).

## References

1. Johnson DV, Al-Khayri JM, Jain SM (2015) Introduction: date production status and prospects in Asia and Europe. In: Al-Khayri JM, Jain SM, Johnson DV (eds) Date palm genetic resources and utilization. Vol 2: Asia and Europe. Springer, Dordrecht, pp 1–16

2. El Hadrami A, Al-Khayri JM (2012) Socioeconomic and traditional importance of date palm. Emir J Food Agric 24:371–385

3. Ozeker E (1999) Phenolic compounds and their importance. Anadolu J AARI 9 (2):114–124

4. Mazid M, Khan TA, Mohammad F (2011) Role of secondary metabolites in defense mechanisms of plants. Biol Med 3(2):232–249

5. Biglari F, Al-Karkhi AFM, Easa AM (2008) Antioxidant activity and phenolic content of various date palm (*Phoenix dactylifera* L.) fruits from Iran. Food Chem 107:1636–1641

6. J'Aiti F, Verdeil JL, El Hadrami I (2009) Effect of jasmonic acid on the induction of polyphenoloxidase and peroxidase activities in relation to date palm resistance against *Fusarium oxysporum* f. sp. *albedinis*. Physiol Mol Plant Path 74:84–90

7. El Hadrami A, Daayf F, El Hadrami I (2011) Secondary metabolites of date palm. In: Jain SM, Al-Khayri JM, Johnson DV (eds) Date palm biotechnology. Springer, Dordrecht, pp 653–674

8. Vayalil PK (2012) Date fruits (*Phoenix dactylifera* Linn.): An emerging medicinal food. Crit Rev Food Sci Nutr 52:249–271

9. Andlauer W, Furst P (2003) Special characteristics of non-nutrient food constituents of plants - phytochemicals. Introductory lecture. Int J Vitam Nutr Res 73:55–62

10. McCue PP, Shetty K (2004) Inhibitory effects of rosmarinic acid extracts on porcine pancreatic amylase in vitro. Asia Pac J Clin Nutr 13:101–106

11. Murthy HN, Lee EJ, Paek KY (2014) Production of secondary metabolites from cell and organ cultures: strategies and approaches for biomass improvement and metabolite accumulation. Plant Cell Tissue Organ Cult 118 (1):1–16

12. Taha HS, Bekheet SA, El-Bahr MK (2012) A new concept for production and scaling up of bioactive compounds from Egyptian date palm (Zaghlool) cultivar using bioreactor. Emir J Food Agric 24:425–433

13. Al-Khayri JM (2012) Determination of the date palm cell suspension growth curve, optimum plating efficiency, and influence of liquid medium on somatic embryogenesis. Emir J Food Agric 24(5):444–455

14. Al-Khayri JM, Al-Bahrany AM (2012) Effect of abscisic acid and polyethylene glycol on the synchronization of somatic embryo

development in date palm (*Phoenix dactylifera* L.) Biotechnology 11(6):318–325

15. Al-Bahrany AM, Al-Khayri JM (2012) In vitro responses of date palm cell suspensions under osmotic stress induced by sodium, potassium and calcium salts at different exposure durations. Am J Plant Physiol 7(3):120–134

16. Taha HS, Abdel-El-Kawy AM, Abd-El-Kareem FM, El-Shabrawi HM (2010) Implement of DMSO for enhancement and production of phenolic and peroxides compounds in suspension cultures of Egyptian date palm (Zaghlool and Samany) cultivars. J Biotech Biochem 1:1–10

17. Shehata WF, Aldaej MI, Alturki SM, Ghazzawy HS (2014) Effect of ammonium nitrate on antioxidants production of date palm (*Phoenix dactylifera* L.) in vitro. Biotechnology 13 (3):116–125

18. Murashige T, Skoog F (1962) A revised medium for rapid growth and bioassays with tobacco tissue cultures. Physiol Plant 15:473–497

# Chapter 27

# In Vitro Assessment of Abiotic Stress in Date Palm: Salinity and Drought

## Jameel M. Al-Khayri, Poornananda M. Naik, and Hussain A. Alwael

## Abstract


Date palm is one of the major crops growing in regions where abiotic stress conditions are extreme. Abiotic stress affects plant growth, development, physiology, and biochemical processes. This chapter describes a protocol to evaluate the response of date palm cultures to abiotic stresses. Tolerance to salinity stress is assessed using calcium chloride ($CaCl_2$), potassium chloride ($KCl$), and sodium chloride ($NaCl$) at 11.96, 12.06, and 9.45 g/L, respectively (equivalent to 0.8 MPa osmotic potential), with different exposure durations (1–12 days). Polyethylene glycol (PEG 8000) is tested at 0–30% (w/v) to assess tolerance to drought stress. Techniques are described to define the effects of these stress agents on the growth of callus and cell suspension cultures, water content, proline accumulation, and $Na^+$ and $K^+$ content ratio, in addition to the technique used for determining the median lethal dose ($LD_{50}$) for PEG (29.5%) and $NaCl$ (11.54 g/L). This protocol will be useful for future studies of in vitro selection of tolerant cell lines.

**Key words** Abiotic stress, Callus, Cell suspension, Drought, In vitro stress, Ion content, PEG, Proline, Salinity


## 1 Introduction

Salinity and drought are the major adverse environmental conditions, particularly in the arid and semiarid regions of the Middle East and North Africa where date palm is predominantly grown [1, 2]. Salinity and drought stresses are the most significant type of abiotic stress, which adversely affect the plant growth, development, physiological, and metabolic processes [3–5]. The cells of whole plants behave similarly as in vitro cultured cells subjected to salinity and drought conditions [6]. The drought tolerance levels in species are totally different when compared to whole plants and cell culture [7]. Cell cultures provide genetically and morphologically uniform cells and also uniform environmental conditions for studying salt and drought stress responses by eliminating complications arising from the tissues of whole plants and variable field conditions [8].


Jameel M. Al-Khayri et al. (eds.), *Date Palm Biotechnology Protocols Volume 1: Tissue Culture Applications*,
Methods in Molecular Biology, vol. 1637, DOI 10.1007/978-1-4939-7156-5_27, © Springer Science+Business Media LLC 2017

The long juvenile phase of date palm strongly affects conventional breeding techniques [9]. Plant tissue culture techniques are alternative tools to overcome these breeding limitations. Date palm in vitro regeneration was successfully demonstrated by a number of researchers [10–14]. To understand the effect of salinity and drought on the behavior of date palm in vitro culture, several studies have been conducted [8, 15–19].

The date palm in vitro cultures which possess great tolerance to abiotic stress will play an important role in breeding programs. A few studies related to date palm response to in vitro salt and drought stresses are recounted below. Al-Bahrany and Al-Khayri [15] reported that exposure of $CaCl_2$, KCl, and NaCl resulted in the reduction of callus dry weight, callus water content, and $Na^+/K^+$ ratio. In another article, Al-Khayri [16] noted an increase in proline accumulation in callus in response to increased salinity. Bekheet [17] showed that as mannitol or PEG concentration increased in culture medium, growth parameters like survival, fresh weight, and osmotic tolerance ratio decreased. Al-Khayri and Al-Bahrany [8] observed increasing PEG concentration associated with a reduction in the callus fresh weight, relative growth rate, index of tolerance, and water content and elevated the endogenous free proline content.

This chapter describes the procedures for the establishment of date palm explant leading to callus induction and proliferation for use in cell suspension culture, in addition to the methods employed for the determination of callus weight, relative growth rate, water content, proline content, $Na^+/K^+$ concentrations, and ratio in response to exposure to salinity and drought stresses.

## 2 Materials

### 2.1 Plant Materials and Sterilization

1. Explant source: Date palm cv. Barhee offshoots.

2. 70% Ethanol.

3. Disinfectant solution: 1.6% (w/v) sodium hypochlorite solution (30% v/v Clorox, commercial bleach) containing three drops of Tween 20 per 100 mL solution.

4. Antioxidant solution: Ascorbic acid and citric acid, 150 mg/L each.

### 2.2 Culture Medium

1. Basal culture medium: the stock solutions and final medium compositions of modified Murashige and Skoog (MS) [20] medium (MS stock I, II, III, and IV) are listed in Table 1 (*see* **Note 1**).

2. Hormone stock solutions: 2,4-dichlorophenoxyacetic acid (2,4-D) (5 mg/mL), naphthaleneacetic acid (NAA)

**Table 1**

Components of MS basal medium [19] and additives used for date palm in vitro culture stages. Hormones, agar, and activated charcoal are added according to the culture stage as shown in Table 2

| Medium composition | Stock concentration (mg/500 mL) | Final concentration in culture medium (mg/L) |
| --- | --- | --- |
| *Stock I: Major inorganic nutrients (10× stock) use 50 mL to prepare 1 L of medium* | | |
| $NH_4NO_3$ | 16,500 | 1650 |
| $KNO_3$ | 19,000 | 1900 |
| $CaCl_2 \cdot 2H_2O$ | 4400 | 440 |
| $MgSO_4 \cdot 2H_2O$ | 3700 | 370 |
| $KH_2PO_4$ | 1700 | 170 |
| $NaH_2PO_4 \cdot H_2O$ | 1700 | 170 |
| *Stock II: Minor inorganic nutrients (100× stock) use 5 mL to prepare 1 L of medium* | | |
| KI | 83 | 0.83 |
| $H_3BO_3$ | 620 | 6.2 |
| $MnSO_4 \cdot 2H_2O$ | 2230 | 22.3 |
| $ZnSO_4 \cdot 7H_2O$ | 860 | 8.6 |
| $Na_2.MoO_4 \cdot 2H_2O$ | 25 | 0.25 |
| $CuSO_4 \cdot 5H_2O$ | 2.5 | 0.025 |
| $CoCl_2 \cdot 6H_2O$ | 2.5 | 0.025 |
| *Stock III: Iron Source (100× stock) use 5 mL to prepare 1 L of medium* | | |
| $FeSO_4 \cdot 7H_2O$ | 2780 | 27.8 |
| $Na_2EDTA \cdot 2H_2O$ | 3730 | 37.3 |

(continued)

**Table 1**
**(continued)**

| Medium composition | Stock concentration (mg/500 mL) | Final concentration in culture medium (mg/L) |
| --- | --- | --- |
| *Stock IV: Vitamins (100× stock) use 5 mL to prepare 1 L of medium* | | |
| Myoinositol | 12,500 | 125 |
| Nicotinic acid | 100 | 1 |
| Pyridoxine HCl | 100 | 1 |
| Thiamine HCl | 100 | 1 |
| Glycine | 200 | 2 |
| Calcium pentothenate | 100 | 1 |
| Biotin | 100 | 1 |
| *Carbon source* | | |
| Sucrose | – | 30,000 |
| *Antioxidants* | | |
| Glutamine | – | 200 |
| Citric acid | – | 100 |
| Ascorbic acid | – | 100 |

**Table 2**
**Different culture stages and their corresponding hormonal and activated charcoal additives supplemented to the MS medium listed in Table 1**

| Culture stage | Hormones, agar, and activated charcoal additives | | | | | |
| | 2,4-D | 2ip | NAA | Agar | Activated charcoal | CaCl$_2$, KCl, NaCl, and PEG |
|---|---|---|---|---|---|---|
| Culture initiation (CI) | 100 mg/L | 3 mg/L | – | 7 g/L | 1.5 g/L | – |
| Callus proliferation (CP) | – | 30 mg/L | 10 mg/L | 7 g/L | 1.5 g/L | – |
| Embryogenic callus (EC) | – | 6 mg/L | 10 mg/L | 7 g/L | 1.5 g/L | – |
| Callus maintenance (CM) | – | 1.5 mg/L | 10 mg/L | 7 g/L | – | – |
| Cell suspension (CS) | – | 1.5 mg/L | 10 mg/L | – | – | Salts: 11.96, 12.06, 9.45 g/L PEG: 0, 5, 10, 15, 20, 25, 30% |

(1 mg/mL), and 2-isopentenyladenine (2iP) (1 mg/mL) (*see* **Note 2**).

3. Additives to MS medium for various culture stages: culture initiation (CI), callus proliferation (CP), embryogenic callus (EC), callus maintenance (CM), cell suspension (CS) (Table 2).

4. Solutions to adjust pH: 1 N and 0.1 N of each KOH and HCl.

*2.3 Salinity and Drought Experiment*

1. Salinity stress agent: calcium chloride (CaCl$_2$), potassium chloride (KCl), and sodium chloride (NaCl).

2. Drought stress agent: polyethylene glycol (PEG 8000).

*2.4 Equipment*

1. Instruments: autoclave, weighing balance, refrigerator, microwave oven, hot air oven, water bath gyratory shaker, UV/VIS spectrophotometer, and flame photometer.

2. Glassware: 125-mL conical flask, 1000-mL conical flask, 200-mL tissue culture jar, 500-mL measuring cylinder, 1000-mL measuring cylinder, and digestion flask.

3. Tools: Whatman Grade 2 filter paper and Buchner funnel.

4. Supplies: nitric acid (HNO$_3$), sulfuric acid (H$_2$SO$_4$), sulfosalicylic acid, toluene, glacial acetic acid, and acid ninhydrin.

## 3 Methods

*3.1 Culture Medium Preparation*

1. MS stocks: weigh the components of MS stocks I, II, III, and IV individually by using electronic weighing balance, dissolve

one by one components of each stock in 400-mL distilled water, and make up the volume to 500 mL by using 500-mL measuring cylinder. Transfer the stocks into reagent bottles, and store the solutions in a refrigerator at 4 °C until use.

2. To prepare the different culture media, take 1000-mL flask and add MS stocks, carbon source, and antioxidants as mentioned in Table 1. In addition, add different hormonal combinations and activated charcoal contents as described in Table 2 for each culture stage. Make up the medium volume 800 mL using distilled water, and adjust the medium to pH 5.8 using the solution 1 N NaOH and 1 N HCl. Finally make up the medium volume of 1000 mL using distilled water and 1000-mL measuring cylinder. Heat the medium to dissolve the agar in a microwave oven.

3. Dispense 40 mL media in a 200-mL tissue culture jar with screw cap, and 50 mL media in a 125-mL culture flask for cell suspension, capped using aluminum foil and rubber bands. Autoclave all the media for 15 min at 121 °C and 1.1 kg/cm$^2$. Store the media in a room temperature, maximum of 2 weeks.

### 3.2 Preparation of Explant

1. Separate 3–4-year-old offshoots from mother trees (Fig. 1a). Remove the outer leaves exposing the shoot-tip region (Fig. 1b).

2. Surface sterilize the shoot-tip tissue for 1 min in 70% ethanol, and then 15 min in disinfectant solution (Fig. 1c). Rinse the tissue in sterilized distilled water four times.

3. Place the tissue in a beaker containing a chilled sterile antioxidant solution to prevent oxidation-induced browning (Fig. 1d).

4. Carefully trim the tissue surrounding the shoot tip, using forceps and scalpel, until about 1–1.5 cm$^3$ of the terminal remains (Fig. 1e). Excise the tip and section longitudinally into 4–8 small sections, which serve as the explants (Fig. 1f).

### 3.3 Callus Induction and Maintenance

1. For callus culture initiation, place shoot tips on the surface of solid CI medium (Fig. 2a; Table 2).

2. Incubate cultures at 24 ± 3 °C for 12 weeks in the dark. Explants initially swell during the first 4 weeks of culture (Fig. 2b). Subculture the callus cultures to a fresh medium at 3-week intervals. Transfer callus to the CP medium and maintain for an additional 3 weeks (Fig. 2c; Table 2).

3. Transfer the callus from CP medium onto EC medium, and maintain for 9 weeks by subculturing it at 3-week intervals (Fig. 2d; Table 2).

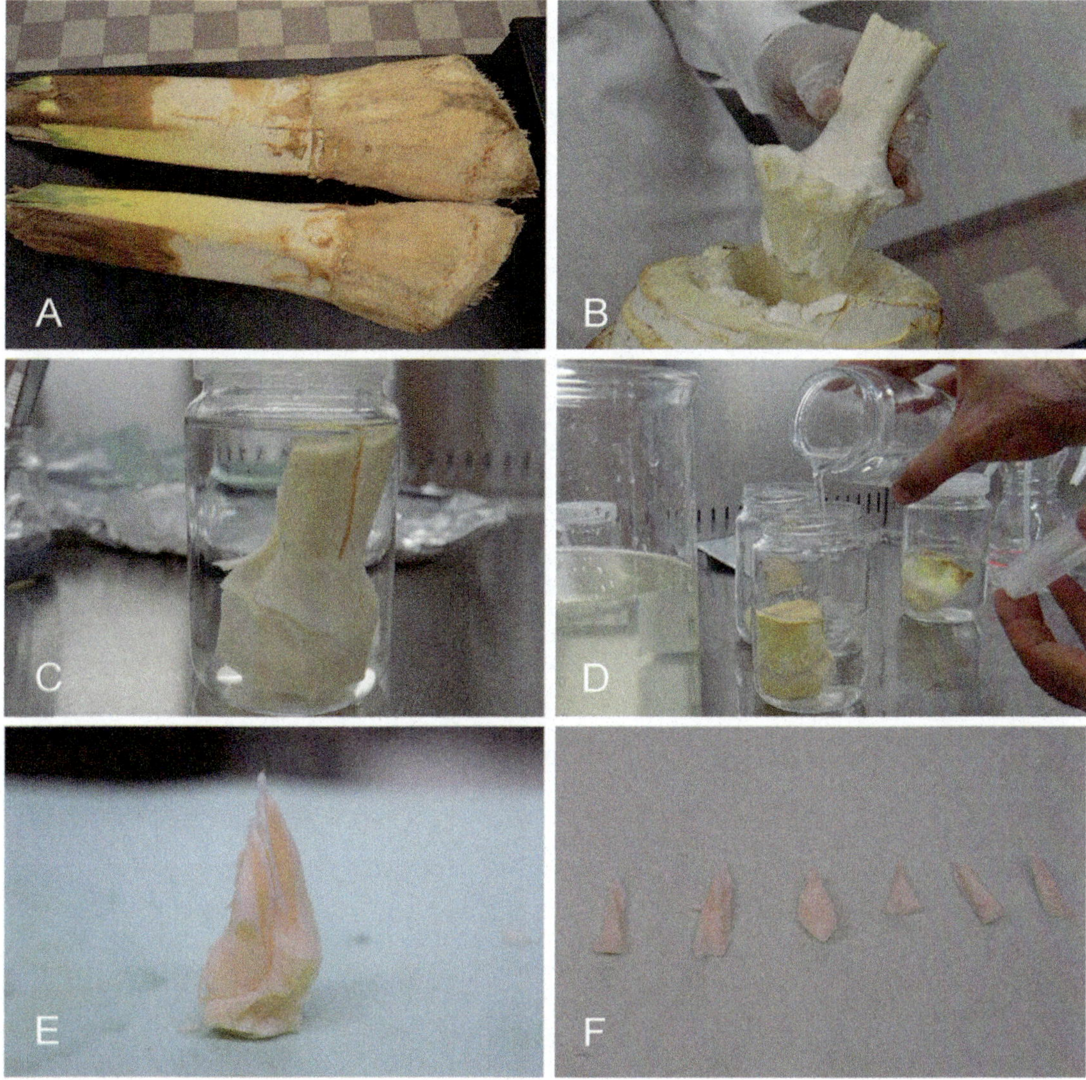

**Fig. 1** Stages of date palm explant preparation: (**a**) 3–4-year-old offshoots, (**b**) shoot-tip region, (**c**) shoot-tip region in disinfectant solution, (**d**) shoot-tip region in the antioxidant solution, (**e**) prominent shoot-tip portion, (**f**) explants isolated from the shoot-tip region

4. Subculture embryogenic callus collected from the EC medium onto the CM medium (Fig. 2e; Table 2), and maintain cultures for 6 months by subculturing at 3-week intervals to obtain sufficient amount of callus as desired.

***3.4 Cell Suspension Experiment for Salinity and Drought Stress Studies***

1. Culture actively growing embryogenic callus in 125-mL conical flask containing 50 mL of CS medium (Fig. 2f; Table 2). Use 1 or 2 g of callus within 2 weeks of subculture to ensure active growth (*see* **Notes 3** and **4**).

**Fig. 2** Stages of date palm cell suspension culture: **(a)** Explants on culture initiation medium, **(b)** expansion of explant after 4 weeks, **(c)** callus induction on callus induction medium, **(d)** callus formation in the proliferation medium, **(e)** callus in the maintenance medium, **(f)** cell suspension culture

2. Place the culture flasks on a gyratory shaker, set at 150 rpm, under a 16-h photoperiod of cool-white florescent light ($40\ \mu mol/m^2/s$) at $24 \pm 2\ °C$ (*see* **Notes 5** and **6**).

3. Filter the culture through Whatman Grade 2 filter paper fitted in a Buchner funnel. Weigh the residue and score it as callus fresh weight (*see* **Note 7**).

4. Dry the callus samples of known fresh weight in an oven set at 65 °C for 48 h and reweigh. Determine the difference between the weights (*see* **Note 8**).

5. The water content is expressed as a percentage of callus fresh weight (*see* **Note 9**).

6. Calculate relative growth rate (RGR) based on the fresh weight using the following formula: $RGR = \frac{[\ln\,(\text{finalweight}) - \ln\,(\text{initialweight})]}{\text{growthperiod}}$ (*see* **Note 10**).

*3.4.1   Determination of Median Lethal Dose (LD$_{50}$)*

1. After scoring callus fresh weight, convert the data into percentage using the following formula: $\text{Callus}\% = \frac{\text{Sample F.W.Value}}{\text{Control F.W.Value}} \times 100$.

2. Create a bar chart using Excel by selecting the data and choosing "Bar" from "Chart" group on the "Insert" tab.

3. Add a trend line to the chart by checking the "Trend line" box from "Chart Element" button (*see* **Note 11**).

4. Draw a line to the right from 50% on the y-axis until it intercepts with the added trend line; draw another line downwards from the interception to determine LD$_{50}$ value (Fig. 3a, b).

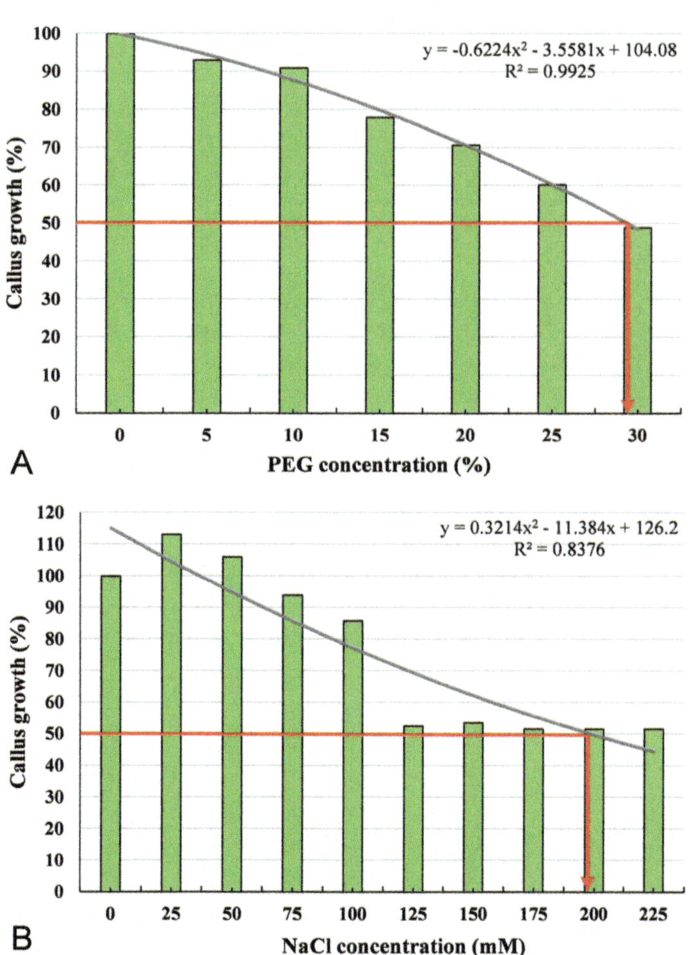


**Fig. 3** Determination of median lethal dose (LD$_{50}$) for date palm in vitro culture: **(a)** PEG LD$_{50}$ = 29.5%, based on data from [8]. **(b)** NaCl LD$_{50}$ = 197.5 mM (or 11.54 g/L), based on data from [16]

*3.4.2 Extraction and Estimation of Proline*

1. Homogenize 500-mg fresh callus per sample in 10 mL 3% (w/v) aqueous sulphosalicylic acid and filter the homogenate through Whatman Grade 2 filter paper.

2. Mix the 2 mL filtrate in a test tube with 2-mL acid ninhydrin and 2-mL glacial acetic acid. Incubate the mixture for 1 h at 100 °C in a water bath (*see* **Note 12**).

3. Keep the reaction mixture in an ice bath, and extract with 4 mL toluene. Mix vigorously and stir 15–20 s.

4. Aspirate the chromophore phase from the aqueous phase, and warm to room temperature.

5. Check the absorbance at 520 nm using UV/VIS spectrophotometer. Use toluene for blank (*see* **Note 13** and **14**).

*3.4.3 Extraction and Estimation of $Na^+$ and $K^+$*

1. Place 500-mg fresh callus in a digestion flask and add 8 mL concentrated $HNO_3$ and 2 mL of $H_2SO_4$.

2. Heat the flasks gently over a hot plate until the solution turns colorless.

3. Cool the digestion and dilute to 50 mL with distilled water.

4. Determine $Na^+$ and $K^+$ contents using flame photometer (*see* **Note 15**).

---

## 4   Notes

1. Prepare MS medium (Table 1) stock solutions of major, minor inorganic nutrients, iron source, vitamins (organic supplements) separately; add stock solutions separately while preparing the media to avoid precipitation. Store stock solutions at 4 °C for no longer than 2–3 months to avoid contamination and precipitation. The chemical composition of stock solutions is stable at 4 °C, but thawing and freezing negatively affects the stability of the stock solutions.

2. Prepare the sock solutions of auxins by dissolving 2,4-D and NAA in 95% ethanol or 1 N NaOH; and 2iP cytokinin using 1 N HCl, make the required volume by adding double-distilled water. Store in the refrigerator at 4 °C for 1 month.

3. In the salinity experiment, add 1 g embryogenic callus per flask with salt solution; $CaCl_2$, KCl, and NaCl at 11.96, 12.06, and 9.45 g/L, respectively. This is equivalent to osmotic potential of 0.8 MPa (−8 bars). Add 0.5 g of embryogenic callus per flask when using NaCl at 1.5, 3, 4.4, 5.84, 7.3, 8.8, 10.22, 11.7, and 13.15 g/L (25, 50, 75, 100, 125, 150, 175, 200, and 225 mM, respectively) to evaluate $LD_{50}$.

4. Add 2 g embryogenic callus per flask in the drought experiment with PEG concentrations of 0, 5, 10, 15, 20, 25, and 30%.

5. To evaluate the callus response in the salinity experiment, expose the cultures to different time duration: 1, 3, 6, 9, and 12 days.

6. In the drought experiment, harvest callus cultures at the end of the second week.

7. Increasing drought stress results in the progressive reduction of growth as expressed in date palm callus fresh weight (Fig. 4a).

8. Exposure to salt stress results in the reduction of callus dry weight as compared to the control (Fig. 5a).

9. Callus water content decreases in response to the increase of PEG concentration and extending exposure duration regardless of the salt type used (Figs. 4b and 5b).

10. Steady reduction in RGR is seen in response to increasing PEG concentration when compared to the control (Fig. 4c).

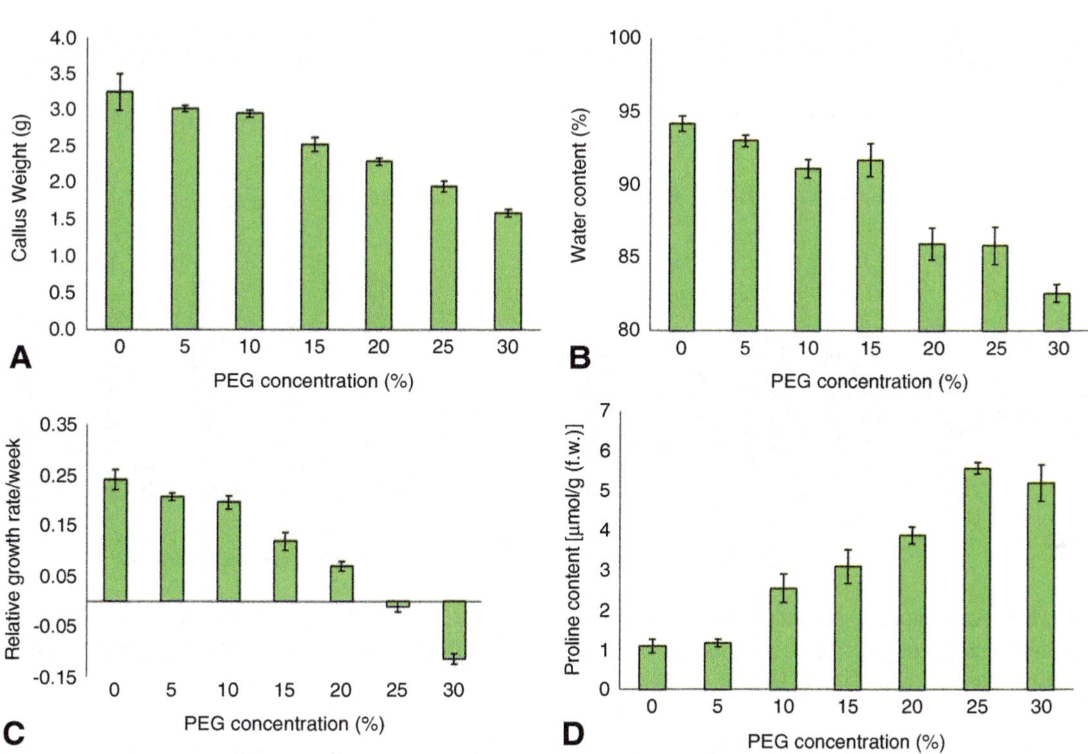

**Fig. 4** Effect of polyethylene glycol concentration on **(a)** callus weight, **(b)** relative growth rate, **(c)** water content, **(d)** proline content of date palm cell suspension culture Source: Based on data from [8]

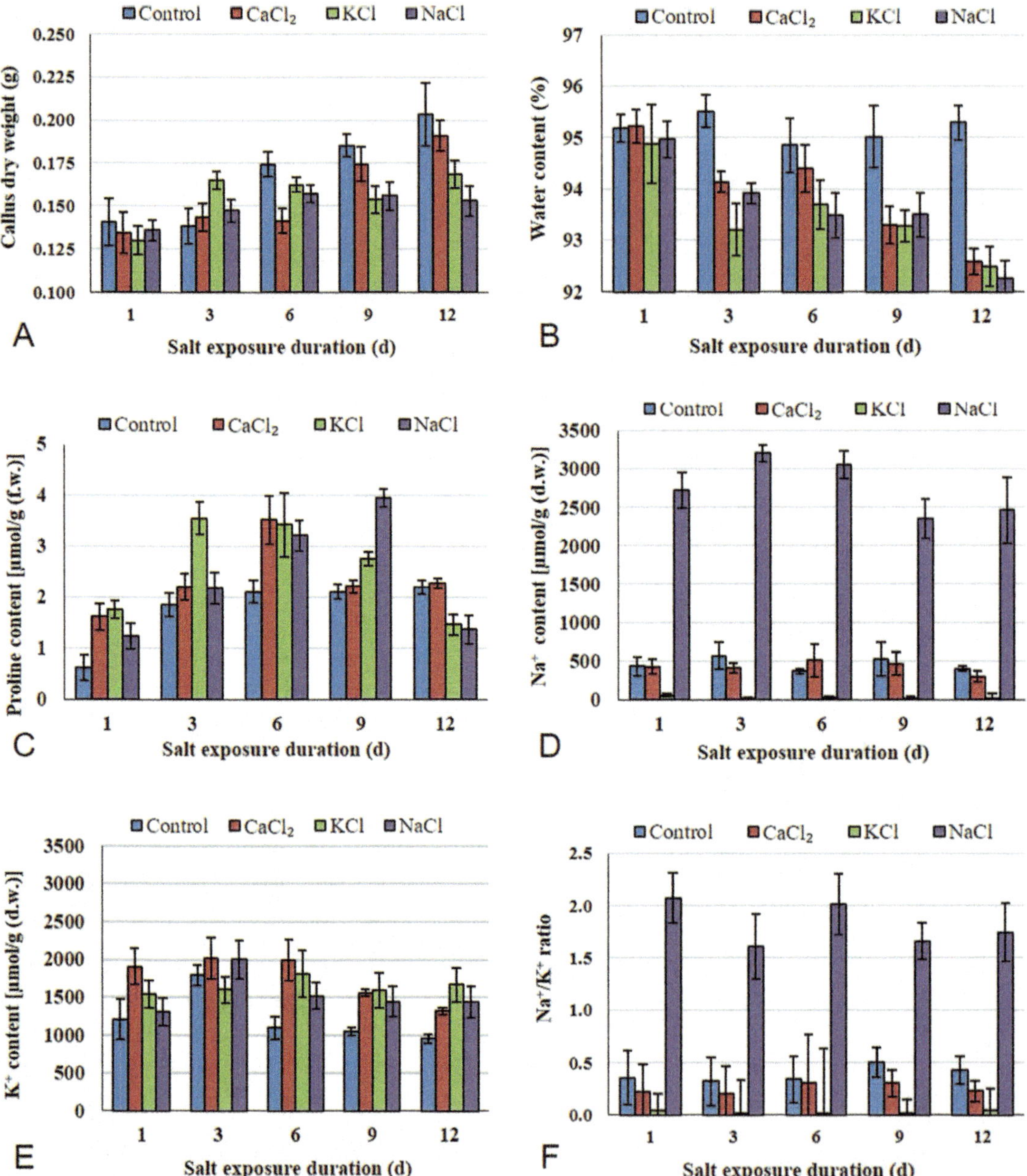

**Fig. 5** Effect of salts on **(a)** callus dry weight, **(b)** water content, **(c)** proline content, **(d)** $Na^+$ content, **(e)** $K^+$ content, and **(f)** $Na^+/K^+$ ratio at different duration (1, 3, 6, 9, and 12 days) of date palm cell suspension culture Source: Based on data from [15]

11. Choose the right trend line type (linear, logarithmic, exponential, polynomial, or power) based on the R-squared value; it is most accurate when the value is at or near 1. You can show $R^2$ value from trend line options by clicking the trend line twice and checking "Display R-squared vale on chart" button.

12. Acid ninhydrin preparation: Warm 1.25 g ninhydrin in 30-mL glacial acetic acid and 20-mL 6 M phosphoric acid with agitations until dissolved, store at 4 °C. Reagent remain stable for 24 h.

13. The proline concentration is determined from a standard curve and calculated on a fresh weight basis as following:

$$\mu moles\, proline/g\, fresh\, weight\, material$$

$$= \frac{(\mu g\, proline/mL \times mL\, toluene)/115.5\, \mu g/\mu mole}{g\, sample/5}$$

14. Accumulation of proline is positively related to PEG concentration (Fig. 4d). In the salt experiment, increasing the exposure duration up to 6 days caused increase in the proline content compared to the control, but extending the exposure duration of KCl and $CaCl_2$ to 9 days caused reduction in the proline content (Fig. 5c).

15. By increasing duration of NaCl exposure, higher $Na^+$ content is observed as compared to control (Fig. 5d). There is a significant increase in $K^+$ content by increasing KCl exposure duration (Fig. 5e). Increasing salt exposure duration to NaCl resulted in higher $Na^+/K^+$ ratio when compared to the control (Fig. 5f).

## References


1. Zohary D, Hopf M (2000) Domestication of plants in the old world: the origin and spread of cultivated plants in West Asia, Europe and Nile Valley, 3rd edn. Oxford University Press, New York

2. Johnson DV (2011) Introduction: date palm biotechnology from theory to practice. In: Jain SM, Al-Khayri JM, Johnson DV (eds) Date palm biotechnology. Springer, Dordrecht, pp 1–11

3. Amirjani MR (2010) Effect of salinity stress on growth, mineral composition, proline content, antioxidant enzymes of soybean. Am J Plant Physiol 5:350–360

4. Naik PM, Al-Khayri JM (2016) Abiotic and biotic elicitors-role in secondary metabolites production through in vitro culture of medicinal plants. In: Shanker AK, Shankar C (eds) Abiotic and biotic stress in plants - recent advances and future perspectives. InTech, Croatia, pp 247–277

5. Naik PM, Al-Khayri JM (2016) Impact of abiotic elicitors on in vitro production of plant secondary metabolites: a review. J Adv Res Biotech 1(2):7

6. Attree SM, Moore D, Sawhney VK, Fowke LC (1991) Enhanced maturation and desiccation tolerance of white spruce [*Picea glauca* (Moench) Voss] somatic embryos: effects of a non-plasmolysing water stress and abscisic acid. Ann Bot 68:519–552

7. Santos-Diaz MS, Ochoa-Alejo N (1994) Effect of water stress on growth, osmotic potential and solute accumulation in cell cultures from chili pepper (a mesophyte) and creosote bush (a xerophyte). Plant Sci 96:21–29

8. Al-Khayri JM, Al-Bahrany AM (2004) Growth, water content, and proline accumulation in drought-stressed callus of date palm. Biol Plant 48:105–108

9. El Hadrami I, El Hadrami A (2009) Breeding date palm. In: Jain SM, Priyadarshan PM (eds) Breeding plantation tree crops. Springer, New York, pp 191–216

10. Al-Khayri JM (2003) *In vitro* germination responses of date palm: effect of auxin concentrations and MS salts. Curr Sci 84:680–683

11. Al-Khayri JM (2007) Date palm *Phoenix dactylifera* L. micropropagation. In: Jain SM, Haggman H (eds) Protocols for micropropagation

of woody trees and fruits. Springer, Berlin, pp 509–526

12. Othmani A, Bayoudh C, Drira N, Marrakchi M, Trifi M (2009) Somatic embryogenesis and plant regeneration in date palm *Phoenix dactylifera* L., cv Boufeggous is significantly improved by fine chopping and partial desiccation of embryogenic callus. Plant Cell Tissue Organ Cult 97:71–79

13. Al-Khayri JM (2011) Basal salt requirements differ according to culture stage and cultivar in date palm somatic embryogenesis. Am J Biochem Biotech 7:32–42

14. Khan S, Tabassum BB (2012) Direct shoot regeneration system for date palm (*Phoenix dactylifera* L.) cv. Dhakki as a means of micropropagation. Pak J Bot 44(6):1965–1971

15. Al-Bahrany AM, Al-Khayri JM (2012) In vitro responses of date palm cell suspensions under osmotic stress induced by sodium, potassium and calcium salts at different exposure durations. Am J Plant Physiol 7:120–134

16. Al-Khayri JM (2002) Growth, proline accumulation, and ion content in sodium chloride-stressed callus of date palm. In Vitro Cell Dev Biol Plant 38:79–82

17. Bekheet SA (2015) Effect of cryopreservation on salt and drought tolerance of date palm cultured in vitro. Sci Agric 9(3):142–149

18. Al-Khayri JM, Ibraheem Y (2014) In vitro selection of abiotic stress tolerant date palm (*Phoenix dactylifera* L.): a review. Emir J Food Agri 26(11):921–933

19. Helaly MN, El-Hosieny HA, El-Sarkassy NM, Fuller MP (2017) Growth, lipid peroxidation, organic solutes, and anti-oxidative enzyme content in drought-stressed date palm embryogenic callus suspension induced by polyethylene glycol. In Vitro Cell Dev Biol Plant 53:133–141

20. Murashige T, Skoog F (1962) A revised medium for rapid growth and bioassays with tobacco tissue cultures. Physiol Plant 15:473–497

# INDEX




Jameel M. Al-Khayri et al. (eds.), *Date Palm Biotechnology Protocols Volume 1: Tissue Culture Applications*,
Methods in Molecular Biology, vol. 1637, DOI 10.1007/978-1-4939-7156-5, © Springer Science+Business Media LLC 2017

CPSIA information can be obtained
at www.ICGtesting.com
Printed in the USA
LVOW05*0145110817

544592LV00005B/21/P